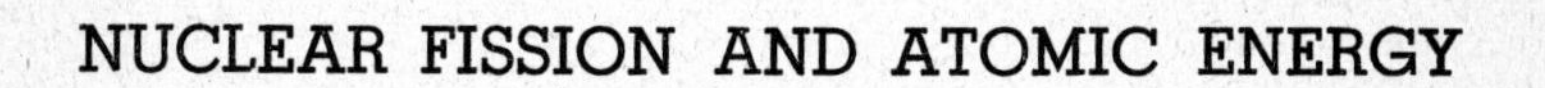

NUCLEAR FISSION AND ATOMIC ENERGY

NUCLEAR FISSION
and
ATOMIC ENERGY

by

WILLIAM E. STEPHENS (*editor*)

and

PARK HAYS MILLER, JR.
KNUT KRIEGER
MARGARET N. LEWIS
SIMON PASTERNACK
BERNARD GOODMAN
WALTER E. MEYERHOF
BERNARD SERIN
ROBERT H. VOUGHT

Members of the Staff of the University of Pennsylvania

THE SCIENCE PRESS
Lancaster, Pa.
1948

PRINTED BY
THE DARWIN PRESS
NEW BEDFORD, MASS.

CONTENTS

FOREWORD

Free and unrestricted research in nuclear physics ceased abruptly in 1941. Activity in the field went underground and certain aspects were the subject of intense study and investigation in secret under the forced draft of military urgency and unlimited support. It emerged on August 6, 1945, with the most destructive explosion that has ever been produced by man. The same dramatic event answered affirmatively the outstanding question which had engaged nuclear physicists previously: Is a self-sustaining nuclear chain reaction possible? The successful culmination of the work of the Manhattan District in the explosion of the bombs over Japan punctuated the end of the war and announced the scientific fact that nuclear chain reactions could be brought about.

With the cessation of hostilities nuclear physicists have returned from a wide variety of war research assignments to this, their chosen field. The obvious first step in resuming programs of fundamental research has been a review of the literature and a taking of scientific stock in the light of available information. At the University of Pennsylvania a series of seminars was conducted by Dr. Stephens and the staff of the Department of Physics resident in the autumn of 1945, for the purpose of reviewing all freely available information and reorienting the interests of the research group. The extensive examination of all the pertinent literature at their disposal and the careful study of its implications in the light of present common scientific knowledge has been of great value in the planning of a research program at the University of Pennsylvania. It is in the hope that the efforts of this group may serve a much broader purpose in assisting their scientific colleagues elsewhere to resume their research programs or enable them to enter their field of fundamental investigation that these seminar notes have been edited for publication.

Unfortunately this book perforce marks a departure from traditional scientific publications, a departure which it is hoped is only a temporary result of abnormal post-war conditions. The authors of this book, in common with authors of reviews in other branches of physical science, have dealt only with information that is available to all. But unlike authors of pre-war treatises they are aware that there exists a body of pertinent knowledge inaccessible to them. To avoid any possible imputation of inadvertent breach of security they have been at pains not to discuss these topics with any persons in possession of classified knowledge concerning them. Though a more complete book on the subject might be written by men who have participated in the atomic bomb project, such persons are at present legally precluded from such an undertaking. The very ignorance of the authors of this book thus enables them to contribute their special training to the writing of it as a contribution to the advancement of knowledge in the best scientific tradition. There is nothing herein that any physicist, be he American, English, Russian, French, Indian or Chinese, could not already know if he himself had taken the time to rework the excellent report of Dr. H. D. Smyth and the recent literature of physics with nuclear fission in mind.

Nuclear physics involves a considerable number of concepts which are not familiar through common experience. This fact, together with the necessarily mathematical nature of the theoretical analyses and the formulation of results, renders the subject a peculiarly difficult one for persons untrained in this science. This situation undoubtedly has led to much popular misunderstanding in the matter of security and has contributed to loose and hasty thinking about supposed secrets which Nature is thought to have shared with this country alone. In a sense the fact that this book could be written by physicists having access to no material not freely available to scientists the world over makes it clear that Nature is the only possible guardian of her own secrets.

A sincere effort is being made to bring about a popular clarification of these matters in order that the advancement of knowledge may not be legally interdicted in this country to the great detriment of our national well being. The achievements of generations of free scientists, which chiefly differentiate our life from that of the dark ages, are the best arguments for the preservation of freedom of intellectual inquiry. This book documents the thesis that the understanding of natural phenomena, of which nuclear fission is no exceptional instance, can be gained by any trained and inquiring mind. Nature will not be a party to man's attempt at discrimination between nations, races or individuals. If the publication of this book contributes to the general appreciation of this fact it will have performed an important additional service in clearing away erroneous conceptions and in promoting a wiser and more constructive approach to current national problems.

April 15, 1946

G. P. Harnwell
Chairman, Department of Physics,
University of Pennsylvania

EDITOR'S PREFACE

THIS book originated in a series of seminars on nuclear fission held in the Physics Department of the University of Pennsylvania in the fall of 1945. These seminars reviewed the known facts of nuclear fission as published in the literature. The notes were mimeographed from week to week and formed a record of the discussions. Because of the interest expressed by other physicists, we have been persuaded to publish the material. We hope that this book will enable physicists who, like most of the authors, are not specialists in nuclear physics, to obtain a semiquantitative understanding of the phenomena concerned.

The persons who presented the seminars and wrote the various chapters have had no connection with the Manhattan District project. All were on the staff of the University of Pennsylvania. The seminar speakers and their topics in order of presentation were:

MARGARET N. LEWIS........................*"Transuranic" Elements*
WILLIAM E. STEPHENS..............................*Fission Fragments*
ROBERT H. VOUGHT.................................*Fission Products*
BERNARD SERIN ....*Secondary Neutrons*
BERNARD GOODMAN .... *Theoretical Considerations*
WALTER E. MEYERHOF............................*Isotope Separation*
SIMON PASTERNACK ....*Controlled Chain Reactions—Piles*
KNUT KRIEGER .... *Chemistry of Plutonium*
PARK HAYS MILLER, JR.....................................*Atom Bomb*
WILLIAM E. STEPHENS...............*Potentialities of Fission Technique*

We wish to thank Prof. L. A. Turner, Prof. J. A. Wheeler, and Dr. T. Lauritsen for consenting to the reproduction of their work, and to acknowledge our indebtedness to the excellent review article by Professor Turner, *Rev. Mod. Phys. 12,* 1 (1940), to the classic paper by Bohr and Wheeler, *Phys. Rev. 56,* 426 (1939), and to the comprehensive report of H. D. Smyth, *Rev. Mod. Phys. 17,* 351 (1945).

April 15, 1946 WILLIAM E. STEPHENS.

Owing to difficulties of publication, the appearance of this book has been delayed far beyond the date originally hoped for. Subsequent to the preparation of manuscript and its initial distribution in mimeographed form, much information on details of the fission process, fission products, and heavy nuclei was released and published in scientific periodicals. However, the main ideas contained in this book were not essentially changed thereby, although some details need to be corrected and many details added. Consequently, we have added a supplementary bibliography of publications that came to our attention after the manuscript was prepared and before October, 1947.

November 15, 1947 W. E. S.

CHAPTER 1

DISCOVERY OF FISSION

The explosions at Hiroshima and Nagasaki demonstrated to the public for the first time that successful nuclear chain reactions could be produced by man. The discovery of the fission of the uranium and thorium nuclei in 1939 and the subsequent observations which showed that several secondary neutrons were emitted in the process had given evidence that such a reaction might be produced. The story of this discovery of fission and the recognition of the tremendous energy released in the process forms one of the most fascinating chapters in physics.

It was the discovery of the neutron by Chadwick in 1932 that prompted Fermi to search for transuranic elements in the products of uranium and thorium bombarded by neutrons, thereby starting the chain of events leading to the discovery of fission. The complicated processes which occurred were not understood at first, and the years which followed the publication of Fermi's original paper found several groups of workers in different countries trying to understand what had happened. This problem was finally resolved by the observation by Hahn and Strassmann of the existence of a light element among the products of uranium bombarded by neutrons and by the insight of Meitner and Frisch into the tremendous energy evolved in the process. The existence of fission was immediately confirmed in the laboratories of several countries.

The years of confusion.—In his paper in 1934, Fermi[1] reported that uranium and thorium had been bombarded by neutrons and suggested that elements of atomic number greater than 92 had been produced. Because of the general instability of the heavy nuclei it was hoped that bombarding them with neutrons might give rise to successive transformations with the consequent production of transuranic elements. The activities of irradiated uranium and thorium were, of course, very complicated and the exact nature of the processes involved was not evident. The first effort to disentangle the half-life curves gave four activities for the products from uranium, and two from thorium. One of the activities of uranium was attributed to element 93. The chemical separations for element 93 were based on the assumption that, since uran-

[1] E. Fermi, *Nature* **133**, 898 (1934).

ium is in group VI of the periodic table, element 93 would lie in group VII and would be chemically similar to rhenium, masurium and manganese. This, as we shall see later, was a misleading assumption. By use of the radioactive isotopes 90 UX_2, 91 UX_1, 92 U, 88 $MsTh_1$ and 89 $MsTh_2$, Fermi showed that at least part of the newly formed active elements did not behave like any of the elements from 88 to 92. Since elements 86 (radon, a gas) and 87 (Eka Cs) could also be ruled out, he concluded that if any known reaction had taken place the new element must lie near the original uranium, but it must be on the other side of uranium. It must be one of the looked for transuranic elements! Noddack[2] in 1934 pointed out that the methods used did not disprove the possible existence of lighter elements among the bombardment products. Unfortunately, this idea was not followed up.

In the four years after the work of Fermi, many papers appeared which described the efforts of several groups to unravel the mystery. A review of this work is included in Turner's[3] article on nuclear fission. Among the investigators in this field were the group in Italy; Hahn, Strassmann and Meitner in Berlin; a group in Paris; several workers in Zurich; and others in this country. They separated the products chemically, measured the half-lives of the activities and studied some of the radiations. It is possible in retrospect to see how close several of these groups were to the discovery of fission.

Curie and Savitch[4] (1937-38), in their experiments with the products of neutron bombardment, used a copper absorber to eliminate beta rays of energy less than 2 Mev. Several half-lives were found, one of which—a 3.5 hour activity—resembled lanthanum. The reason that this was not recognized as lanthanum is explained in the following quotation from Turner[4a]. The "experiment indicated that the 3.5 hr stuff tended to concentrate in the portion first precipitated (i.e., not with lanthanum). This result prevented Curie and Savitch from being confronted with the perfect chemical identity of La and $R_{3.5\ hr.}$ It may be that the presence of the recently discovered active yttrium, also a fission product and also of a half-life of 3.5 hr,[5] was responsible for the observed fractionation." Hahn and Meitner[6] had suggested that

[2] I. Noddack, *Zeits. f. Angew. Chimie* **37**, 653 (1934).
[3] L. A. Turner, *Rev. Mod. Phys.* **12**, 1 (1940).
[4] Curie & Savitch, *J. de Phys.* (7) **8**, 385 (1937); Curie & Savitch, *J. de Phys.* (7) **9**, 355 (1938).
[4a] L. A. Turner, *Rev. Mod. Phys.* **12**, 1 (1940).
[5] C. Lieber, *Naturwiss.* **27**, 421 (1939).
[6] Hahn and Meitner, *Naturwiss.* **23**, 320 (1935).

a short-lived activity in the products of thorium bombarded by fast neutrons might be radium produced by an (n, α) reaction. Braun, Preiswerk and Sherrer[7] used an ionization chamber and a linear amplifier to look for these alpha particles, and reported finding alpha particles with energies greater than 9 Mev. However, since they measured energy by range they were not able to separate the fission fragments from the numerous natural alpha particles because the ranges are comparable.

Von Droste[8] (1938), also tried this experiment with uranium and thorium. The use of thin foils to eliminate the natural particles probably prevented him from getting the fission fragments in the ionization chamber and observing the large bursts of ionization that they produce.

The discovery of fission.—At the end of 1938, Meitner left Berlin because of the threatening storm cloud of Nazism, but she took with her the information gained in the Berlin laboratory. The indisputable evidence which was published by Hahn and Strassmann[9] in the first days of 1939 gave proof of the existence of an isotope of barium among the products of uranium bombarded by neutrons. This evidence was interpreted by Meitner and Frisch[10] to mean that the heavy uranium nucleus had divided into two light elements which separated with kinetic energies of the order of magnitude of 100 Mev. This was immediately and independently established by the experimental work of Joliot and Frisch. The evidence for the existence of barium was found in experiments extending the earlier work of the group in Berlin when Meitner was there. In studying the products of neutron irradiated uranium, four activities had been found that could be attributed only to isotopes of radium or barium. Some of the separated product, called Ra IV in the earlier work, was added to a solution containing barium and a small amount of Th X or Ms Th_1, both being used as tracer isotopes of radium. By means of fractional precipitations and crystallizations to separate radium isotopes from barium they found that Ra IV separated, not with radium, but with barium and was consequently an isotope of barium. The other three activities which had also been attributed to isotopes of radium could be reasonably inferred to be isotopes of barium, and the four daughter products which had been attributed to $_{89}$Ac would be really isotopes of lanthanum. The product form-

[7] Braun, Preiswerk and Scherrer, *Nature* **140**, 682 (1937).
[8] G. Von Droste, *Zeits. für Physik* **110**, 84 (1938).
[9] Hahn and Strassmann, *Naturwiss.* **27**, **11** (1939).
[10] Meitner and Frisch, *Nature* **143**, 239 (1939).

erly called Ac II (2.5 hr activity) was added to a solution containing lanthanum and $_{89}$Ms Th_2. The "Ac II" separated with the lanthanum and not with the $_{89}$Ms Th_2, thus giving additional evidence of the presence of barium.

Bohr was informed of the fact that barium had been found among the products of uranium bombarded with neutrons, and of the calculation of Meitner and Frisch that the uranium nucleus fissioned into two light nuclei with the release of about 200 Mev of energy. Shortly afterward Bohr came to this country to spend some time at Princeton. Immediately upon his arrival in this country on January 16, 1939 he informed his former student, J. A. Wheeler, of this idea; the news was spread by word of mouth to other physicists, including Fermi who was at Columbia. Experiments were undertaken at Columbia to find the fission fragments.

A discussion of the experimental results of Hahn and Strassmann and the hypothesis of uranium fission was given by Bohr and Fermi before the Fifth Washington Conference on Theoretical Physics on January 26, 1939. This created great excitement among physicists and in the popular press. Before the meeting adjourned on January 28, Roberts, Meyer and Hafstad[12] were able to demonstrate the existence of fission by the large pulses of ionization produced by the fragments in an ionization chamber.

Meanwhile Frisch[13] in Copenhagen had obtained physical evidence in support of the hypothesis of fission by observing the ionization produced by the recoil fragments. These observations were cabled to Bohr, then in this country, and were published in *Nature* in a letter dated January 16. Frisch used a uranium-lined ionization chamber connected to a linear amplifier. When a radium-beryllium source of neutrons was placed near the chamber, large pulses of ionization were observed. Surrounding the neutron source with paraffin increased the number of pulses by a factor of two. Experiments with thorium in place of uranium gave similar results, except that paraffin did not enhance the effect.

Joliot's[14] results, obtained independently and almost simultaneously, were published in the *Comptes Rendus* of January 30, 1939, and also reported the presence of the highly energetic recoil particles which emerged from the irradiated uranium and collected on a nearby plate. These were detected by their radioactivity.

[12] Roberts, Meyer and Hafstad, *Phys. Rev.* **55**, 416 (1939).
[13] O. R. Frisch, *Nature* **143**, 276 (1939).
[14] F. Joliot, *Comptes Rendus* **208**, 341 (1939).

Dunning[15] and his co-workers at Columbia, who had been told of the fission hypothesis before the Washington meeting, demonstrated the existence of these high energy particles on January 25,[16] after Fermi had left for Washington.

Other investigators who confirmed the fission hypothesis by demonstrating the large pulses of ionization from the fission fragments were Green and Alvarez[17] of California and Fowler and Dodson[18] of Johns Hopkins.

An independent method of showing the fission of uranium was used by Abelson,[19] who studied the X rays from a 72 hour activity. By critical absorption measurements these were shown to be the K X rays of iodine. The 72 hour period was shown to be due to tellurium, and the daughter substance, which was separated quantitatively, was shown to be 2.5 hr iodine. Similar results were obtained independently by Feather and Bretscher.[20]

The discovery of nuclear fission aroused so much interest among physicists that almost 100 papers were published about the subject within a year. In January 1940 the review article by L. A. Turner[21] summarized the information and gave a bibliography covering the work until almost the end of 1939.

The volume of published material on the subject of uranium fission fell off in 1940. This, we learn from Smyth (3.2),[22] was the result of a voluntary censorship system instituted by nuclear physicists through the National Research Council. Not until August 6, 1945 did the world know the outcome of the experiments on nuclear chain reactions. Much of the dramatic story of what went on behind the scenes has been told to us by Smyth. We must wait for further publication for the complete picture of what advances were made in nuclear physics and other branches of physics related to the successful solution of the chain reaction experiments.

[15] Anderson, Booth, Dunning, Fermi, Glasoe and Slack, *Phys. Rev.* **35**, 511 (1939).
[16] Date given by: Roberts, Meyer and Hafstad, *Phys. Rev.* **55**, 416 (1939).
[17] Green and Alvarez, *Phys. Rev.* **55**, 417 (1939).
[18] Fowler and Dodson, *Phys. Rev.* **55**, 417 (1939).
[19] P. Abelson, *Phys. Rev.* **55**, 418 (1939).
[20] Feather and Bretscher, *Nature* **143**, 516 (1939).
[21] L. A. Turner, *Rev. Mod. Phys.* **12**, 1 (1940).
[22] H. D. Smyth, "Atomic Energy for Military Purposes", Princeton University Press (1945). (Since we shall refer to this report quite frequently we shall abbreviate the reference by the word Smyth followed by the paragraph number.) Also *Rev. Mod. Phys.* **17**, 351 (1945).

CHAPTER 2

PRODUCTION OF FISSION

Although the early work on fission was done with neutrons, it was soon realized that the disturbance of uranium and thorium leading to fission might be produced by other nuclear agents. This chapter describes the various methods by which fission was attempted, the attendant success and, where possible, observed cross sections and energy thresholds. Since experimental data for the new nuclei, plutonium and neptunium, are unavailable, the problem of theoretically estimating the cross section for fission of these nuclei is left to chapter 8. For the same reason, the fission properties of the rare isotopes of uranium, 234 and 233, also will have to be estimated.

Slow neutrons.—Since the 238 isotope of uranium is so abundant, it is expected that most of the effects in uranium are due to that isotope. Bohr and Wheeler[23] first pointed out that the slow neutron effect in uranium probably should be ascribed to the rare isotope U^{235}. This was observed experimentally at Columbia[24] by the use of isolated uranium isotopes, separated by mass spectrometer methods. A value of 3×10^{24} cm^2 is given for the cross section for fission by slow neutrons on normal uranium.[25] Since the U^{235} is present to only one part in 140 in normal uranium, the cross section for slow neutron capture by the single isotope would be about 420×10^{-24} cm^2. Although no experimental data are available, theoretical considerations (see page 114) indicate that U^{233} and Pu^{239} will also fission with slow neutrons with similar cross sections.

It is expected on theoretical grounds that this slow neutron induced fission follows the $1/v$ law and has no pronounced resonances (see page 115). Experimentally this $1/v$ dependence has been verified by the Columbia group.[26]

Fast neutrons.—Probably most of the experimental work on fission has been done with medium fast neutrons. The D(d,n)

[23] Bohr and Wheeler, *Phys. Rev.* **56**, 426 (1939).
[24] Nier, Booth, Dunning and Grosse, *Phys. Rev.* **57**, 546, 748 (1940); Kingdon, Pollock, Booth and Dunning, *Phys. Rev.* **57**, 749 (1940).
[25] Dunning, Booth and Slack, *Phys. Rev.* **56**, 800 (1939).
[26] Anderson, Booth, Dunning, Fermi, Glasoe and Slack, *Phys. Rev.* **55**, 511 (1939).

and Be (d,n) reactions are convenient sources of neutrons for this purpose, giving approximately 2 Mev monochromatic energy and 6 Mev maximum energy neutrons respectively. Photo neutrons from the Be $(\mathrm{Ra}\gamma, n)$ reaction and neutrons from the $C(d,n)$ reaction give about 1 Mev and 0.5 Mev neutrons, respectively. The $\mathrm{Li}(d,n)$ reaction gives neutrons up to 17 Mev. By comparing yields from these different sources (see table 3), crude thresholds and yield curves can often be determined. More elaborate methods are necessary, however, for precise work.

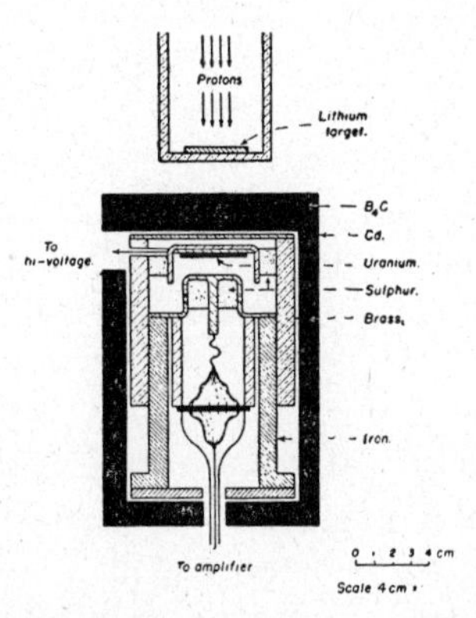

Fig. 1. Experimental arrangement for using variable maximum energy neutrons to observe fission thresholds. (Westinghouse Research Laboratories.)

To determine threshold values of neutron energies necessary for fission, neutrons from the $\mathrm{Li}(p,n)$ reaction can be utilized. Monochromatic protons of several million volts energy are directed onto a thick target of lithium. Figure 1 shows an experimental arrangement. The neutrons coming off the lithium target in the forward direction have an energy E_N determined by the (p,n) threshold energy E_t and the energy E_ρ of the protons.

$$E_N = \left(\frac{\sqrt{E\rho}}{8} + \frac{7}{8}\sqrt{E\rho - E_t} \right)^2$$

E_t for lithium has been measured[27] accurately to be 1.85 ± 0.02 Mev. If a thick target of lithium is used there will be only neutrons of lower energy. Consequently, if the energy of the protons is increased until fission is observed, the value of E_N necessary for fission can be calculated. This gives a maximum

[27] Haxby, Shoupp, Stephens and Wells, *Phys. Rev.* 58, 1035 (1940).

value of the fission threshold since extrapolation is uncertain. However, the thresholds seem quite definite. Such threshold curves for fast neutron induced fission of thorium and uranium are given in figures 2 and 3. The uranium curve has a background of neutron induced fission from U^{235} even though the number of slow neutrons was minimized by cadmium and boron shields.

For measuring yield curves, this integral method is not suitable. By using thin targets of lithium, however, an essentially monochromatic variable energy source of neutrons can be obtained. Work along these lines was started at Westinghouse Research Laboratory in 1940, and according to Smyth (12.44) this tech-

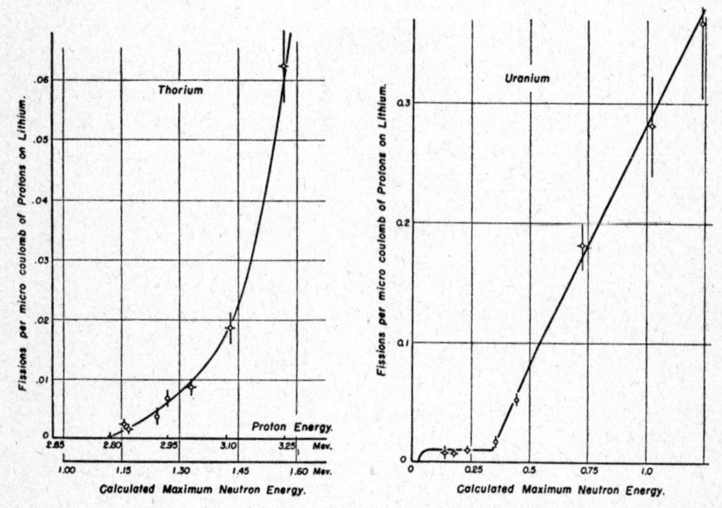

Fig. 2. Threshold curve for fast neutron induced fission in thorium. (Westinghouse Research Laboratories.)

Fig. 3. Threshold curve for fast neutron induced fission in uranium. The background probably is due to fission of U^{235}. (Westinghouse Research Laboratories.)

nique has been developed to the point where good yield curves can now be obtained within the range of neutron energies from 3 kv to 2 Mev.

The threshold values for fast neutron induced fission were observed at Westinghouse[29] as 0.35±0.1 Mev for uranium and 1.1±0.1 Mev for thorium.

[29] Haxby, Shoupp, Stephens, Wells and Goldhaber, *Phys. Rev.* 57, 1088 (1940); 58, 199 (1940).

The cross section for fission of uranium by neutrons of intermediate energies was given as 0.006×10^{-24} cm^2 at 0.5 Mev and 0.012×10^{-24} cm^2 at 1.0 Mev, by M. A. Tuve at the Princeton meeting of the American Physical Society in 1939. These values seem low compared to the value of 0.5×10^{-24} cm^2 reported by the Princeton group for neutrons of 2.4 Mev on uranium.[30] Ladenburg and his co-workers also give a value of 0.1×10^{-24} cm^2 for the cross section for fission of thorium for the same D(d,n) neutrons of about 2.4 Mev. They find these cross sections to be constant within 10 per cent between 2.1 and 3.1 Mev neutron energy. The Italian group[31] has measured the cross section for higher energy neutrons and infers that the uranium cross section remains roughly constant after 2 Mev until perhaps 8 Mev. It starts rising then, and reaches a value about 40 per cent higher at about 10 Mev. Thorium behaves similarly. Figure 4 roughly indicates the yield curves as interpolated between these observed values.

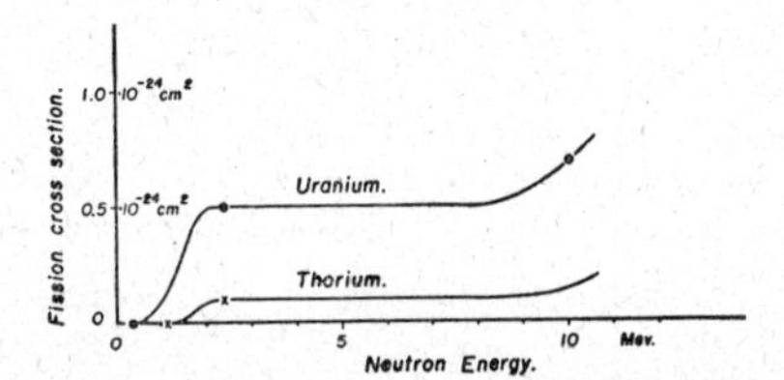

Fig. 4. Yield curves for fast neutron induced fission in uranium and thorium.

Radioactive alpha particles make observation of fission recoils difficult in other cases. However, at Columbia it was found that protactinium fissions with fast neutrons, with a cross section about thirty times that of thorium.[32] Since protactinium fission was not observed with Be(Raγ,n) photo neutrons, but was observed with D(d,n) neutrons, the threshold was estimated to be about 1 Mev neutron energy.

Ionium has been reported to fission with neutrons.[33] The cross section was about 2.7 times that for thorium [33a] using Be(d,n) neutrons with a 6.7 Mev deuteron source.

[30] Ladenburg, Kanner, Barschall and Van Voorhis, *Phys. Rev.* **56, 168** (1939).
[31] Ageno, Amaldi, Bocciarelli, Cacciapuoti and Trabacchi, *Phys. Rev.* **60,** 67 (1941).
[32] Grosse, Booth and Dunning, *Phys. Rev.* **56**, 382 (1939).
[33] Jentschke, Prankl and Hernegger, *Nature* 28, 315 (1940).
[33a] Curie and Joliot, *Ann. de Phys.* **19, 107 (1944)**.

Many other elements have been investigated, but no fission observed. Roberts[34] and his collaborators, using fast neutrons, found no fission in bismuth, rubidium, thallium, mercury, gold, platinum, tungsten, tin or silver although they estimated they could have detected one thousandth of the thorium effect.

Photo fission.—The fission of uranium and thorium by gamma rays, first observed at Westinghouse,[35] was considerably weaker than neutron induced fission. To eliminate the possibility that the observed fission was caused by photoneutrons or (p,n) neutrons, the absorption curve of the fission producing radiation was measured to be 0.53 per cm of lead, which agrees with that of the 6.3 Mev gamma ray used. Fission was observed with both the $F(p,\gamma)$ and $Li(p,\gamma)$ gamma ray, whose energies are 6.3 Mev and 17 Mev, respectively. The cross sections for the 6.3 Mev gamma ray were measured to be 3.5×10^{-27} for uranium and 1.7×10^{-27} cm^2 for thorium.[36] The errors were estimated at about 30 per cent. The threshold values for photofission have been determined with X rays from a betatron.[37] As electron energy is increased, the maximum energy of the X rays is increased also until photofission is observed. The values reported are 5.76 ± 0.1 Mev and 6.21 ± 0.15 Mev for U^{238} and Th^{232}, respectively. As the electron energy is increased, a peak is reached at 1.7 Mev above the threshold where the yield from uranium is 1.8 times that of thorium. Above this energy, the thorium yield remains roughly constant while the uranium yield keeps rising until at 13 Mev it is 8.6 times that of thorium. Using a 100 Mev betatron, the yields rise rapidly to about 20 Mev and then drop off up to 100 Mev. In this range the uranium yield is about twice that of thorium, giving a maximum of about 260 fissions per roentgen from a 4 mg/cm^2 layer of uranium oxide.[38] No fission recoil fragments were observed in a balanced ionization chamber lined with lead, thallium, bismuth, tungsten, gold or samarium, which was exposed to 20r of 100 Mev betatron X rays. An effect of one thousandth of that produced in thorium could have been detected.

Charged particle induced fission.—It is difficult to produce fission by charged particles since these heavy nuclei have such a

[34] Roberts, Meyer and Hafstad, *Phys. Rev.* **55**, 416 (1939).
[35] Haxby, Shoupp, Stephens and Wells, *Phys. Rev.* **58**, 92 (1940).
[36] Haxby, Shoupp, Stephens and Wells, *Phys. Rev.* **59**, 57 (1941).
[37] H. W. Koch, "Thresholds of Photo Fission", Univ. of Illinois Thesis (1944).
[38] G. S. Klaiber, *Bull. Am. Phys. Soc.* **21**, 15 (1946).

large coulomb repulsion. However, fission has been produced by energetic charged particles in both uranium and thorium.

Dessauer and Hafner[39] bombarded thick uranium and thorium targets with 6.9 Mev protons from a cyclotron. They detected fission by catching the recoil fission fragments on neighboring plates. One plate collected fission fragments from the front surface of the target, giving the proton induced fission yield plus any neutron induced fission. A second catcher plate collected fission fragments from the back of the target, measuring only the neutron induced fission. The difference gave the true proton induced fission yield. The yields were similar for uranium and thorium. The threshold was around 5.8 Mev. Gant found fission produced by bombardment of uranium and thorium [40] with deuterons of over 8 Mev. Jacobsen and Lassen measured the yield curve and obtained cross sections of $2.2 \pm 0.1 \times 10^{-26}$ cm^2 and $1.5 \pm 0.7 \times 10^{-26}$ cm^2, for uranium and thorium, respectively, at 9 Mev deuteron energy.[41] This ratio of uranium to thorium cross sections was checked by Krishnan and Banks.[42]

Table 1

FISSION THRESHOLDS

Compound Nucleus	Threshold Energy for exciting fission		Cause of excitation	Target
$_{90}Th^{232}$	6.21 ±0.15 Mev	(1)	γ	Th^{232}
$_{90}Th^{233}$	1.1 ±0.1	(2)	*n*	Th^{232}
$_{91}Pa^{232}$	~ 1	(3)	*n*	Pa^{231}
$_{91}Pa^{233}$	< 6.9	(4)	*p*	Th^{232}
$_{91}Pa^{234}$	~ 8	(5)	*d*	Th^{232}
$_{92}U^{236}$	< 0	(6)	slow *n*	U^{235}
$_{92}U^{238}$	5.76 ±0.1	(1)	γ	U^{238}
$_{92}U^{239}$	0.35 ±0.1	(7)	*n*	U^{238}
$_{93}Np^{239}$	< 6.9	(4)	*p*	U^{238}
$_{93}Np^{240}$	~ 8	(5)	*d*	U^{238}

(1) H. W. Koch, Univ. of Illinois Thesis, 1944: "Threshold of Photofission"
(2) Haxby, Shoupp, Stephens, Wells and Goldhaber, *Phys. Rev.* **57**, 1088 (1940).
(3) Grosse, Booth and Dunning, *Phys. Rev.* **56**, 382 (1939).
(4) Dessauer and Hafner, *Phys. Rev.* **59**, 840 (1941).
(5) Jacobsen and Lassen, *Phys. Rev.* **59**, 1043 (1941).
(6) Nier, Booth, Dunning and Grosse, *Phys. Rev.* **57**, 546 (1940).
(7) Haxby, Shoupp, Stephens and Wells, *Phys. Rev.* **58**, 199 (1940).

[39] Dessauer and Hafner, *Phys. Rev.* **59**, 840 (1941).
[40] D. H. T. Gant, *Nature* **144**, 707 (1939).
[41] Jacobsen and Lassen, *Phys. Rev.* **58**, 867 (1940); *Phys. Rev.* **59**, 1043 (1941).
[42] Krishnan and Banks, *Nature* **145**, 860 (1940).

Fermi and Segrè[43] bombarded a thick ammonium uranate target for about one minute with several milliamperes of 32 Mev alpha particles from the Berkeley cyclotron. They detected that fissions had been produced by observing the radioactivity of the fission products.

Natural fission.—Various observations have been attributed to natural fission in uranium and estimates of half-life have been made.[44] The best estimate, however, seems to be based on Seaborg's observation [45] that Pu^{239} is found in a concentration of one part in 10^{14} in pitchblende. If this is in equilibrium then it must be produced at the rate it is decaying. Its production can be estimated by assuming with Seaborg that it is produced by the capture in U^{238} of secondary neutrons from the natural fission of U^{235}.

$$U^{235} \xrightarrow[\text{spontaneous}]{\lambda_1} \text{fragments} + \text{several neutrons}$$

$$n + U^{238} \longrightarrow U^{239} \longrightarrow Np^{239} + \beta$$

5 volt resonance or slow neutrons

$$\longrightarrow Pu^{239} + \beta$$

$$_{94}Pu^{239} \xrightarrow[2.4 \times 10^4 \text{ yr}]{\lambda_2} {}_{92}U^{235} + \alpha \qquad \text{(see chapter 6)}$$

Assuming the U^{238} to be one tenth of the pitchblende and that one neutron per natural fission is captured as indicated, the activity of $Pu = \lambda_2 N_{Pu}$ = production of $Pu = \lambda_1 N_{U^{235}}$.

$$\lambda_1 = \lambda_2 \frac{N_{Pu}}{N_{U^{235}}} = \lambda_2 \frac{1}{10^{11}}$$

Hence the half-life of U^{235} for spontaneous fission is 10^{11} times the alpha particle half-life of Pu^{239}, or about 10^{15} years. This is a lower limit since, depending on the impurities in the pitchblende, fewer neutrons may be captured in uranium. We have also assumed that the 235 isotope of uranium is the most unstable towards natural fission. The theoretical estimate of the half-life of this process (10^{15} years) is discussed in chapter 8.

[43] Fermi and Segrè, *Phys. Rev.* 59, 680 (1941).
[44] Thibaud and Moussa, *Comptes Rendus* 208, 562, 744 (1939); Flerov and Petrjak, *Phys. Rev.* 58, 89 (1940); *J. Phys. USSR* 3, 275 (1940).
[45] G. T. Seaborg, *Chem. and Eng. News* 23, 2192 (1945).

Production of compound nucleus in fission.—Theoretical considerations (see page 92) indicate that fission produced by particles is preceded by the formation of a compound nucleus, as in other transmutations. This compound nucleus then breaks up into the resultant fragments. The life of a compound nucleus, sufficiently excited so that it will fission, is expected to be extremely

Table 2

FISSION PRODUCTION CROSS SECTIONS
(in units of 10^{-24} cm^2)

Cause	Target Nucleus				
	Uranium		Thorium 232	Prot-actinium 231	Ionium 230
	238	235			
neutrons slow	no	420 (1) thermal energy	no	no	
fast	0.5 (2) at 2.4 Mev	yes	0.1 (2) at 2.4 Mev	3 (3)	0.3 (4)
gamma rays	0.0035 (5) at 6.2 Mev		0.0017 (5) at 6.2 Mev		
protons	yes (6)		yes (6) similar to uranium		
deuterons	0.22 (7) at 9 Mev		0.015 (7) at 9 Mev		
alpha particles	yes (8)				

(1) Nier, Booth, Dunning and Grosse, *Phys. Rev.* **57**, 748 (1940).
(2) Ladenburg, Kanner, Barschall and Van Voorhis, *Phys. Rev.* **56**, 168 (1939)
(3) Grosse, Booth and Dunning, *Phys. Rev.* **56**, 382 (1939).
(4) Curie and Joliot, *Ann. de Physique* **19**, 107 (1944).
(5) Haxby, Shoupp, Stephens and Wells, *Phys. Rev.* **59**, 57 (1941).
(6) Dessauer and Hafner, *Phys. Rev.* **59**, 840 (1941).
(7) Jacobsen and Lassen, *Phys. Rev.* **59**, 1043 (1941).
(8) Fermi and Segrè, *Phys. Rev.* **59**, 680 (1941).

Table 3

NEUTRON SOURCES OF DIFFERENT ENERGIES

Reaction photo neutrons	Q reaction energy	E_n maximum neutron energy for RaC gamma rays		Comments
D^2 (γ,*n*)	–2.18 Mev	0.02 Mev		
Be^9 (γ,*n*)	–1.63	0.5		
(*d,n*) reactions		for deuterons of 1 Mev	for deuterons of 16 Mev	
C^{12} (*d,n*)	—.28	0.7(0°) 0.58(90°)	13.2 Mev(0°)	possible weak higher energy neutrons
D^2 (*d,n*)	3.18	4.05 3.1	10.6	monochromatic
Be^9 (*d,n*)	4.2	5.3	16.7	max. energy
N^{14} (*d,n*)	5.1	5.9	18.3	possible weak higher energy
F^{19} (*d,n*)	10.8	11.0	23.6	max. energy
B^{11} (*d,n*)	13.5	13.7	25.8	max. energy
Li^7 (*d,n*)	15.0	15.5	26.6	max. energy
(*p,n*) reactions		for protons of 1.85 Mev	for protons of 4.0 Mev	
Li^7 (*p,n*)	—1.62	0.03 (0°)	2.75 (0°)	Probably monochromatic if thin target is used
Be^9 (*p,n*)	—1.83	protons of 2.03 Mev 0.03 (0°)	2.3 (0.°)	
(α, *n*) reaction		for Po α's	RaC′ α's	
Be^9 (α,*n*)	5.8	11 Mev	13.7 Mev	strongest intensity ½-1½ Mev

short. Experiments to measure this delay in fission have succeeded in putting an upper limit of 5×10^{-13} sec on the half-life of this compound nucleus.

Feather[46] collected the recoiling fission fragments produced in a thin uranium foil by fast neutrons. He observed 19 per cent more recoil activity on the collector on the forward side (relative to the neutron velocity) than on the backward side. He interpreted this as showing that the fission breakup occurred while the uranium nucleus still had appreciable velocity of recoil left from the initial neutron impact. Since the time estimated as necessary for the struck uranium atom to lose its momentum is 5×10^{-13} seconds, this is an upper limit on the time in which the fission occurred.

Since it is the compound nucleus which fissions, the bombarding particle not only transfers its kinetic energy to the transmutation energy, but changes the target nucleus into the compound nucleus. This adds to the excitation energy of the compound nucleus the binding energy of the bombarding particle (with respect to the target nucleus) and, consequently, the threshold energies measured can be simply interpreted only in the case of photofission. In the other cases the observed threshold energy must be added to the binding energy and then applied to the compound nucleus. Table 1 summarizes some of these data. Binding energies and more complete data are given in chapter 7.

The other results discussed in this chapter are summarized in table 2 which gives the observed production of fission for various nuclei and agents.

[46] N. Feather, *Nature* **143,** 1027 (1939).

CHAPTER 3

FISSION FRAGMENTS

Enormous energy of recoil fragments.—Meitner and Frisch[47] realized that if fission of uranium were to occur, the energy released in the process would be large and should consequently give rise to high energy recoil fragments. These are easy to detect, and very soon Frisch,[48] and within a few days Joliot,[49] detected these fission recoil particles.

A crude but simple picture accounts for the energy release. At the instant of fission, two highly positively charged groups of nuclear particles break apart. They have typical charge numbers of $Z_1 = +54$ and $Z_2 = +38$, and are initially at a distance d, of about 1.5×10^{-12} cm apart (center to center). The energy of mutual electrostatic coulomb repulsion, E, is simply

$$E = \frac{Z_1 Z_2 e^2}{d} \approx 200 \text{ Mev.}$$

This energy divided between the recoiling particles endows them each with roughly 100 Mev. Not all this energy goes into kinetic energy, and the uneven splitting makes one particle heavier than the other, thus reducing its share of the energy. We shall see that experimental observations are consistent with this picture.

This relatively great energy makes the recoil fragments easy to detect and observe. The total ionization produced is more than ten times that of the most energetic alpha particle. Consequently, ionization chambers easily detect individual fission particles. Cloud chambers show dense tracks when recoil particles pass. Photographic plates also show recoil particle tracks. The recoil obviously tears fission fragments loose from the uranium surface. They can then be collected on a neighboring sheet and detected by their radioactivity.

All these methods have been used to detect fission and to measure the range, energy, energy loss and other characteristics of the fission fragments.

[47] Meitner and Frisch, *Nature* **143**, 239 (1939).
[48] O. R. Frisch, *Nature* **143**, 852 (1939).
[49] F. Joliot, *Comptes Rendus* **208**, 341 (1939).

Range of recoil fragments.—Many early fission experiments utilized the relatively simple method of collecting the recoil fragments and observing their radioactivity. Joliot,[50] Thibaud[51] and McMillan[52] used this method to measure the range of the recoil fragments. McMillan stacked thin aluminum foils (0.57 mg/cm^2 thick) and laid them on top of a uranium slab. On irradiation with neutrons, the fission fragments recoiled into the stack of foils. By measuring the radioactivity in each foil separately he obtained a rough absorption curve. The activity curve dropped to background activity at a depth in the stack corresponding to a range of 2.2 ± 0.2 cm air equivalent.

Joliot found a range of about 10 microns in UO_2, whereas Thibaud observed a range of about 5 microns in aluminum.

At Columbia[53] the range in air was measured by the use of a flat, thin ionization chamber. The source of fission recoils was fixed relative to the ionization chamber, but the pressure of the air could be varied to get a curve showing the number of fragments penetrating different equivalent thicknesses of standard air. They found evidence for two groups of recoils of ranges 2.2 ±0.1 cm and about 1.5 cm. Range values given by Haxel[54] are 1.8 ±0.24 and 1.5 ±0.2 cm.

Several investigators[55] observed cloud chamber tracks of fission recoils soon after the discovery of fission. Corson and Thornton estimated a range of about 3 cm. However, the most accurate work has been done by the Danish group[56] in Bohr's laboratory. They observed two groups of recoil tracks with ranges of 2.5 ±0.2 cm and 1.9 ±0.2 cm.

These short ranges are obviously due to the high charge of the fragments, which results in a large energy loss in passing through matter. They also explain why fission fragments were not detected in previous searches for high energy alpha particles from the postulated reaction U (n,α). A thin absorber was ordinarily placed over the uranium to cut out the natural alpha particles, which have an energy of 4.13 Mev and a range of 2.37 cm in air. This absorber automatically also cut out the fission fragments.

[50] F. Joliot, *Comptes Rendus* **208**, 341 (1939); *J. de Phys.* (7) **10**, 388 (1939).
[51] Thibaud and Moussa, *Comptes Rendus* **208**, 652 (1939).
[52] E. McMillan, *Phys. Rev.* **55**, 510 (1939).
[53] Booth, Dunning and Glasoe, *Phys. Rev.* **55**, 982 (1939).
[54] O. Haxel, *Zeits. für Phys.* **112**, 681 (1939).
[55] F. Joliot, *Comptes Rendus* **9**, 647 (1939); W. Perfilov, *C. R. Acad. Sci. USSR* **23**, 896 (1939); Corson and Thornton, *Phys. Rev.* **55**, 409 (1939).
[56] Bøggild, Brostrøm, T. Lauritsen, "Cloud Chamber Studies of Fission Fragment Tracks," *Det. Kgl. Danske Vid. Selsk.* **18**, 4 (1940).

Fission recoils have also been detected in photographic plates placed close to uranium and exposed to neutrons.[57]

Energy of recoil fragments.—The most direct way to measure the energy released in the fission process is to measure the heat produced in a calorimeter by a known number of fissions. Henderson[58] did this by measuring simultaneously the rise in temperature of a calorimeter filled with 13.36 gm of metallic uranium and the number of fissions in an adjacent ionization chamber containing a thin layer of uranium (54 micrograms) when the apparatus was exposed to neutrons. The "atomic powers" observed were about 40 microwatts. After making appropriate corrections and calculations, Henderson obtained a value of 177 ± 2 Mev for the energy per fission. This excludes energy emitted in the form of hard gamma rays and neutrinos but includes soft radiation and short-lived beta emission energy. Consequently, the kinetic energy of the recoiling particles should be somewhat less than this.

A more direct method of measuring recoil energies is to observe the ionization produced by the recoils, which can be readily compared to the ionization produced by alpha particles of known energy. This has been done by many investigators and the following results are apparently the most carefully measured:

	E_1	E_2	E
Haxel [59]	52 Mev	74 Mev	
Booth, Dunning and Slack [60]	50	80	
Jentschke and Prankl [61]	61	98	
Kanner and Barschall [62]	65	97	159

The quantities E_1 and E_2 are ionization energies (assuming the mean energy per ion pair is the same for fission recoil particles as for alpha particles) of the most numerous particles in each of the two groups of particles usually observed. Jentschke and Prankl's values agree well with those of Kanner and Barschall, whose results seem quite reliable. Kanner and Barschall also measured both recoils simultaneously and got a total ionization energy, E, of 159 Mev. Since 12 Mev is a reasonable average energy expended in beta ray emission by the radioactivity recoil particles, this total ionization energy plus reasonable beta ray energy is 171 Mev, which compares well with Henderson's calorimeter value of 177 Mev per fission.

[57] Myssowsky and Idanoff, *Nature* **143**, 794 (1939); Lark-Horovitz and Miller, *Phys. Rev.* **59**, 941 (1941).
[58] M. C. Henderson, *Phys. Rev.* **56**, 103 (1939); **58**, 200, 774 (1940).
[59] O. Haxel, *Zeits. für Phys.* **112**, 681 (1939).
[60] Booth, Dunning and Slack, *Phys. Rev.* **55**, 981 (1939).
[61] Jentschke and Prankl, *Naturwiss.* **27**, 134 (1939).
[62] Kanner and Barschall, *Phys. Rev.* **57**, 372 (1940).

The half width of the peaks corresponding to the two groups of recoils is about 18 Mev, so that the groups overlap. The half width of the peak for total ionization is 30 Mev. This is not entirely experimental error but is a result of the statistical fashion in which the fission breakup occurs. However, the number of particles in each group is the same, indicating that in each fission one recoil is of one group while the other recoil is of the second group. Consequently, we can calculate the masses of the most probable fragments by assuming conservation of momentum and by taking Kanner and Barschall's values of E_1 and E_2. These masses are 96 and 143, in agreement with the observed fission product masses listed in table 4.

Lark-Horovitz[63] has investigated the fission recoils, using the high energy neutrons from Li (d,n) in a cyclotron. His preliminary results suggest three groups of fragments with energies of 52, 86 and 110 Mev and some recoils with energies as high as 135 Mev. Total energy measurements gave 172 Mev, with some values up to 200 Mev. These effects were interpreted by Lark-Horovitz as evidence for (a) asymmetric fission, 52 and 110 Mev peaks, (b) symmetric fission, 86 Mev peak, and (c) triple fission, total energies of 200 Mev. Other evidence in fission product identification confirms the increasing probability of symmetric fission at higher neutron energies.

Energy loss of recoil fragments.—Much theoretical work has gone into the interpretation of experimental results to give information on the mechanisms and rate of energy loss in matter of these

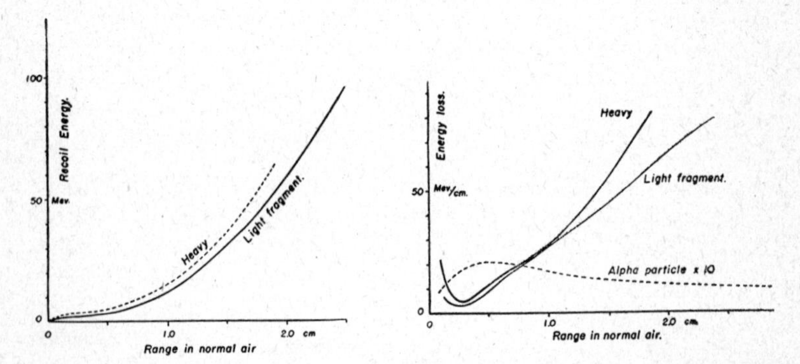

Fig. 7. Range energy curves for typical fission fragments. (Calculated from data of Bøggild, Brostrøm and T. Lauritsen.)

Fig. 8. Energy loss curve for typical fission fragments. (Calculated from Fig. 7.)

[63] Lark-Horovitz and Schreiber, *Phys. Rev.* **60,** 156 (1941).

heavy, highly charged fragments. The principal work has been done in Bohr's laboratory[64] with a statistical analysis of their beautiful cloud chamber pictures of fission recoil tracks (see figures 5 and 6).

The tracks are easily distinguishable from alpha particle tracks by their heavy ionization and the number of heavily ionized "branches" due to collisions with gas nuclei. By counting the branchings as a function of residual range and using appropriate formulae, it is possible to show the existence of two groups of recoils and to calculate the range-energy curve near the end. This curve can be extended to the initial conditions and an energy loss curve deduced. Such curves are shown in figures 7 and 8.

An important consideration in interpreting the energy loss is the effective nuclear charge of the fragment. The recoils are not entirely stripped of their electrons; those electrons remain whose binding energies exceed the kinetic energy of electrons in the material relative to the recoiling nucleus. Lamb[65] estimated the initial charge to be $+17e$ for the light fragment and $+13e$ for the heavy one. A measurement of the radius of curvature of fission recoils in a magnetic field has been made by Lassen.[66] He identified the group to which the particles belong by their ionization energy in the ionization chamber detector. The low energy group of recoils contains the fragments of heavy mass. Since the recoils have equal momenta, the $H\rho$ (magnetic field times the radius of curvature) should be proportional to the reciprocal of the effective charge. The low energy, heavy mass group was found to have slightly lower $H\rho$ and consequently larger charge. But our previous consideration indicated that the lighter mass had the higher velocity and hence a larger charge. Lassen suggests that the recoil particles were not in equilibrium with matter in the $H\rho$ measurements. In any case, Lassen calculates the light group to have a charge of $+20e$ and the heavy group $+22e$.

In connection with this question of charge Lamb[67] has suggested that the difference in energy loss and consequent range of the two groups of fragments might be primarily due to a difference

[64] Theoretical articles: N. Bohr, *Phys. Rev.* 58, 654 (1940); **59**, 270 (1941); W. E. Lamb, *Phys. Rev.* 58, 696 (1940), 59, 687 (1941); Knipp and Teller, *Phys. Rev.* **59**, 659 (1941).
Experimental articles: Bøggild, Brostrøm and T. Lauritsen, *Phys. Rev.* 59, 275 (1941); Bohr, Bøggild and T. Lauritsen, *Phys. Rev.* **58**, 839 (1940); Bøggild, *Phys. Rev.* **60**, 827 (1941); Brostrøm, *Phys. Rev.* **58**, 651, **59**, 275 (1940).

[65] W. E. Lamb, *Phys. Rev.* **59**, 687 (1941).

[66] N. O. Lassen, *Phys. Rev.* **68**, 142 (1945).

[67] W. E. Lamb, *Phys. Rev.* **59**, 687 (1941).

in shell structure which will cause the fragments to pick up electrons differently. He calculates the range of typical fragments in a 160 Mev fission to illustrate that there is no simple relation between mass and range.

Fragment	Range
$_{36}Kr^{87}$	1.4 cm
$_{42}Mo^{100}$	0.74
$_{50}Sr^{124}$	0.89
$_{52}Ba^{143}$	0.9

The high charge not only produces heavy ionization, which is predominantly responsible for the loss of energy in the first part of the range, but also makes nuclear collisions highly probable at relatively high velocities, so that near the end of the range such collisions will produce appreciable curvature of path and will be mainly responsible for the stopping effect.

CHAPTER 4

FISSION PRODUCTS

When fission occurs in one of the heavier elements of the periodic table the fragments into which it splits are atoms of the elements occurring in the middle region of the table. These atoms are at first highly unstable and change by radioactive disintegration into the stable nuclei observed in nature. It is the purpose of this chapter to describe qualitatively some of the processes by which these transformations occur and to indicate some of the methods available for determining what nuclei are produced and the probability of their production.

A variety of questions require information on the type of nuclei produced by fission and on the radiations they emit. Because of the possibility that these products may absorb many of the neutrons needed to produce fission, it might be difficult to realize a chain reacting pile that could operate for a reasonable period of time (see page 150). There is also the question of the effects on personnel of the radioactivity of the products in chain reacting piles. Information of this type would be needed to determine the duration and intensity of radiations in regions that have experienced large scale exposure to fissioning materials and their products. In the production and use of radioactive tracers described on page 242 knowledge of the efficiency of production and half-lives of the obtainable products would of course be necessary. The distribution in mass of fission products is useful for checking theories of the mechanism of fission as explained on page 89.

Immediately following the work of Hahn and Strassmann,[68] which definitely identified fission as such, a large number of papers appeared in the literature in which their observations were confirmed and more products identified. Seaborg compiled a table of isotopes that includes all the products that had been observed prior to June 1, 1944.[69] This table also includes the types of radioactivity observed, their half-lives, energies and methods of measurement; also given are the types of fission that produced these products and a complete bibliography of references. E. Segrè has

[68] Hahn and Strassmann, *Naturwiss.* **27**, **11** (1939).

[69] G. T. Seaborg, *Rev. Mod. Phys.* **16**, 1 (1944). He does not report results of government-sponsored research.

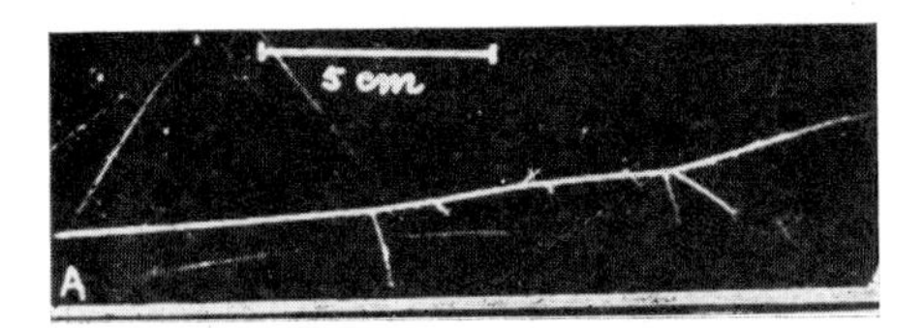

Fig. 5. Cloud chamber picture of a fission fragment recoiling in hydrogen. The large probability of nuclear collisions is clearly shown by the numerous branches. The range is amplified by the use of low pressure of hydrogen. The cloud chamber was filled with hydrogen gas plus vapor of one third alcohol and two thirds water, making a total pressure of 13 cm of mercury. (Bøggild, Brostrøm and T. Lauritsen, *Det. Kgl. Danske Vid. Selshab, Math-pys. Medd.* XVIII 4 (1940))

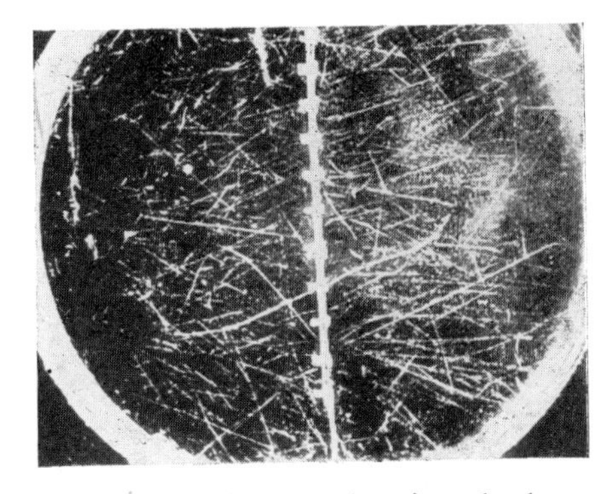

Fig. 6. Cloud chamber picture showing both recoil fragment tracks of a fission originating on the foil. The cloud chamber was filled with argon and water vapor to a pressure of 20 cm and the mica foil was 1.2 mg/cm^2 thick. (Bøggild, Brostrøm and T. Lauritsen).

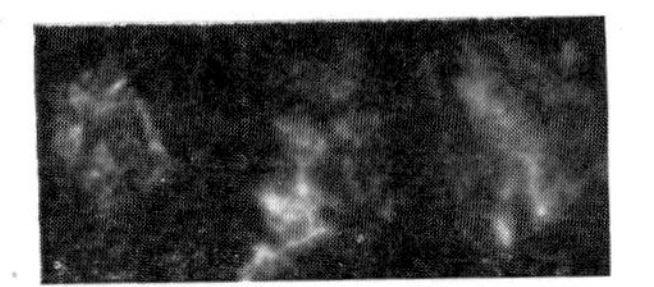
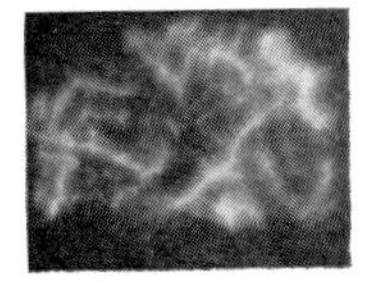

Fig. 50. Autoradiograph of slabs of silicon with small amounts of aluminum impurity. Each slab was irradiated with fast neutrons and placed on film. The exposed streaks are where the aluminum was changed to radioactive sodium which reveals its presence by the emission of beta rays. (See page 248.)

prepared a chart of all the known isotopes, using the results of more recent unreported research.[70] This chart gives no references.

Nature of fission fragments.—The previous chapter describes the process of fission itself, and it will be discussed in more detail in chapters 7 and 8. Immediately after the fissioning nucleus breaks up, we have two particles flying apart with enormous energies of approximately 80 Mev each. The sum of the charges on these fragments is probably equal to the charge on the original nucleus. The sum of their masses is two or three mass units less than the mass of the original owing to the instantaneous emission of neutrons upon fission (see page 44). Initially, as we have seen roughly 15 electrons are missing from the electron shell, but as the particles slow down they acquire more electrons until they come to rest and are complete atoms.

However, the nuclei of these new atoms are highly unstable. The ratio of neutrons to protons in ${}_{92}U^{235}$ is about 1.5, and it is reasonable to expect that the direct products of fission would have approximately the same ratio, not allowing for the instantaneous neutrons emitted. In a typical reaction in which the protons split 40 to 52 we would get

$$ {}_{92}U^{235} + {}_{0}n^{1} \rightarrow {}_{40}Zr^{97} + {}_{52}Te^{137} + {}_{0}n^{1} + {}_{0}n^{1} \qquad (1) $$

The heaviest known stable isotope of zirconium is ${}_{40}Zr^{96}$, and for tellurium it is ${}_{52}Te^{130}$. Perhaps a more realistic viewpoint, anticipating the actual decay process, is to say that if a nucleus of mass 97 is to be stable it must have at least 42 protons, and a nucleus of mass 137 requires 56 protons. These nuclei would be stable ${}_{42}Mo^{97}$ and ${}_{56}Ba^{137}$. In either case it is evident that the neutron-proton ratio of the direct fission products must be reduced by some transformation until a stable nucleus is reached. Two processes are available: beta emission and neutron emission.

Decay mechanisms.—The most important decay mechanism is the emission of beta rays, when a neutron in the nucleus is converted into a proton, and an electron is ejected. These electrons, or beta rays, do not all have the same energy but may have any energies from zero up to some maximum characteristic of the particular nucleus. An important feature of beta rays is that, like other radioactive processes, their rate of decay follows an exponential law. The half-life of this decay is a characteristic property of the nucleus and may be measured to aid in identifying the product. Evidently a beta ray emission decreases the number of neu-

[70] Revised May 15, 1945.

trons by unity and increases the number of protons a like amount.

The second process is neutron emission. This is much less frequent and occurs only about once for every 500 beta emissions. Since this process is not used in identifying fission products, it is merely mentioned here. It will be discussed in chapter 5.

Fission products would not be expected to disintegrate by K-electron capture or positron emission, since both processes increase the neutron-proton ratio rather than decrease it. Similarly, alpha particles which would decrease the number of neutrons and protons by equal amounts, and hence increase the ratio slightly, have not been observed.

An effect observed to accompany most other types of radioactive disintegration is the emission of gamma rays. This does not change the neutron-proton ratio but is merely a mechanism by which a nucleus can emit the excess energy resulting when a particle emission leaves the resultant nucleus in an excited state. The gamma ray is usually observed so soon after the particle emission that it appears as if it were part of the same process. Usually the particle is ejected from the parent nucleus and the daughter nucleus then emits the gamma ray. If the nuclear transition that produces the gamma ray is strongly forbidden, the probability for the emission will be small and a measurable half-life may be observed. A nucleus that has a measurable half-life for this gamma radiation is said to be in a *metastable* state. A nucleus in such a state is said to be *isomeric* with respect to the ground state into which it can fall by emitting a gamma ray. Hence, *isomers* are nuclei that have the same number of neutrons and protons but different energy states. The transition that produces the gamma ray is an *isomeric transition.* The fact that gamma rays have line spectra of immeasurable line breadth is evidence of well-defined nuclear energy levels.

Secondary effects due to the emission of gamma rays can also be used to identify fission products. Frequently a gamma ray leaving a nucleus ejects a photoelectron from one of the X ray levels of the atom in which it originated. This process is called *internal conversion.* Internally converted electrons have a discrete energy spectrum, since they are produced from a definite energy level in the atom by gamma rays of definite energy. A further consequence of internal conversion is that when the vacancy in the X ray level is filled, characteristic X rays are produced, and these can be used to identify the element.

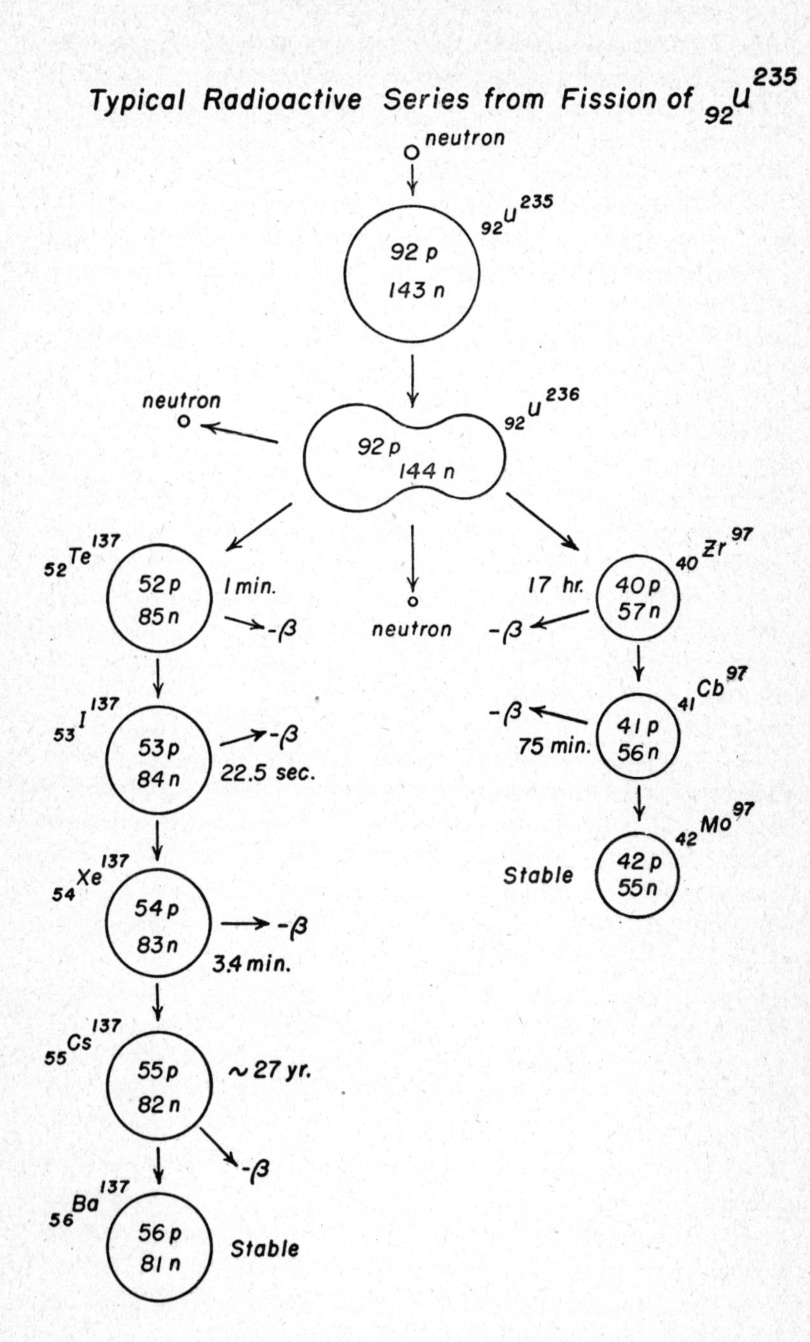

Fig. 9. Typical radioactive series from fission of U^{235}.

Table 4

Method of production	Th-n			Th-n U-n	Th-n U-n	U-n		U-n	Th-n U-n	U-n	U-n	U-n	U-γ U-n Th-n	U-n	U-n	U-n		U-n		U-n Th-n	
Atomic No. \ Mass.	80	81	82	83	84	85	86	87	88	89	90	91	92	93	94	95	96	97	98	99	100
34Se	S		S	A β 30m																	
35Br	A * 4.4 h; A β 18m 2.0	S		A β 140m 1.05	A β 33m 5.3	A β 3m		B β 50s													
36Kr	S		S	A * 113 m; S	S	A β 4.6h 1.0	S	B β 74m 4.0; B β >5y 0.8	A β 3h 2.5	A β 2.7m	B β Short	B β 5.7s	D β 1.5s		D β Short	B β 1.5s					
37Rb						S		A β $6.3 \cdot 10^{10}$y 0.13	A β 17.5m 5.1	A β 15m 3.8	B β Short	B β Short	D β 80s		D β Short	B β Short					
38Sr					S		S	A * 2.7 h; S	S	A β 55d 1.52	B β ~30y 0.65	B β 9.7h 1.3–3.2	D β 2.7h		D β 2m	B β 7m					
39Y										S	A β 60h 2.2	B * 50m; B β 57d 1.6	D β 3.5h 3.5		D β 20m	B β 11.5h 3.4					
40Zr											S	S	S		S	B β 65d .42–1.0	S	B β 17h 2.2			
41Cb														S		B * 9.0 h; B β 35d 0.15		B β 75m 1.4			
42Mo													S		S	S	S	S	S	B β 67h 1.5	S
43																				B * 6.6h	
44Rh																	S		S	S	S
Branching ratio.																		6.1			

Table 4 (Cont.)

Method of production		U-n	U-n	U-n		U-n						U-n Th-n	U-n Th-n			U-n		U-n	U-n	
Atomic No. \ Mass.	100	101	102	103	104	105	106	107	108	109	110	111	112	113	114	115	116	117	118	119
42Mo	S	B β 14.6m 1.8	D β 12m			B β Short														
43		B β 14m 1.1	D β <1m			D β Short														
44Ru	S	S	S		S	B β 4h 1.5						D β 4m 4								
45Rh				S		A β 35h 0.6						D β 24m 1.2								
46Pd			S		S	S	S		S		S	A β 26m 3.5	A β 21h							
47Ag								S		S		A β 7.5d 0.8, 1.0	A β 3.2h 3.6							
48Cd							S		S		S	S	S	S	S	A β 2.33 d 1.25 / A β 43 d 0.95 –1.7	S	A β 2.83h	D * 48.7m	
49In														S		A * 4.5h / S		A β 117m 1.95	M	
50Sn													S		S	S	S	S	S	S

Table 4 (Cont.)

Method of production	U-n	U-n	U-n	Th-n?	U-α U-n	U-α Th-n U-n	U-α U-n Th-n		U-n Th-n	U-α U-n	U-n	U-n	U-n Th-n	U-n Th-n	U-n Th-n U-γ	U-n Th-n U-γ	U-n	U-n	U-n		U-n
Mass / Atomic No.	127	128	129	130	131	132	133	134	135	136	137	138	139	140	141	142	143	144	145	146	147
50 Sn	D β 10d 2.6; D β 6.2h 0.8	D β 70m 0.7 2.8; D β ~20 m																			
51 Sb	A β 80h 1.2	D β 60m	A β 4.2h			A β <10m	D β <10m			D β 5m 1.3											
52 Te	A * 90d; A β 9.3h 0.6	S	A * 32d; A β 72m 1.8	S	A * 30h; A β 25m	D β 43m	D β 60m		A β <2m	D β 77h ~0.3	D β ~1m										
53 I	S		M	A β 12.6h 0.61, 1.03	A β 8d 0.6	D β 54m 1.7	A β 22h 1.4		A β 6.7h 1.4	D β 2.4h 2.2	D β 22.5s 2.1	E 1.8m									
54 Xe		S	S	S	S	S	D β 5.5d 0.42	S	A* β 15.6 m; A β 9.4h 1.0	S	D 68m; D β 3.4m 4	D β 17m 1.1	A β 41s 2.1	A β 16s	A β 1.7s			D β Short			
55 Cs							S		M		D β ~27y	D β 33m 2.6	A β 7m	D β 40s	A β Short	D β Short		D Short			
56 Ba				S		S		S	S	S	S	S	D 3m; A β 86m 2.3	A β 12.5d 1.05	A β 18m	B β 6m	D β Short	E β <1m	D β Short		
57 La													S	A β 40h 1.45	A β 3.5h 2.9	E β <30m	B β 74m	D β Short	D β Short		
58 Ce										S		S		B * 140d; S	A β 30d 0.6	S	B β 36h 1.36	D β 1.8h	D β ~15m		D β 275d 0.348
59 Pr															S		B β 13.5d 0.95	D β 4.5h 3.2	D β 24.6m 3		D β 17m 3.07
60 Nd																S	S	S	S	S	E β 47h 0.95
61																					E β 2.7h
62 Sm																		S			S
Branching ratio.	0.18		0.34		1.6	12	7.6		9(?)	5.2			6.4	8.4							

To illustrate some of the processes described, figure 9 depicts a possible result of neutron-induced fission of $_{92}U^{235}$, which produces the reaction described in equation 1. The disintegration series shown are two that have actually been observed. This does not necessarily mean that these series *do* result from the same fission process, but they have been observed and the numbers of neutrons and protons are such that it is a *possible* reaction. The isomerism of $_{54}Xe^{137}$ found in transmutation experiments [71] and indicated in table 4 has not been reported among fission products.

Table 4 lists all the series of this type that have been reported, including the latest values as given on Segrè's chart. Isotopes of a particular element occur in horizontal rows; radioactive series of isobars of a particular mass appear in vertical columns. Each square contains the half-life for beta ray emission and the maximum energy of these beta rays expressed in Mev. The squares divided by a vertical line give information for two isomeric states of the nucleus and the times indicated in squares marked by asterisks are the half-lives of gamma rays in the isomeric transitions. The letters in the upper left corner of each square have the following meanings:

A—isotope certain (mass number and element certain)
B—isotope probable, element certain
C—one of few isotopes
D—element certain
E—element probable
F—insufficient evidence

An M in a square indicates that the nucleus is believed to belong to the series but that its activity has never been reported. The stable nuclei are marked by S.

Early X ray identification.—The identification of fission products first reported in this country was made by Abelson[72] using the characteristic X rays produced by internal conversion as described above. The X rays had been observed a year earlier coming from a neutron-irradiated sample of uranium. Their absorption by copper had been studied and yielded results that might be expected from a transuranic element by internal conversion. The discrepancy between expected and observed values was attributed to poor geometry. However, after Hahn and Strassmann reported fission, the experiment was repeated with a stronger sample and several different absorbing materials were used. It was found that the

[71] E. P. Clancy, *Phys. Rev.* **60**, 87 (1941).
[72] P. Abelson, *Phys. Rev.* **55**, 418 (1939).

absorption increased with the atomic number of the absorber up to $_{50}$Sn, after which it dropped; in fact, tin showed two absorption coefficients. Figure 10 shows a schematic diagram of the absorption coefficients of indium ($Z=49$) and tin ($Z=50$), and superimposed on it are the iodine K_α and K_β lines. It is seen that the K_α and K_β radiations would both be strongly absorbed in indium, whereas only the K_β would be appreciably absorbed in tin. This would produce just the effect observed, and so the X ray emitting substance was identified as iodine. Using the same method Feather and Bretscher[73] arrived at the same conclusion independently.

The complete process consists of emission of a beta ray of 77 hour half-life by a tellurium atom, changing it to iodine; the iodine nucleus then emits a gamma ray that is internally converted, thus producing the X ray.

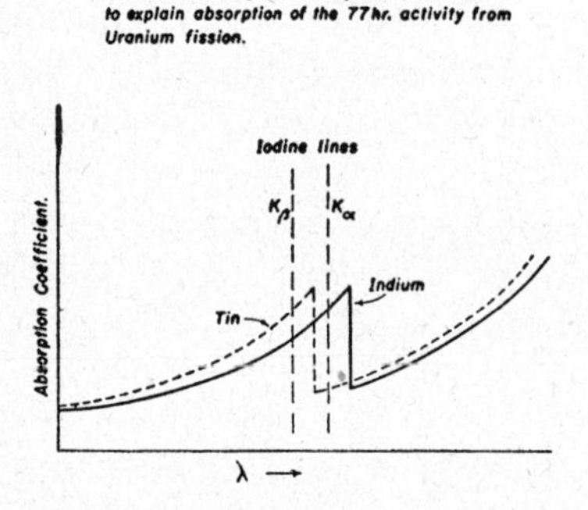

Fig. 10. Schematic diagram of X-ray characteristics to explain absorption of the 77-hour activity from uranium fission.

The procedures most widely used in identifying fission products involve various combinations of chemical separations and measurements of radioactive constants. We shall describe the processes and phenomena first and then proceed to indicate how they are used to obtain the desired information.

Chemical separation methods.—If a uranium compound is irradiated with a beam of neutrons from a cyclotron for about an hour the fissions will produce minute amounts (less than micromicrograms) of many products, each of which decays with a characteristic half-life to some daughter product. We desire to know which isotopes are present, from what isotopes they were

[73] Feather and Bretscher, *Nature* **143**, **516** (1943).

produced and into what they will decay. There are three principal desiderata to be considered for the chemical separations. Evidently they must be specific for one element. We also desire that they may be performed quickly, especially for investigation of short half-lives of direct fission products. Lastly, the methods should be semiquantitative, particularly for the determination of branching ratios (see page 40).

The first requirement can be satisfied by fairly standard methods of oxidation and reduction, precipitation and extraction. By reasonable compromise all three can be satisfied sufficiently for many identifications. Details of methods used would take too much space to include here, but they can be found in several papers on fission products.[74] However, it is interesting to point out a few methods that are particularly applicable to this work.

One common practice is the use of carriers. The separation of such minute quantities of elements as are produced by fission is not easily accomplished by ordinary chemical means. However, added extra chemicals (generally additional quantities of the substance to be separated) will often "carry" the interesting small amounts along with them in a reaction (see page 231).

The recoil that a nucleus experiences when disintegration or radiation occurs provides a method of separating the different types of disintegrating nuclei of the same element. The nuclei to be separated are used in the preparation of a compound that can be separated by the Szilard-Chalmers method.[75] When one of the nuclei to be separated ejects a photon or particle, the recoil is frequently sufficient to break the chemical bond that holds the atom in the compound. If the compound used is such that interchange among the freed atoms and those still in compound cannot occur, the freed atoms can then be separated and their disintegration characteristics measured without interference from other products.

There are two methods of particular interest for studying xenon and its parents and daughters. The first uses "emanating samples" described by Langsdorf and Segrè.[76] Iodine and bromine are first distilled with carriers from the irradiated uranium solution into a dilute sulphite solution. Iodine alone is oxidized to the free state by adding ferric chloride and then extracted with carbon tetrachloride. The iodine may then be transferred to an

[74] P. Abelson, *Phys. Rev.* **56**, 1 (1939); Hahn and Strassmann, *Naturwiss.* **27**, 89 (1939); Grosse and Booth, *Phys. Rev.* **57**, 664 (1940); Glasoe and Steigman, *Phys. Rev.* **58**, 1 (1940); Anderson, Fermi and Grosse, *Phys. Rev.* **59**, 52 (1941).

[75] Szilard and Chalmers, *Nature* **134**, 462 (1934).

[76] Langsdorf and Segrè, *Phys. Rev.* **57**, 105 (1940).

emanating sample by shaking silica gel impregnated with silver nitrate in the solution, drying it, and placing it in an evacuated chamber. The iodine produces xenon by beta decay, and this xenon, emanating from the silica gel, can then be collected in another chamber where its activity can be measured without contamination due to other radiations. In the other method,[77] the radioactive xenon is passed into or through a chamber consisting of a grounded metal cylinder around a central electrode at a high negative potential. When the xenon disintegrates by beta emission it becomes a positively charged cesium ion which is attracted to the negative electrode and deposited as cesium. This electrode can then be removed quickly and the radiations from the cesium measured. These methods are useful also for the study of krypton and its genetically related isobars.

Radioactivity measurements.—Standard procedures are known for determining characteristics of disintegration processes and only brief mention of a few of them will be made here. More complete discussions can be found in several sources.[78]

The beta particles emitted in radioactive decay are charged particles of sufficient energy to produce ionization in the medium through which they pass. One convenient method of detecting and measuring this ionization is by means of a gas-filled chamber, in which an electric field is used to collect the ions produced. (This chamber should have a suitable window for admitting the particles.) The amount of charge collected on one insulated electrode in the ionization chamber can be measured with an electrometer, or an electronic circuit can be used to amplify the current. The amount of ionization is a measure of the rate of occurrence of disintegrations, which is called the *activity* of the source. Another method is to expose a charged electroscope to the radiation whose intensity is to be measured.[79] The rate of discharge is then a measure of the activity.

The energies of the particles can be measured by observing the radii of curvature of their paths in a magnetic field, either by photographing their tracks in a Wilson cloud chamber or by using a beta-ray spectrograph.[80]

[77] Glasoe and Steigman, *Phys. Rev.* **58,** 1 (1940). Originally used by Rutherford, *Phil. Mag.* **49,** 161 (1900).
[78] Livingstone and Bethe, *Rev. Mod. Phys.* **9,** 256 (1937); F. Rasetti, "Elements of Nuclear Physics", Chapter 1, Prentice-Hall (1936); O. Glasser, "*Medical Physics*", p. 643, Year Book Publishers (1944).
[79] Lauritsen and Lauritsen, *Rev. Sci. Inst.* **8,** 438 (1937).
[80] Plesset, Harnwell and Seidl, *Phys. Rev.* **13,** 451 (1942).

Also useful are the results of beta-ray absorption experiments, which are performed by inserting varying thicknesses of absorbers between the source and the measuring device. The transmitted intensity is plotted as a function of absorber thickness. The incident electrons have an energy distribution that would not be expected to give a simple result in such an experiment and since electrons are so light, their scattering in matter is very complicated. However, with the essential requirements of "good geometry" these effects combine to give an almost exactly exponential absorption up to a certain limit of thickness.

Elementary theory of beta disintegration.—A theory of the mechanism of beta disintegration, which accounts for the phenomena observed when activity and energy measurements are made, was developed by Fermi[81] and a summary of that theory will be given here. The distribution in energy of beta particles is found to be of the type shown in figure 11. Significant features of this curve are: (1) a continuous distribution up to some maximum energy, ϵ_0, exists; (2) the most probable energy of a particle is approximately one-third of this maximum; and (3) the curve approaches zero at high energies with a small slope.

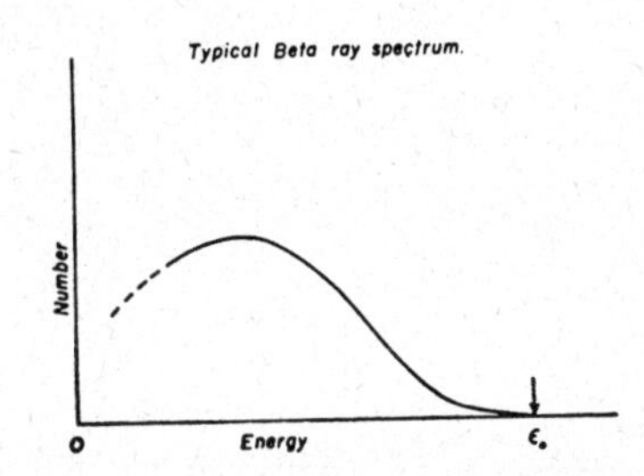

Fig. 11. Typical beta ray spectrum.

The first of these facts leads to a postulation of the neutrino. The parent and daughter nuclei involved in a beta disintegration have discrete energy levels and yet the beta particles emitted give a continuous spectrum. If we keep the law of conservation of energy we must postulate another particle, the neutrino, as yet undetected, which carries off some of the energy (and spin and angular momentum). Thus we can interrupt the emission of a beta particle as a process in which one of the neutrons in the nucleus is transformed into a proton, an electron and a neutrino, according to the reaction

$$ {}_0n^1 \longrightarrow {}_{+1}p^1 + {}_{-1}\beta^0 + {}_0\mu^0 \qquad (2)$$

The sum of the energies of the electron and neutrino emitted is the

[81] E. Fermi, *Zeits. für Phys.* 88, 161 (1934).

maximum of the energy distribution, ϵ_0. To solve the problem of beta disintegration one must calculate the probability of this reaction.

A new type of interaction was postulated by Fermi which would produce the transformation indicated in equation 2 and the probability calculation is analogous to the determination of optical transition probabilities among electron levels of the atom. Integrating a suitable combination of operators and wave functions [82] over the coordinates of the nucleus, we get the probability that the transformation will occur, resulting in emission of a beta particle with energy between ϵ and $\epsilon + d\epsilon$. This function is of the same general nature as that shown in figure 11 except that the most probable energy is $\frac{1}{2}\ \epsilon_0$. Konopinski[83] has made a thorough analysis of many experiments on beta emission and indicates that the actual distribution is probably more nearly that given by Fermi's function. He attributes the shape of the experimental curves to the use of thick samples and other spurious effects. He concludes that Fermi's original theory is to be preferred to the modifications suggested by Konopinski and Uhlenbeck,[84] but that the Gamow-Teller[85] selection rules give better agreement with experimentally observed intensities than do the Fermi rules.

The third characteristic of the energy distribution curves requires that the mass of the neutrino be very nearly zero; for it can be shown from the equations derived for transition probabilities that if the mass of the neutrino is comparable with that of the electron the upper end of the energy distribution curve should approach zero with a vertical tangent. As this is not usually observed, it is concluded that the neutrino mass is small. In fact, from a measurement of the C^{13} (p,n) N^{13} reaction threshold and the known positron energy in the decay of N^{13} to C^{13}, Haxby, Shoupp, Stephens and Wells[86] conclude that the neutrino mass is less than one-tenth of the mass of the electron.

Decay curve characteristics.—If the probability derived above is integrated over all energies we get the total probability of decay per unit time. This is frequently called the *disintegration constant*, λ. If there are N of these atoms, the rate of decay, or activity will be simply λN. Activities are measured in units of the *curie*, which is the activity observed for 1 gram of radium, 3.71×10^{10} disintegrations per second. Letting the original number be N_0, we may integrate to find the number at any time to be $N_0 e^{-\lambda t}$, so

[82] Bethe and Bacher, *Rev. Mod. Phys.* 8, 82 (1936).
[83] E. J. Konopinski, *Rev. Mod. Phys.* 15, 209 (1943).
[84] Konopinski and Uhlenbeck, *Phys. Rev.* 48, 7 (1935).
[85] Gamow and Teller, *Phys. Rev.* 49, 895 (1936).
[86] Haxby, Shoupp, Stephens and Wells, *Phys. Rev.* 58, 1035 (1940).

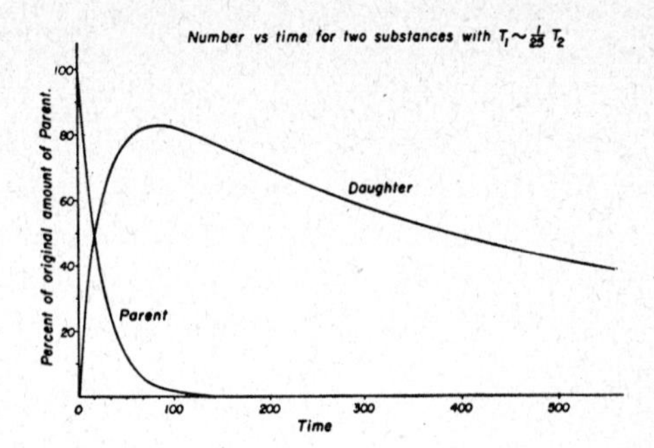

Fig. 12. Number *vs.* time for two radioactive substances with $T_1 \sim 1/25\ T_2$.

that the rate of decay $-dN/dt$, at any time is $\lambda N_0 e^{-\lambda t}$. This gives an exponential curve for the activity which is exactly what is observed if we start with a nucleus that decays by one beta emission to a stable daughter substance. Activities are identified by their *half-life,* which is the time required for one-half of the original number to disintegrate. Evidently if the logarithm of the activity is plotted against time we get a straight line of slope $-\lambda$, and the half-life T is

$$T = \log_e 2/\lambda = -\frac{\log_e 2}{slope} = -\frac{0.693}{slope}$$

The measurement of half-lives of the many products resulting from fission is never so easy as indicated above where one disintegration results in a stable nucleus. Since a direct fission product undergoes an average of about four disintegrations before arriving

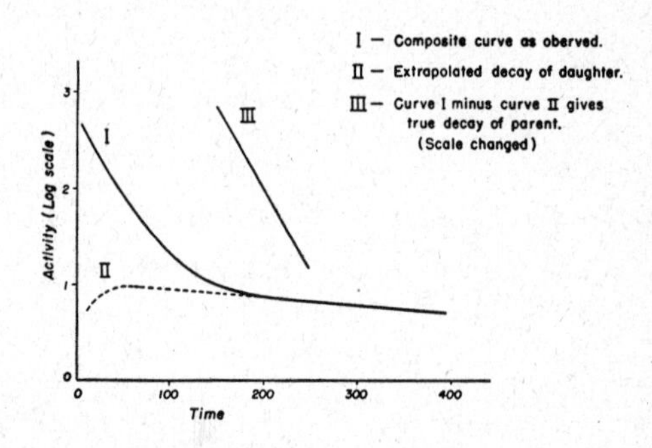

Fig. 13. Activity curve for fig. 12. Activity *vs.* time for decay of two radioactive substances with $T_1 \sim 1/25\ T_2$. (Meitner, Hahn and Strassmann.)

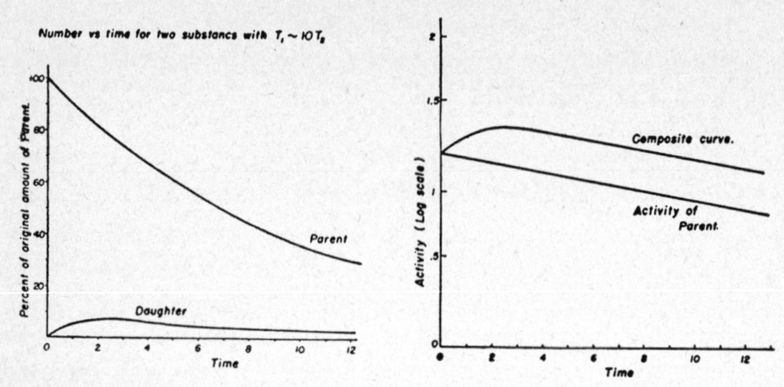

Fig. 14. Number *vs.* time for two radioactive substances with $T_1 \sim 10T_2$.

Fig. 15. Activity curve for fig. 14. Activity *vs.* time for decay of two radioactive substances with $T_1 \sim 10T_2$.

at a stable nucleus, and any one element may occur in as many as six or more disintegrating series, it is evident that a chemical separation of a particular element would still result in a mixture of confusing activities. Consider first the relatively simple case of one nucleus that decays into a daughter substance, which in turn disintegrates to a stable element. If the parent half-life (T_1) is very short compared to that of its daughter (T_2), the parent disappears in a short time, producing the daughter substance which then decays with its own half-life. Figure 12 shows the number of atoms of each substance as a function of time and in figure 13 the logarithm of the activity is plotted against time. There would be no difficulty in measuring each of these half-lives. However, suppose $T_1 > T_2$, so that the daughter disintegrates almost as fast as it is formed. The corresponding curves in this case are shown in figures 14 and 15, from which it is seen that even in this simple case only an estimate of the daughter activity can be obtained without separating it from the parent.

To measure half-lives and establish parent-daughter relationships in the complicated situations that could occur, it is evident that special techniques must be employed. Actually, these techniques are merely applications of partial knowledge already obtained. To indicate how some of these determinations are made, a brief account of a few of the actual early experiments will be given. It will he helpful to our understanding of why certain procedures were useful and necessary, to consult the table of activities, table 4, although undoubtedly the original determination of these methods resulted from a certain amount of trial and error.

Classification of activities.—Abelson[87] used some interesting methods in classifying 16 activities resulting from fission and we shall discuss first the series

$$_{51}\text{Sb} \xrightarrow{\textit{5 min.}} {}_{52}\text{Te} \xrightarrow{\textit{77 hr.}} {}_{53}\text{I} \xrightarrow{\textit{2.4 hr.}} {}_{54}\text{Xe}\ (\textit{stable}).$$

It was found that if tellurium was separated a week after irradiation of a uranium compound, curve I in figure 16 was obtained for the activity. This is the type described in the preceding section for which the daughter half-life is considerably shorter than that of the parent. When iodine was separated from the tellurium precipitates, the half-life measured was 2.4 hours with no contaminating activities appearing (curve II, figure 16). From the values of the half-lives, table 5, we see that after one week only the 77-hour, 32-day and 90-day tellurium activities would be present in appreciable quantities. As the daughter product of the 90-day tellurium is stable and none has been detected for the 32-day substance, we can understand the appearance of only the 2.4-hour activity when the iodine was separated.

Meitner, Hahn and Strassmann[88] had reported 66 hours for the half-life of the activity here reported as 77 hours. This result could have arisen from the mixture of the 77-hour and 30-hour isotopes and a method of periodic separations of iodine was used to clear up the confusion. The iodine separations were made at periodic intervals after precipitation of the tellurium. The activity of the 2.4-hour iodine was then measured immediately after each separation, and this activity was a measure of the amount of tellurium

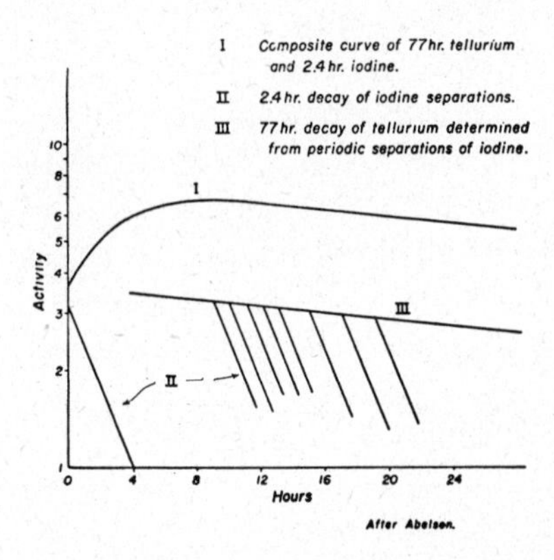

Fig. 16. Activity curves for radioactive decay of tellurium and iodine fractions. (Abelson.)

[87] P. Abelson, *Phys. Rev.* **56**, 1 (1939).
[88] Meitner, Hahn and Strassmann, *Zeitz. für Phys.* **106**, 249 (1947).

Table 5
Data for estimating half-life of parent of 77-hour tellurium.

	Time after start of irradiation	Activity
Conclusion of irradiation	30 min.	
1st precipitation	40 min.	3.4 div/sec
2d precipitation	50 min.	0.25 div/sec
3d precipitation	60 min.	0.07 div/sec
4th precipitation	70 min.	0.02 div/sec

that had decayed and hence indicated the activity of the tellurium as a function of time. The results confirmed the 77-hour value (curve III, figure 16).

Abelson also proved the report of a 59-minute parent of the 77-hour tellurium to be in error. Tellurium precipitates from an irradiated uranium sample were made at 10-minute intervals and immediate determinations of the 77-hour activity were made by measuring both the beta activity and the X ray intensity. The results, in table 5, indicate a 5-minute parent, although it is not necessarily antimony. The same decay characteristics would be obtained if a 5-minute tin decayed to a very short-lived antimony, which then went to tellurium.

The 2.4-hour iodine was thought to give stable xenon because a fast iodine separation from the 77-hour tellurium gave a strictly exponential decay. The half-life of any daughter would be less than 1 minute or greater than 100 days. Thus we have a complete account of the genetic relationships among the elements of this radioactive series. Unreported research has apparently confirmed these observations, for mass number 136 has been assigned to the series and the end product is given as ${}_{54}Xe^{136}$ on Segrè's chart.

Abelson's work on the series of mass number 129 is also interesting as an illustration of the Szilard-Chalmers separation method and of the value of comparing observations with those of nuclear transmutation experiments. The series of isobars is short:

$$ {}_{51}Sb^{129} \xrightarrow{4.2\,hr} {}_{52}Te^{129} \xrightarrow{72\,min} {}_{53}I^{129} $$

The upper isomer of Te^{129} has not been reported among fission products, although it does exist. The iodine nucleus is not observed in nature, but its mode of decay has not been determined.

An irradiated sample of uranium was allowed to age for 6 hours, at which time antimony was precipitated. Consulting table 4, we see that this would give a mixture of isotopes of masses 127.

128 and 129. However, owing to the small percentage abundance of mass 127 and its long half-life (see branching ratios, page 41 and table 4), its effect on the 4.2-hour activity would be negligible. The effect of the 128 isotope is not negligible. However, the tellurium in the 128 series is not radioactive, so that if tellurium is now separated from the antimony precipitate, only the 72-minute activity of $_{52}Te^{129}$ is measured. This also permits an accurate determination of the 4.2-hour antimony half-life by a series of periodic separations of tellurium. Just as in the case illustrated in figure 16, periodic separations of tellurium and measurements of the initial activities give a measure of the rate of disintegration, hence of the half-life of its parent antimony.

The determination of the mass number of the series as 129 resulted from the fact that the 72-minute half-life observed for tellurium was the same as that reported for the lower isomer of Te^{129} produced by the reaction Te^{128} (d,p) Te^{129}. To confirm the identification, beta-ray absorption curves were taken of the activities. The tellurium produced by the deuteron bombardment contained both the upper and lower isomers of Te^{129}. In order to measure the 72-minute activity of the lower isomer it first had to be separated from the mixture. This was accomplished by the Szilard-Chalmers method described on page 31. The recoil produced by the 0.1 Mev gamma ray was sufficient to release the lower isomer from compound and appreciable intensity was attained. When the resulting absorption curves from the two sources were compared, they agreed over a factor of 100 in intensity and so the mass number of the series is assigned as 129.

Products from various types of fission.—Products of fission of U^{235} by slow (i.e., thermal) neutrons are the most thoroughly investigated at present. The resulting series arrange themselves largely into two main groups; the masses in the heavy group range from about 127 to 147 and in the light group the range is about 105 to 80. However, it is not expected that for each series in the heavy group there is a corresponding series in the light group such that the sum of their masses is constant. One reason for this is the variation in the number of neutrons released at the time of fission.

At the head of each column of table 4 the reaction that produced the observed series is indicated.[89] The notation U-*n* refers to thermal neutrons except for the series of masses 111 and 112. These series resulted from bombarding uranium or thorium with

[89] These data are taken from Seaborg, *Rev. Mod. Phys.* **16**, 1 (1944).

neutrons of energies greater than 10 Mev,[90] and correspond to symmetric fission in contrast to the asymmetric fission produced by slow neutrons. Lark-Horovitz and Schreiber [91] used fast neutrons and obtained a peak in the curve of number of fissions versus energy of fragments, which they attribute to symmetric fission. They also noted that the asymmetric peaks extend to higher and lower energies, indicating greater numbers of highly asymmetric fissions.

Langer and Stephens [92] reported that the ratio of the initial activities of strontium and barium produced by gamma ray fission is the same as for slow neutron fission. Several iodine activities have been reported from bombardment of uranium with alpha particles, but further search has apparently not been made. Identification of the products from fission induced by protons and deuterons have not yet been reported.

Branching ratios.—In addition to knowing *what* isotopes are produced, and the genetic relationships among them, it would be desirable to know *how much* of each isotope is produced. This would give more specific knowledge as to the intensity of certain types of radiations and would also allow more accurate calculations of the seriousness of poisoning caused by the products generated in piles. Any theory of the fission mechanism will predict a general trend in the distribution of products, and one useful check of its validity would be a comparison with quantitative measurements of the type to be described. Anderson, Fermi and Grosse[93] started a systematic study of this problem, but only one set of data on slow neutron fission of uranium was reported.

The object is to determine the percentage of fissions that result in a particular radioactive series of the type discussed above. This percentage is called the *branching ratio* of that series. It is assumed that none of the radioactive series branches into others, on the basis that disintegrations observed are almost exclusively by beta emission. Then a determination of the percentage of fissions producing any element in that series is its branching ratio, R. An idealized way of determining this number is to irradiate uranium for such a long time that eventually the numbers of atoms of all members of the series become constant. Then the activities of all the members would be equal, since the rate of production of any member equals its rate of decay and this activity is just equal to R times the rate of fission production. Thus a measurement of this

[90] Segrè and Seaborg, *Phys. Rev.* 59, 212 (1941); Nishina, Yasaki, Kimura and Ikawa, *Phys. Rev.* 59, 677 (1941).
[91] Lark-Horovitz and Schreiber, *Phys. Rev.* 60, 156 (1941).
[92] Langer and Stephens, *Phys. Rev.* 58, 759 (1940).
[93] Anderson, Fermi and Grosse, *Phys. Rev.* 59, 52 (1941).

rate of fissioning and of any one activity in the series would give the branching ratio. It is seen that in the actual experiment a correction will be necessary for the finite time of irradiation and that this correction will involve knowing the half-lives of all elements preceding the one on which the measurement is made. However, if one of these half-lives is not known accurately, it is relatively unimportant provided it is short and not preceded by any long-lived ancestors. This means that the lack of precise information on direct fission products is not too serious.

To determine the rate of fission production, the uranium solution used in the experiment was replaced by a solution of manganese sulphate of such strength that the manganese atoms absorbed approximately as many neutrons as the uranium. Then the activity of these manganese atoms was measured and multiplied by the ratio of cross section for fission to cross section for capture by manganese, to obtain the rate of fission production. Only one measurement of this type was necessary for uranium solutions of the same composition, for it could be used to calibrate a gold foil that was then used to monitor the neutron intensity in subsequent irradiations.

A further correction was required to account for the variation in absorption of different beta rays with energy. In most activity measurements we desire only to know how a particular activity varies with time, but here we must know the absolute magnitude of the activity and hence allowance must be made for absorption in the counting arrangement used. This is sometimes difficult to determine accurately because of the difficulty in separating the desired activity from that of its products and this accounts for the doubtful figure indicated for the series with mass number 135.

The results of this experiment are shown in the bottom row of the table of fission products, table 4, where the value given is the R for that series. The values for the series with masses of 131, 132, 133, 135 and 136 were determined from separations of iodine. Procedures similar to that described on page 30 were used to isolate each activity from the others. Columns 127 and 129 were obtained by separation of antimony; columns 139 and 140 by measurements of barium. The only series in the light group, mass number 97, was determined from the zirconium.

It would be expected that the sums of the branching ratios in each group, the heavy and light, would be approximately 100, whereas the reported values total only about 50. Measurements on other series will, of course, bring this value up, but Anderson, Fermi and Grosse suggest that their measurement of the rate of fission production may be in error. Smyth (8.17) reports that the most abundant fission product constitutes a little less than 10 per cent of the total, which probably means that the branching ratio for that series is nearly 20. This is appreciably larger than the value of 12 for the series of mass 132, which Anderson, Fermi and Grosse found to have the largest branching ratio.

CHAPTER 5

SECONDARY NEUTRONS

The nuclei that result from the fission of uranium have an excess of neutrons over the stable nuclei of the same atomic number. This excess may be relieved, as we have already seen, by the emission of beta particles—at each emission a neutron changing into a proton and the nucleus increasing its atomic number by unity. At the time of the discovery of fission, it occurred to many physicists that the excess could also be relieved by the emission of neutrons at the instant of fission or else by the emission of neutrons after a very short time from the highly excited fragment nuclei produced in fission. It was suggested also that a fragment nucleus may have to go through several beta emissions before a nucleus is reached that could reduce its energy by emitting a neutron. The latter possibility would result in the delayed emission of neutrons, the delay depending on the half-lives of the previous beta decay periods. A nucleus that can emit a neutron probably does so with such an extremely short half-life that this does not contribute to the delay.

Delayed neutrons.—The observation of the delayed emission of neutrons accompanying fission was reported first. Roberts, Meyer and Wang[94] exposed uranium to thermal neutrons from a Li (d,n) source covered with paraffin. The source was then turned off and the uranium placed in front of a boron-lined ionization chamber surrounded by paraffin. Neutrons were observed to come from the uranium for as long as 1½ minutes after the exposure. The neutrons actually decayed with a half-life of 12.5 ± 3 seconds. It was thought possible that the neutrons were produced by gamma rays from the fission products and a hard gamma ray activity of about the same half-life was observed. However, surrounding activated uranium by more inert uranium did not increase the number of secondary neutrons emitted, and exposing uranium to gamma rays from the Li(p,γ) reaction did not produce uranium that emitted delayed neutrons. Thus it was concluded that the delayed neutrons resulted from the primary slow neutrons which

[94] Roberts, Meyer and Wang, *Phys. Rev.* 55, 510 (1939); Roberts, Hafstad, Meyer and Wang, *Phys. Rev.* 55, 664 (1939).

produced fission in the uranium. From a calibration of the ionization chamber with a standard source, the cross section for this process was determined to be 4 x 10^{-26} cm^2. A study of recoils in a cloud chamber showed that the energy of the secondary neutrons did not exceed 1Mev. Delayed neutrons could also be produced by exposing uranium to fast neutrons, but no secondary neutrons came from uranium exposed to the neutrons of intermediate energy from the C(*d,n*) reaction. It is to be recalled that fission is produced also by fast and slow neutrons but not by neutrons of this same intermediate energy. Delayed neutrons were observed to come also from thorium activated by fast neutrons.

Booth, Dunning and Slack[95] found another half-life, 45 seconds, for delayed neutrons from uranium. At equilibrium the total number of delayed neutrons emitted per minute was 1/60 of the number of fissions per minute. If 3 x 10^{-24} cm^2 is taken as the cross section for fission in natural uranium, the above result indicates that the cross section for the production of delayed neutrons by thermal neutrons is 5 x 10^{-26} cm^2, in good agreement with the value given above.

Brostrøm, Koch and Lauritsen[96] found delayed neutrons with half-lives of 3 seconds and 0.1 — 0.3 seconds.

In 1942, Snell, Nedzel and Ibser[97] reexamined this phenomenon. The method of investigation was similar to the method used by Roberts, Meyer and Wang discussed at the beginning of this section. Delayed neutron periods of 57 ± 3 seconds, 24 ± 2 seconds, 7 seconds and 2.5 seconds were observed. The relative intensities of these activities activated to saturation were respectively 0.135, 1.0, 1.2 and 1.2. No activity period longer than 57 seconds was observed. In a separate experiment it was noted that 1.0 ± 0.2 per cent of the neutrons were delayed by at least 0.01 seconds and that approximately 0.07 per cent were delayed by at least 1 minute.

Instantaneous neutrons.—The emission of neutrons immediately accompanying the fission of uranium was reported by von Halban, Joliot and Kowarski.[98] Nothing in the experimental procedure indicated directly that the observed neutrons were emitted instantaneously. However, more than one neutron was observed for each fission and neutron energies of at least 2 Mev were observed, so it was assumed that these neutrons did not come from the same source as the delayed neutrons that were pre-

[95] Booth, Dunning and Slack, *Phys. Rev.* **55**, 876 (1939).
[96] Brostrøm, Koch and T. Lauritsen, *Nature* **144**, 830 (1939).
[97] Smyth (Appendix 3).
[98] Von Halban, Joliot and Kowarski, *Nature* **143**, 470, 680 (1939); Anderson, Fermi and Hanstein, *Phys. Rev.* **55**, 797 (1939).

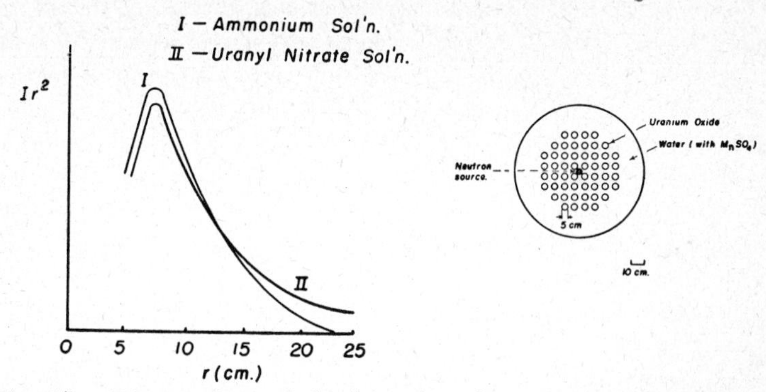

Fig. 17. The product of the intensity of neutrons and the square of the distance from the source is shown plotted as a function of the distance from the neutron source for a uranyl nitrate solution and an ammonium nitrate solution. (Von Halban, Joliot and Kowarski.)

Fig. 18. Horizontal section through the center of the cylindrical tank containing the manganese sulphate solution. The photoneutron source is at the center surriunded by the cans of uranium oxide. The cans are 60 cm high, and concentration of H too U is 17 to 1. (Anderson, Fermi and Szilard.)

viously observed but were neutrons that immediately accompanied the fission process. In the experiment a Raγ—Be source of neutrons was placed in a large vessel containing a 1.6-molar solution of uranyl nitrate and the intensity I of neutrons at various distances r (along one radius) from the source was measured by determining the activity induced in a dysprosium detector placed at r. It was assumed that the neutron distribution was spherically symmetric. A plot was made of Ir^2 vs. r as shown in figure 17. It had been shown that the area under such a curve is proportional to Qt[99], where Q is the rate of production of neutrons and t is the mean life of a neutron before capture. The experiment was then repeated with the container filled with 1.6-molar solution of ammonium nitrate; this solution differed only by 2 per cent in hydrogen concentration from the uranium solution. It was to be expected that the area under the Ir^2 vs. r curve for the uranium solution would be smaller than the area under the curve for the ammonium–nitrate solution, since the mean life t for uranium is smaller. This is because uranium has a greater total cross section for the absorption of thermal neutrons and, more important, a resonance for the capture of neutrons of 5 ev energy. The resonance would result in the capture of neutrons before they reached thermal energies. However, the area under the uranium curve

[99] Amaldi and Fermi, *Phys. Rev.* **50**, 899 (1936); Amaldi, Hafstad and Tuve, *Phys. Rev.* **51**, 896 (1937).

was 5 percent greater than area under the curve for the ammonium nitrate solution. Frisch, von Halban and Koch[1] showed that the introduction of hydrogen or nuclei that merely act to capture neutrons changes the Ir^2 vs. r curve of a solution in such a way that the new curve may be made to coincide with the original curve by multiplying the ordinates by a suitable factor and the abscissae by another factor; this cannot be done for the two curves shown in figure 17. Furthermore, the energy of the primary neutrons was too small to produce an $(n, 2n)$ reaction. Thus it was concluded that the increased area was attributable to an increase of Q, that is, to an increase in the rate of production of neutrons attributable to the fission reaction. The number of secondary neutrons was calculated to be 3.5 ± 0.7 per fission.

A similar experiment with a better geometrical arrangement was performed by Anderson, Fermi and Szilard.[2] Cylindrical cans containing uranium oxide were placed around a central source of photoneutrons in a cylindrical tank containing a 10 percent solution of manganese sulphate as shown in figure 18. A total mass of about 200 kg. of uranium oxide was used. The average neutron density in the solution was determined by first mixing the solution and then measuring the activity of the manganese in a small sample. Alternate measurements were taken with the cans filled with uranium oxide and with empty cans. The activity was 10 percent greater with the uranium oxide present than without it. It was therefore concluded that more neutrons were produced by uranium than were absorbed by uranium. A further experiment determined that 1.5 neutrons were emitted per thermal neutron absorbed in natural uranium.

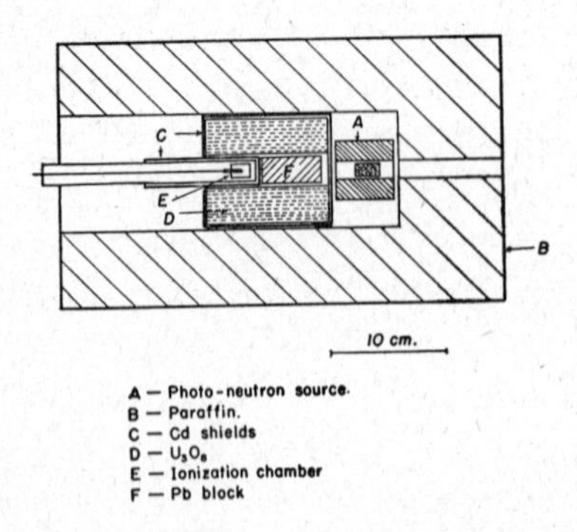

Fig. 19. Arrangement for observing the fast neutrons emitted by uranium. (Szilard and Zinn.)

[1] Frisch, von Halban and Koch, *Danske Vid. Selsk.* **15**, 19 (1938).
[2] Anderson, Fermi and Szilard, *Phys. Rev.* **56**, 284 (1939).

Turner[3] reexamined the data from a similar experiment performed by von Halban, Joliot, Kowarski and Perrin[4] and found that the results indicated that 1.7 neutrons were emitted per thermal neutron absorbed.

Zinn and Szilard[5] used a different technique to determine the number of neutrons produced per fission. Their experimental setup is shown in figure 19. The helium-filled ionization chamber is kept covered with the cadmium shield to keep thermal neutrons out of the chamber. The lead block shields the chamber from gamma rays from the photoneutron source. The cadmium shield around the uranium oxide may be removed, thereby exposing the uranium to thermal neutrons coming from the paraffin. Fifty pulses per minute were observed when the oxide was exposed to thermal neutrons and only five pulses per minute were observed with the cadmium shield in place. Assuming the collision cross section of helium for neutrons to be 3.5×10^{-24} cm^2 and taking into account the solid angle and the size and pressure of the ionization chamber, they calculated the total number of neutrons coming from the uranium oxide. In order to find the number of neutrons produced per fission, the total number of fissions occurring in the uranium oxide had to be determined; this was done with the aid of a separate experiment. The uranium oxide was removed and the helium ionization chamber was replaced by an ionization chamber lined with a thick layer of uranium oxide. When exposed to thermal neutrons the new chamber gave 45 fissions per minute. From the range of fission fragments in uranium oxide, the mass of uranium oxide producing the 45 fissions per minute was calculated; this permitted the total number of fissions occurring in the original mass of uranium oxide to be calculated. It was found that about 2 neutrons were produced per fission. The neutrons appeared in less than 1 second. The greatest error in the experiment was introduced by the fact that the collision cross section for helium has resonances.[6] The experiment was repeated with essentially the same apparatus, except that hydrogen recoils were used to determine the number of neutrons produced per fission. This experiment gave 2.3 as the number of neutrons produced per fission, or 1.4 neutrons for each thermal neutron absorbed, in good agreement with the number 1.5 obtained by Anderson, Fermi and

[3] L. A. Turner, *Phys. Rev.* **57**, 334 (1940).
[4] Von Halban, Joliot, Kowarski and Perrin, *J. de Phys.* (7) **10**, 428 (1939).
[5] Zinn and Szilard, *Phys. Rev.* **55**, 799 (1939); **56**, 619 (1939).
[6] Staub and Stephens, *Phys. Rev.* **55**, 131 (1939).

Szilard. Secondary neutrons were observed to accompany fission induced by gamma rays.[7]

Energy of instantaneous neutrons.—In an ingenious experiment Dodé, von Halban, Joliot and Kowarski[8] demonstrated that secondary neutrons accompanied fission and also obtained a measure of the secondary neutron energy. In the experiment a Raγ-Be neutron source surrounded by crystallized uranium nitrate was placed in a large flask containing carbon disulphide. If fast neutrons were produced in the uranium, they in turn would produce radioactive phosphorus by the S^{32} (n,p) P^{32} reaction. The reaction is endothermic by 0.9 Mev and requires neutrons of at least 2 Mev for a reasonable yield. The primary neutrons do not have enough energy to produce the reaction. After six days, phosphorus was added as a carrier and the phosphorus was separated out by distillation. The isolated phosphorus gave 32 counts per minute, whereas a run without the uranium gave only 5 counts per minute. Thus it was concluded that secondary neutrons of at least 2 Mev energy accompany the fission of uranium.

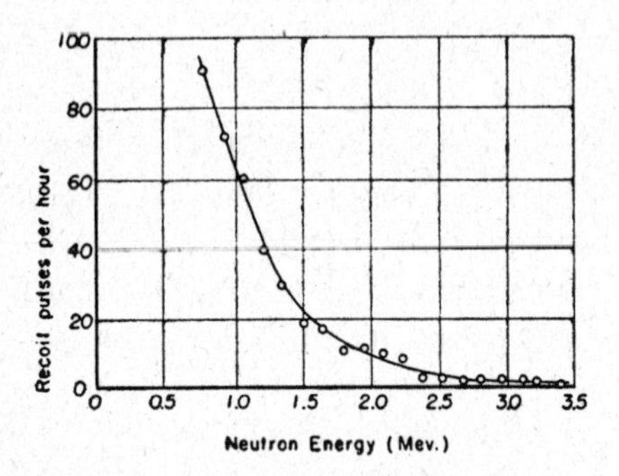

Fig. 20. The energy distribution of secondary neutrons accompanying uranium fission. (Zinn and Szilard.)

Using the arrangement shown in figure 19 Zinn and Szilard[9] determined the energy distribution of secondary neutrons by counting the recoil helium nuclei projected in the helium-filled ionization chamber. Figure 20 shows the energy distribution of the secondary neutrons. The energy of the neutrons did not exceed 3.5 Mev.

Time for the emission of instantaneous neutrons.—Gibbs and Thompson[10] demonstrated that secondary neutrons were emitted

[7] Haxby, Shoupp, Stephens and Wells, *Phys. Rev.* **59**, **57** (1941).
[8] Dodé, von Halban, Joliot and Kowarski, *Comptes Rendus* **208**, 995 (1939).
[9] Zinn and Szilard, *Phys. Rev.* **56**, 619 (1939).
[10] Gibbs and Thompson, *Nature* **144**, 202 (1939).

from uranium at most 0.001 second after the uranium was exposed to thermal neutrons. The ion source of a cyclotron was modulated so as to produce ½ millisecond bursts of neutrons from the D-D reaction at intervals of 5 milliseconds. The neutron source was surrounded by paraffin and a considerable thickness of uranium oxide and the secondary neutrons were detected by a boron-trifluoride ionization chamber shielded by cadmium. The pulses from the chamber were put on an oscilloscope screen along with time markers. Delays greater than 0.001 second could have been detected, but no delay was found.

Theoretical estimates place the time for the instantaneous emission of neutrons at 10^{-14} second (see chapter 8).

CHAPTER 6

HEAVY NUCLEI

The production of new heavy nuclei.—Until 1940 there were elements known only up to atomic number 92 in the periodic table of the elements and even among these there were several gaps because no naturally occurring isotopes had been found. Number 82, lead and one isotope of 83, bismuth, were the heaviest stable elements. All the isotopes of the elements heavier than those were naturally unstable and radioactive. Radioactive isotopes, which fill in the gaps occurring at numbers 43, 61, 85 and 87, had also been produced in the laboratory. Of the heavy radioactive elements the most plentiful are the long-lived elements uranium and thorium. The isotope U^{238} is the parent of the uranium series of natural radioactive isotopes. In this series are found the well known elements radium, radon and polonium. U^{235} is the parent of another radioactive series, called the actinium series, which contains among the daughter products actinium and the rare element 91, protactinium. The other naturally occurring series of radioactive elements starts with thorium, ${}_{90}Th^{232}$.

In May 1940 the first element beyond the classical list was found. This was element 93^{239} found by McMillan and Abelson[11] at the University of California and called neptunium. This isotope is formed by the beta decay of ${}_{92}U^{239}$ produced by radiative neutron capture in ${}_{92}U^{238}$ as shown below:

$${}_{92}U^{238}+{}_{0}n^{1}\rightarrow{}_{92}U^{239}+\gamma,$$

$$\swarrow$$

$${}_{92}U^{239}\xrightarrow[23\text{ min}]{}{}_{93}Np^{239}+{}_{-}\beta$$

Because of the complications introduced by the fission of uranium and the unexpected chemical behavior of neptunium, this new element had not been definitely recognized prior to McMillan's work (see chapter 1).

[11] McMillan and Abelson, *Phys. Rev.* **57**, **1185** **(1940)**.

It was later found[12] that this isotope of neptunium transformed into plutonium, element 94, by beta decay. The Pu^{239} then decays by alpha emission as shown below:

$$
{}_{93}Np^{239} \xrightarrow[2.3\ d]{} {}_{94}Pu^{239} + {}_{-}\beta,
$$

$$
{}_{94}Pu^{239} \xrightarrow[2.4\times 10^{4} yr]{} {}_{92}U^{235} + \alpha .
$$

This important isotope of plutonium has been produced in quantity in the piles. After the chemical properties of plutonium had been established a search was made for plutonium in uranium-bearing ores. Seaborg and Perlman[13] found an alpha activity that they attributed to Pu^{239}. They estimated that it occurs in pitchblende in an amount of 1 part in 10^{14}. It is probably produced by radiative capture by U^{238} of neutrons emitted in the spontaneous fission of uranium (see page 12). However, it was not the first isotope of plutonium discovered or studied chemically. ${}_{94}Pu^{238}$ was the first isotope of this element to be discovered, and it was produced by deuteron bombardment of uranium by Seaborg, McMillan, Wahl and Kennedy late in 1940.[13a] This bombardment led to the reactions below:

$$
{}_{92}U^{238} + {}_{1}H^{2} \longrightarrow {}_{93}Np^{238} + {}_{0}n^{1} + {}_{0}n^{1},
$$

$$
{}_{93}Np^{238} \xrightarrow[2.0\ d]{} {}_{94}Pu^{238} + {}_{-}\beta + 1\ \text{Mev}
$$

$$
{}_{94}Pu^{238} \xrightarrow[50\ yrs]{} {}_{92}U^{234} + \alpha\ 3.9\ \text{cm air}
$$

This isotope was used for the first studies of the chemical properties of plutonium, which were done on the ultramicrochemical scale at a time when only microgram amounts were available. When the Hanford plutonium plant was later put in operation, a step-up factor of 10^{10} from these microchemical experiments to quantity production was achieved.[14]

Another isotope of neptunium, Np^{237}, was discovered in 1942 by Wahl and Seaborg at the University of California. By a

[12] Segrè, Seaborg, Kennedy and Wahl, *Smyth* (4.24).
[13] G. T. Seaborg, *Chem. and Eng. News* **23,** 2190 (1945).
[13a] Seaborg, McMillan, Kennedy and Wahl, *Phys. Rev.* **69,** 366 (1946).
[14] G. T. Seaborg, *Chem. and Eng. News* **23,** 2190 (1945).

$(n, 2n)$ reaction with U^{238} they produced the previously known[15] U^{237} which transforms to Np^{237} by beta emission as shown below: The neptunium is an alpha emitter of very long half-life and starts a long chain of new radioactive elements, the $4n+1$ series, which will be discussed in the next section.

$$_{92}U^{238}+_0n^1 \longrightarrow {}_{92}U^{237}+_0n^1+_0n^1,$$
$$\swarrow$$
$$_{92}U^{237} \xrightarrow[6.8\text{d}]{} {}_{93}Np^{237}+_{-}\beta.$$

Fermi[16] and Meitner, Strassmann and Hahn[17] found that Th^{233}, produced by an (n,γ) reaction in Th^{232}, was also beta active, producing Pa^{233} as shown below:

$$_{90}Th^{232}+_0n^1 \longrightarrow {}_{90}Th^{233}+\gamma,$$
$$\swarrow$$
$$_{90}Th^{233} \xrightarrow[23\text{ min}]{} {}_{91}Pa^{233}+_{-}\beta.$$

The resonance energy of the thorium capture was given as about 2 ev by Meitner.[18] The Pa^{233} decays to U^{233} by beta emission and is undoubtedly also a member of the $4n + 1$ series, as will be discussed in the next section.

Nishina, Yasaki, Kimura and Ikawa[19] reported the production of UY by 15 Mev neutrons on thorium as shown:

$$_{90}Th^{232}+_0n^1 \longrightarrow {}_{90}Th^{231}+_0n^1{}_0n^1.$$

This $_{90}Th^{231}$ is UY, which is beta active with a 24.5 hour half-life.

Protactinium 232, with a 1.6 day half-life for beta emission, has been reported.[20] This might possibly be produced by radiative capture of neutrons by Pa^{231}.

Seaborg[21] has reported the production of elements 95 and 96 as a result of the bombardment of U^{238} and Pu^{239} with very high energy (40 Mev) helium ions by J. G. Hamilton and his group at the University of California. (Element 95 is named "americium" after the Americas or the New World and element 96 is

[15] Nishina, Yasaki, Ezoe, Kimura and Ikawa, *Phys. Rev.* **57**, **1182** (1940); E. McMillan, *Phys. Rev.* **58**, 178 (1940).
[16] E. Fermi, *Proc. Roy. Soc.* A**149**, 522 (1935).
[17] Meitner, Strassmann and Hahn, *Zeit. für Phys.* **109**, 538 (1938).
[18] L. Meitner, *Phys. Rev.* **60**, 58 (1941).
[19] Nishina, Yasaki, Kimura and Ikawa, *Nature* **142**, 874 (1938).
[20] Segrè Chart.
[21] G. T. Seaborg, *Chem. and Eng. News* **23**, 2190 (1945).

named "curium" after Pierre and Marie Curie.[21a]) The identification of the elements was done by Seaborg, James, Morgan and Ghiorso in the Metallurgical Laboratory.[22] Possible reactions to give these elements might be:

$$_{92}U^{238}+_{2}He^{4}\longrightarrow {}_{94}Pu^{241}+{}_{0}n^{1},$$

$$\swarrow$$

$$_{94}Pu^{241}\longrightarrow {}_{95}Am^{241}+{}_{-}\beta,$$

and

$$_{94}Pu^{239}+_{2}He^{4}\longrightarrow {}_{96}Cm^{242}+{}_{0}n^{1},$$

or

$$_{96}Cm^{240}+{}_{0}n^{1}+{}_{0}n^{1}+{}_{0}n^{1}$$

The 4n+1 radioactive series.—Three series of radioactive elements are found in nature and a fourth one among the heavy nuclei has been produced in the laboratory. This last, the $4n+1$ series, starts either from $_{92}U^{237}$ or from $_{90}Th^{233}$, both of which give rise to $_{91}Pa^{233}$. For U^{237} the reaction is

$$_{92}U^{237}\xrightarrow[6.8\text{d}]{}{}_{93}Np^{237}+{}_{-}\beta+0.26\text{ Mev}+\gamma+0.5\text{ Mev},$$

$$\swarrow$$

$$_{93}Np^{237}\xrightarrow[2.25\times 10^{6}\text{ yr}]{}{}_{91}Pa^{233}+\alpha\,.$$

For thorium the reaction is simply

$$_{90}Th^{233}\xrightarrow[23\text{ min}]{}{}_{91}Pa^{233}+{}_{-}\beta+1.6\text{ Mev}.$$

Pa^{233} then decays by the reaction:

$$_{91}Pa^{233}\xrightarrow[27.4\text{ d}]{}{}_{92}U^{233}+{}_{-}\beta+\gamma.$$

$_{92}U^{233}$ is undoubtedly an alpha emitter of long half-life,[23] and continues the $4n+1$ series[24] as shown on following page:

[21a] G. T. Seaborg, *Bull. Am. Phys. Soc.* **21**, 22 (1946).
[22] G. T. Seaborg, *Chem. and Eng. News* **23**, 2190 (1945).
[23] L. A. Turner, *Rev. Mod. Phys.* **17**, 292 (1945).
[24] Segrè Chart.

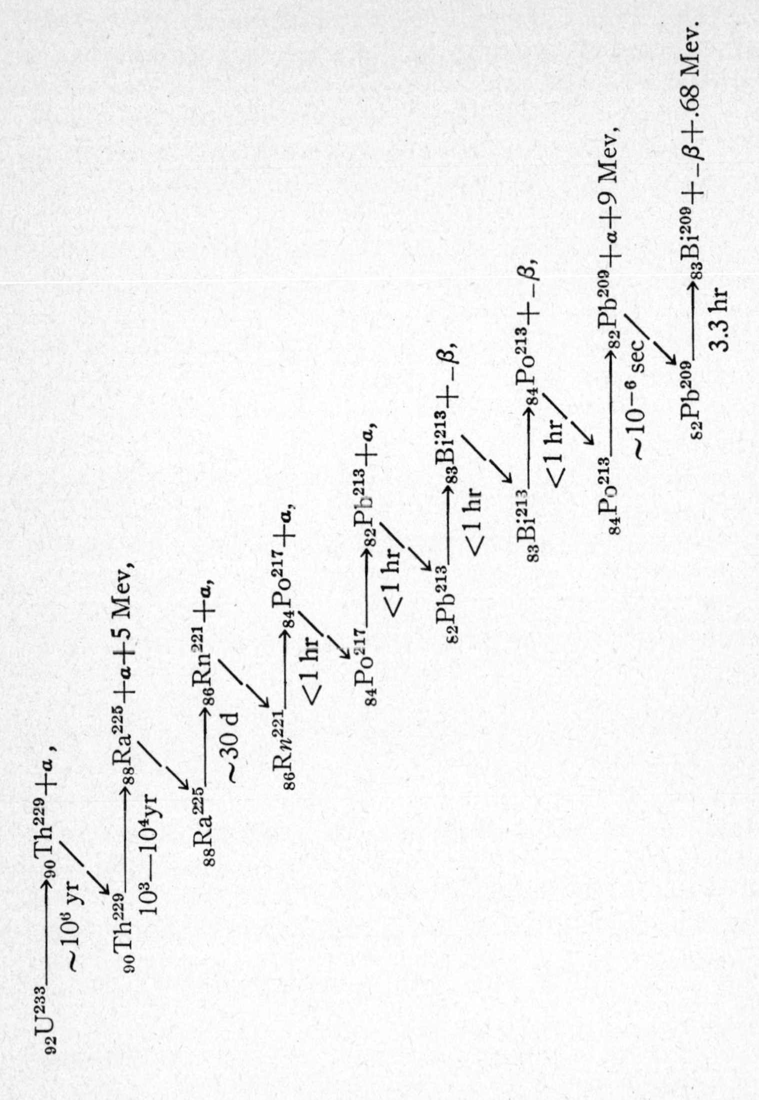

$_{83}Bi^{209}$ is the stable bismuth isotope and the end product of the chain. Another branch from Bi^{213} goes to

$$_{83}Bi^{213} \xrightarrow[<1\ hr]{} {}_{81}Tl^{209} + \alpha$$

$$_{81}Tl^{209} \xrightarrow[<1\ hr]{} {}_{82}Pb^{209} + {}_{-}\beta,$$

$$_{82}Pb^{209} \xrightarrow[3.3\ hr]{} {}_{83}Bi^{209} + {}_{-}\beta + .68 \text{ Mev.}$$

This series is similar to the natural series: the "C" product, $_{83}Bi^{213}$, decays by both alpha and beta particle transformations, as do Th C, Ra C and Ac C.

Predictions[25] of the behavior of this series were correct in several respects. However, the series does not decay through elements 85 and 87 as was suggested, but passes in general through new isotopes of naturally occurring elements.

The missing heavy nuclei.—The reason for the radioactivity of heavy nuclei and the absence of many heavy nuclei in nature has long been a problem confronting nuclear physicists. It has been attributed to a decrease in stability due to the increasing coulomb repulsion of the protons, but its specific details require elaborate consideration. The limits for stability of nuclei against alpha emission and beta emission have been examined by Heisenberg.[26] The general aspects of his treatment give a reasonable account of the known radioactivities although quantitative agreement is not achieved. The discovery of fission provides another mechanism for instability and was discussed by Turner[27] as an explanation of certain missing heavy elements. While these processes may not explain every case, they provide a basis for the discussion of the presence or absence in nature of the very heavy nuclei. The absence of the $4n+1$ chain in nature has been discussed by Turner. His conclusions, together with the deductions that may be drawn from the recent announcement by Seaborg of the production of elements 95 and 96, may suggest the reasons for the absence in nature of this series. The elements that can be considered as starting points for the series are Th^{233}, U^{237}, U^{233} or Np^{237}. Consideration of the possibilities for their presence or absence in nature should show why the other radioactive elements of the series are not found today. The first possibility Th^{233} might be produced by an alpha active U^{237}; however, U^{237} has been found to be beta active with a very short half-life. Any other reaction of the natural radioactive type seems to be equally impossible and therefore Th^{233} probably never existed in nature.

We may next consider the possible parent for U^{237}. A beta active Pa^{237} (produced by an alpha active Np^{241}) might give rise to U^{237}. However, since $_{93}Np^{241}$ is probably beta active and not alpha active, this possibility is also ruled out. An alpha decay from Pu^{241} to produce U^{237} seems unlikely since we have already

[25] L. A. Turner, *Phys. Rev.* **57**, 950 (1940); L. Ponisovsky, *Nature*, **152**, 187 (1943).

[26] W. Heisenberg, *Rapports du Septième Conseil de Physique de l'Institut Internationale de Physique Solvay* (1933). Gauthier-Villars, Paris (1934).

[27] L. Turner, *Phys. Rev.* **57**, 950 (1940); *Rev. Mod. Phys.* **17**, 292 (1945).

Table 5
PREDICTED PROPERTIES OF HEAVY NUCLEI

Z	A	Remarks
92	240	β-active, analogous to Th^{234} (UXI).
92	237	β-active, analogous to Th^{231} (UY).
92	236	β-stable, Est. α half-life $\sim 10^7$ yr.
92	233	β-stable since U^{235} β-stable. Formed from Th^{233} by two successive β disintegrations. Est. α half-life $\sim 10^6$ yr.
92	232	Est. α half-life short.
92	231	Positron emitter since ${}_{91}Pa^{231}$ is β-stable.
93	237	β-stable? (${}_{91}Pa^{231}$ is β-stable.) Est. α half-life$\sim 10^5$ yr.
93	235	Positron emitter since ${}_{92}U^{235}$ is β-stable.
94	244	β-stable ??
94	243	β-active, analogous to Th^{231} (UY).
94	242	β-stable, long α half-life.
94	241	β-stable (?), analogue of U^{235}.
94	240	β-stable.
94	239	β-stable.
94	238	β-stable.
94	237	Probably positron emitter, because of β-stable 93^{237}.
94	236	Probably positron emitter.
95	245	β-active, analogue of ${}_{91}Pa^{233}$.
95	243	?
95	241	Positron ?, or is this stable and ${}_{94}EkaOs^{241}$ β-active?

Taken from: L. A. Turner, *Rev. Mod. Phys.* **17**, 292 (1945).

shown that ${}_{94}Pu^{241}$ is probably beta active and gives rise to 95^{241} (see page 53). Thus U^{237} probably never existed in nature.

The third possible parent of the chain to consider is U^{233}. Aside from the reaction given in the series this might be produced by an alpha or positron emission by ${}_{94}Pu^{237}$ or ${}_{93}Np^{233}$, respectively. Turner's conclusion, however, as seen in table 5, is that ${}_{94}Pu^{237}$ is a positron emitter and not an alpha emitter. If ${}_{94}Pu^{237}$ is an alpha emitter the half-life might be very short by analogy with ${}_{94}Pu^{238}$. Thus there seems to be no long-lived possible parent for U^{233} that would produce it by these two reactions.

The last possibility to consider as a direct parent of the $4n + 1$ series is ${}_{93}Np^{237}$. The reaction given for its production is the beta decay of U^{237}. However, we may consider the possibility of its formation from Am^{241} by alpha decay in competition with a relatively high probability for spontaneous fission. The possible parents for $_{93}Np^{237}$ by alpha decay would be

$$97^{245} \xrightarrow{\alpha} 95^{241} \xrightarrow{\alpha} {}_{93}Np^{237}$$

or

$$96^{245} \xrightarrow{\alpha} 94^{241} \xrightarrow{\beta} 95^{241} \xrightarrow{\alpha} {}_{93}Np^{237},$$

The half-lives for spontaneous fission for 96^{245} would be relatively short and for 97^{245} so short as to assure that none of it would be left in the rocks of the earth. Since the known plutonium half-lives are 10^4 and 50 years for Pu^{239} and Pu^{238}, respectively, it is possible that the alpha half-lives above may be much shorter than those for uranium and thorium. The fact that Bi^{209}, the end product of the chain, does exist may help to support the above ideas and the conjecture that 95^{241} is beta stable and that $_{94}Pu^{241}$ is beta active. The fact that Bi^{209} exists in only a small amount (0.01 percent of Pb^{207}) may show that not much of the parent nonfissioning members of this series were originally produced and that they must have had relatively short half-lives for alpha or beta decay.

The absence of $_{94}Pu^{242}$ may be accounted for by a short alpha half-life or, since $_{94}Pu^{241}$ is beta active, a short beta half-life. The half-lives for spontaneous fission for $_{94}Pu^{242}$ and $_{94}Pu^{241}$ are too long to account for their absence. However, since the alpha half-lives of Pu^{239} and Pu^{238} are short compared to what one would expect for an extrapolation of the Heisenberg nuclear energy surface (see figure 22) and the application of a generalized Geiger Nuttall relation, it is possible that the other isotopes of plutonium may have relatively short alpha half-lives. The half-lives for spontaneous fission are given in table 6. These are taken from Turner's calculations but corrected to an estimated value of 10^{15} years for U^{235} (see page 110).

The absence in nature of U^{236} is another interesting problem. If Pu^{240} is beta stable, as suggested by Turner, but has a short alpha half-life, it may originally have formed U^{236}. Since the estimated half-life for U^{236} is $\sim 10^7$ years or as low as 10^5 years, it may have disappeared if the age of the rocks of the earth is of the order of 2×10^9 years.

Table 6

HALF-LIVES FOR SPONTANEOUS FISSION

Z	Element	A	Half-life
92	U	235	10^{15} years
92	U	233	1.4×10^{14} "
93	Np	237	7.6×10^{13} "
94	Pu	244	6.8×10^{16} "
94	Pu	242	7.7×10^{14} "
94	Pu	241	8.0×10^{13} "
94	Pu	239	1.3×10^{12} "
94	Pu	238	1.6×10^{11} "
95	—	243	6.6×10^{11} "
95	—	241	1.4×10^{10} "

These half-lives are calculated from ratios given by Turner, *Rev. Mod. Phys.* **17**, 292 (1945), on the basis of a half-life of 10^{15} years for U^{235}.

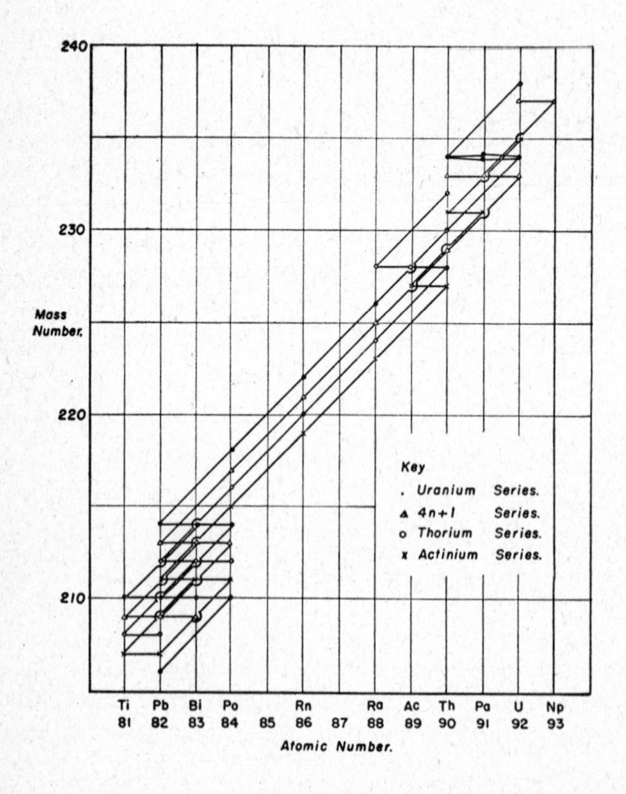

Fig. 21. The radioactive series.

The new data so far released have shown that some of the gaps on the old charts may be filled in a logical and satisfying pattern (see figures 21 and 22). Not enough is known as yet to say whether the energies of the emitted alpha particles as predicted from the Heisenberg nuclear energy surface chart are correct. The two points for which data have been given, Th^{229} and Po^{213}, fit fairly well. The energy predicted from the chart for $_{84}Po^{213}$, the C′ product of the bismuth chain (or should we call it the neptunium chain?), would be about 8.4 Mev, which is a little low compared to 9 Mev given by Segrè. However, this C′ product may give rise to several groups of alpha particles in common with the other C′ products. Th C′ and Ra C′ are classified by Bethe[28]

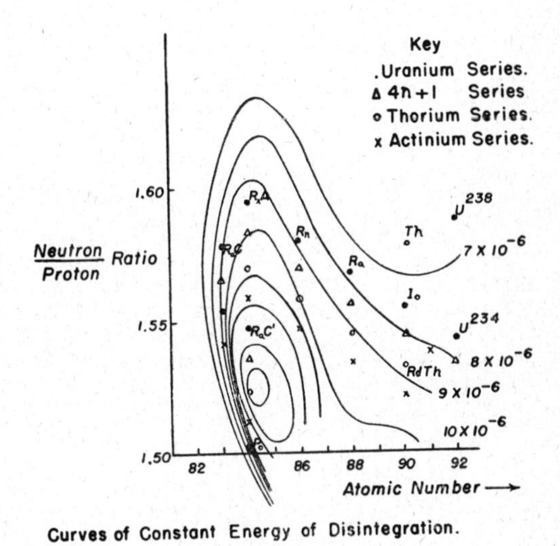

Fig. 22. Curves of constant disintegration energy (Heisenberg). The points for the $4n+1$ series have been added.

as emitting type II alpha spectra, which are interpreted as due to various states of the initial nucleus emitting the alpha particle. Therefore, longer range groups of alpha particles than the normal group may be observed. Consequently, the 9 Mev given on the Segrè chart may represent an admixture of very high energy alpha particles with the main group. The agreement therefore may not be so poor as the figures indicate. For Th^{229}, the 5 Mev predicted from the Heisenberg chart agrees with the data from the Segrè chart.

[28] H. A. Bethe, *Rev. Mod. Phys.* **9**, **69** **(1937)**.

For reference we include table 7, which gives some of the properties of the heavy nuclei. From the table it is apparent that information necessary for a complete interpretation is not yet available. When additional data are obtained the picture of the heavy nuclei may be further clarified.

Table 7 — PROPERTIES OF HEAVY NUCLEI

Atomic Number	Name	Isotope	Relative Abundance	Half-Life	Particle	Energy in Mev	Gamma Ray
96	Curium	242		5 mo.	α		
		240		1 mo.	α		
95	Americum	241		500 yr	α		
94	Plutonium	239		2.4×10^4 yr	α	~ 5.4	
		238		50 yr	α		
93	Neptunium	239		2.3 d	$-\beta$	0.47	0.27
		238		2.0 d	$-\beta$	1.0	γ
		237		2.25×10^6 yr	α		
		236		20 hr	$-\beta$		
		235		8 mo.	K		
		234		4.4 d	K		γ
92	Uranium	239		23 min	$-\beta$		
		238	99.274	4.51×10^9 yr	α	4.13	
		237		6.8 d	$-\beta$	0.26	.05
		235	0.720	8.88×10^8 yr	α	4.52	
		234	0.006	2.33×10^5 yr	α	4.33	
		233			α ?		

Table 7 (Continued)—PROPERTIES OF HEAVY NUCLEI

Atomic Number	Name	Isotope	Relative Abundance	Half-Life	Particle	Energy in Mev	Gamma Ray
91	Protactinium						
	UZ	234		6.7 hr	$-\beta$	0.51	0.7
	UX_2	234		1.14 min	$-\beta$	2.32	0.8
		233		27.4 d	$-\beta$		
		232		1.6 d	$-\beta$		γ
	Protactinium	231		3.2 x 10^4 yr	α	5.049	γ
90	Thorium						
	UX_1	234		24.5 d	$-\beta$	0.13	0.092
		233		23 min	$-\beta$	1.6	none
	Thorium	232		1.39 x 10^{10} yr	α	4.20	
	UY	231		24.6 hr	$-\beta$	0.2	
	Ionium	230		8.3 x 10^4 yr	α	4.66	γ
		229		10^3-10^4 yr	α	5	
	Rd Th	228		1.90 yr	α	5.418	
	Rd Ac	227		18.9 d	α	6.049	
89	Actinium						
	Ms Th_2	228		6.13 hr	$-\beta$	1.55	
					α	4.5	
	Actinium	227		13.5 yr	α	5.0 (99%)	none
					$-\beta$	0.22	

Table 7 (Continued)—PROPERTIES OF HEAVY NUCLEI

Atomic Number	Name	Isotope	Relative Abundance	Half-Life	Particle	Energy in Mev	Gamma Ray
88	Radium						
	Ms Th_1	228		6.7 yr	$-\beta$	0.053	
	Radium	226		1590 yr	α	4.791	0.19
		225		~30 d	α		
	Th X	224		3.64 d	α	5.681	
	Ac X	223		11.2 d	α	5.717	γ
87	Ac K	223		21 min	$-\beta$	1.2	>3
86	Radon	222		3.825 d	α	5.486	
		221		1 hr	α		
	Tn	220		54.5 sec	α	6.282	
	An	219		3.92 sec	α	6.824	
85		218		short	α	6.63	
		216		short	α	7.64	
		211		7.5 hr	α,K	5.94 (α)	
84	Polonium						
	Ra A	218		3.05 min	α	5.998	
					$-\beta$		
		217		<1 hr	α		
	Th A	216		1.58×10^{-1} sec	α	6.774	
					$-\beta$		
	Ac A	215		1.83×10^{-3} sec	α	7.365	

Table 7 (Continued)—PROPERTIES OF HEAVY NUCLEI

Atomic Number	Name	Isotope	Relative Abundance	Half-Life	Particle	Energy in Mev	Gamma Ray
	Ra C′	214		1.5×10^{-4} sec	α	7.680	
		213		$\sim 10^{-6}$ sec	α	9	
	Th C′	212		3×10^{-7} sec	α	8.776	
	Ac C′	211		5×10^{-3} sec	α	7.434	
	Polonium	210		140 d	α	5.298	γ
83	Bismuth						
	Ra C	214		19.7 min	α	5.502	1.8
					$_{-}\beta$	3.15	
		213		1 hr	$\alpha_{-}\beta$		
	Th C	212		60.5 min	α	6.054	
					$_{-}\beta$	2.20	
	Ac C	211		2.16 min	α	6.619	γ
					$_{-}\beta$		
	Ra E	210		5.0 d	$_{-}\beta$	1.17	
	Bismuth	209	100				
		207		6.4 d	K,e^{-}		1.1
82	Lead						
		215		1.6 min	I.T.,e^{-}		
	Ra B	214		26.8 min	$_{-}\beta$	0.65	γ
		213		<1 hr	$_{-}\beta$		

Table 7 (Continued)—PROPERTIES OF HEAVY NUCLEI

Atomic Number	Name	Isotope	Relative Abundance	Half-Life	Particle	Energy in Mev	Gamma Ray
	Th B	212		10.6 hr	$-\beta$	0.36	
	Ac B	211		36.1 min	$-\beta$	0.5	0.8
						1.4	
	Ra D	210		22 yr	$-\beta$	0.025	0.047
		209		3.3 hr	$-\beta$	0.68	
	Lead	208	52.3				
	Lead	207	22.6				
	Lead	206	23.6				
		205		68 min	I.T.,e^-		γ
	Lead	204	1.5				
		203		10.25 min	$+\beta$	1.66	
81	Thallium						
	Ra C″	210		1.32 min	$-\beta$	1.80	
		209		<1 hr	$-\beta$		
	Th C″	208		3.1 min	$-\beta$	1.82	2.62
	Ac C″	207		4.76 min	$-\beta$	1.47	γ
		206		3.5 yr	$-\beta$	0.87	none
	Thallium	205	70.9				
		204		4.23 min	$-\beta$	1.6	none

Table 7 (Continued)—PROPERTIES OF HEAVY NUCLEI

Atomic Number	Name	Isotope	Relative Abundance	Half-Life	Particle	Energy in Mev	Gamma Ray
	Thallium	203	29.1				
		202		11.8 d	K(?),e⁻		0.4
		200(?)		4 min			
				3.8 hr			
		199(?)		44 hr	K(?),e⁻		
		198		10.5 hr	K(?),e⁻		1.0

References:
Segrè, Chart of Isotopes, May, 1945.
G. T. Seaborg, *Chem. and Eng. News* **23**, 2190 (1945).
G. T. Seaborg, *Bull. Am. Phys. Soc.* **21**, 22 (1946).
Seaborg, McMillan, Kennedy and Wahl, *Phys. Rev.* **69**, 366 (1946).

CHAPTER 7

THEORY OF FISSION

In this chapter and the next we shall discuss fission and other nuclear processes in the light of our present ideas about nuclei. Some parts of this discussion will be important for the later chapters on piles and nuclear chain reactions. The most comprehensive treatment of the various aspects of fission is that of Bohr and Wheeler[29] which was published several months after the discovery of fission. Although some work has appeared since then,[30] the general features described by Bohr and Wheeler are still believed to be valid and it is probable that the recent intensive work on the military applications of nuclear energy has not greatly increased our understanding of the fission phenomenon. In fact, it is rather remarkable that the fission of heavy nuclei was not predicted in the two-year period before its discovery, for as we shall see in chapter 8, the liquid drop model of the nucleus had already provided a successful description of many nuclear properties.

The discussion will be divided into two main parts. In this chapter, only energy aspects will be considered. These will include estimates of the energy necessary to produce fission and the energies available from the fission of a nucleus. The following chapter will take up the dynamics of the reaction, i.e., the rate at which the various possible processes occur and the dependence of the fission yield on the competition among these processes.

In order to maintain a certain degree of continuity and to prepare for some of the more quantitative estimates that are essential to the understanding of the experimental results, we include some brief background of the necessary concepts about nuclei. For a much more complete treatment of the properties of nuclei, reference should be made to general reviews of the subject.[31]

This chapter will begin with a review of the fundamental ideas concerning nuclear forces and nuclear radii and a discussion of some methods of determining the latter. These ideas are applied

[29] N. Bohr and J. A. Wheeler, *Phys. Rev.* **56**, 426 (1939).
[30] E.g., R. D. Present and J. N. Knipp, *Phys. Rev.* **57**, 751, 1188 (1940).
[31] H. A. Bethe and R. F. Bacher, "Stationary States of Nuclei", *Rev. Mod. Phys.* **8**, 83 (1936); H. A. Bethe, "Nuclear Dynamics, Theoretical", *Rev. Mod. Phys.* **9**, 69 (1937); See also Gamow, "Atomic Nuclei and Radioactivity", Cambridge (1937).

to the consideration of the mass or total binding energy of the nucleus. A semiempirical expression for the total binding energy enables us to calculate the various interesting energies that play a role in fission. Among these are the fission threshold energies and the energy released in fission. Finally, we will consider the distribution in mass of the fragments and the symmetry of fissioning.

Nuclei—general description.—The most obvious characteristic of nuclei is that they have approximately equal numbers of protons and neutrons. The difference is largest in the heaviest nuclei and even there it is only about 25 percent of the total number of particles. According to our present ideas, this is consistent with the following statements:

(1) The specific nuclear forces (subtracting the Coulomb interaction between protons) are much larger than the electrostatic forces acting on protons in the nucleus.

(2) These nuclear forces are approximately the same for neutrons as for protons.

The quantum states of motion and their energy levels should be similar for both kinds of particle. Because of the Pauli exclusion principle (and the fact that the particle spins are $\frac{1}{2}\hbar$) only two particles of each kind can occupy each of the lowest levels in the nucleus. A nucleus with predominantly one kind of particle would then have a much higher energy than those common in nature and would not exist long (if at all) because of the possibility of transformation to a nucleus of lower energy. The shift toward larger fractions of neutrons in the heavy nuclei is caused by the electrostatic repulsion energy of the protons, which begins to build up rapidly toward the end of the periodic system (proportional to Z^2, the atomic number squared) compared with the slower variation of the nuclear binding energy.

Aside from the very light elements (especially D^2), all nuclei have about the same density and, closely connected with that, they also have nearly the same binding energy per particle, about 8 Mev. This situation is in marked contrast to the electron system about an atom and shows clearly one essential difference in the pictures we must use in describing atoms and nuclei. For example, the Thomas-Fermi statistical method gives a $Z^{7/3}$ behavior for the total binding energy of the orbital electrons as compared with the nearly linear behavior of nuclei. Similarly, if we define the "size" of an atom as that which contains, say, 90 percent of the electron charge, then the Thomas-Fermi method shows that the "radius" of an atom *decreases* as $Z^{-1/3}$. This decrease is a result of the long range Coulomb forces that permit the collective interaction of all the other electron charges with any particular electron. The nucleus, on the other hand, seems to be like a liquid droplet[32] (with a much smaller number of particles), which does have the constant density and linear energy characteristics mentioned above. The nuclear radius varies as $A^{1/3}$, where A is the number of particles in the nucleus.

By extending this sort of consideration, Heisenberg[33] concluded that the specific nuclear forces are short-range attractive forces similar to the familiar exchange interaction between hydrogen atoms in a hydrogen molecule. Two particles interact strongly only when they are close together; and, since this will be true for an appreciable fraction of the time only if the two are in the same quantum state, it follows from the Pauli exclusion principle (and the fact that neutrons and protons have spins of $\frac{1}{2}\hbar$) that at most two protons and two neutrons can comprise a strongly interacting unit. This expresses the saturation character of the nuclear forces and therefore the linearity of the binding energy with atomic weight. To a first approximation there is no tendency for the neutrons and protons to crowd together, since a neutron (or proton) affects only its closest neighbors.

The picture of the random type of motion of the individual particles, with each particle jostling the next one and moving in a rapidly fluctuating field of force, has important consequences in the formal description of nuclei. It is obviously a poor approximation to consider states of motion of one particle in some average field of the others because of the rapid interchange of energy between the particle and the "system." Instead we must consider collective states of motion of the system as a whole, not unlike the modes of motion of a solid lattice. Bethe[34] has emphasized how little progress has been made in the direct quantum mechanical description of the nucleus. However, it is possible to exploit some of the analogies with more familiar systems, such as the liquid droplet, and get a phenomenological description of the heavier nuclei. We shall discuss this more fully in the next chapter, which deals with fission dynamics.

Nuclear radii—Although the nuclear radius is much better defined than an atomic radius, it still essentially lacks definition, chiefly because of the finite range of nuclear forces. Feenberg[35] has estimated the range of the neutron-proton interaction from the mass defects of deuteron and alpha particles and found it to be 2×10^{-13} cm. An interaction of about this range has also been

[32] N. Bohr, *Nature* **137**, 344 (1936).
[33] W. Heisenberg, *Zeits. für Phys.* **77**, 1 (1932).
[34] H. A. Bethe, *Rev. Mod. Phys.* **9**, 69 (1937).
[35] E. Feenberg, *Phys. Rev.* **47**, 850 (1935); **48**, 906 (1935).

measured by Sherr[36] from the scattering of very fast neutrons by protons. There is associated with this finite and nonabrupt range an essential lack of definition of the surface of a nucleus. It follows that the value obtained for the radius of a nucleus, and also the value used in any calculation, must refer to a particular method of measurement and can change by appreciable amounts with the method of definition, experimental or theoretical. Since we shall have occasion to speak of the nuclear radius in at least two different respects, it will be worth while to mention some of the methods of measurement and the spread of values obtained.

Often the quantity $r_0 = RA^{-1/3}$, where A is the integral atomic weight, is used instead of the nuclear radius R to characterize the nucleus. r_0 is a measure of the nuclear density and changes only very little from one nucleus to another. It can also be considered as the radius of the nuclear particles, since $2r_0$ is as close as neutrons and protons approach each other even though there is a large attractive potential between them ($\sim$ 28 Mev). The main methods of determining R (or r_0) are:

(a) By calculation from the natural alpha particle radioactivity of heavy nuclei.
(b) Anomalous scattering of charged particles by nuclei.
(c) Scattering of fast neutrons by nuclei.
(d) Fitting theoretical formulae for mass defects and binding energies to mass spectrograph values.

We shall discuss each method briefly below. It is expected[37] that the radius of a "constant density" nucleus as estimated by method (d) will be smaller than the geometric radius determined from scattering by about $a/2 \sim 10^{-13}$ cm, where a is the range of the nuclear forces. Even for the heavier nuclei the different ways of estimating r_0 give values from 1.2×10^{-13} cm to 1.5×10^{-13} cm. There is also a somewhat questionable estimate[38] that gives the large value $r_0 = 2 \times 10^{-13}$ cm.

(a) ALPHA PARTICLE RADIOACTIVITY: The height of the Coulomb potential barrier and the kinetic energy of the alpha particle determine the probability of escape, or the half-lives of alpha radioactive elements. The height and thickness of the Coulomb barrier are determined by the radius where, and the abruptness with which, the specific nuclear forces become important. Gamow[39] and Bethe[40] estimated R for U^{238} to be 8.9×10^{-13} cm and 12.3×10^{-13} cm, respectively, and obtained values of r_0 for the heavy nuclei of 1.48×10^{-13} cm and 2.05×10^{-13} cm, respectively. Bethe's larger value results from an attempt to apply the many

[36] R. Sherr, *Phys. Rev.* **68**, 240 (1945).
[37] R. D. Present, *Phys. Rev.* **60**, 28 (1941).
[38] H. A. Bethe, *Rev. Mod. Phys.* **9**, 166 (1937).
[39] Gamow, Atomic Nuclei and Radioactivity, Cambridge (1937).
[40] H. A. Bethe, *Rev. Mod. Phys.* **9**, 166 (1937).

body description of the nucleus instead of the familiar potential well picture previously used. According to Bethe, the probability of escape of an alpha particle is the product of the probability of two events: first, that an alpha particle unit will be separated from the rest of the nuclear matter; and second, that the alpha particle will penetrate the potential barrier. In the older theory the first probability was replaced approximately by the oscillation frequency of the alpha particle in the nuclear potential well, $\nu \sim 10^{21}$ per second. This is very much larger than Bethe's estimated separation probability $\Gamma/\hbar \sim 10^{15}$ per second. Bohr and Kalckar,[41] have criticized Bethe's treatment on the grounds that the two events are not so simply separated and it is probable that the value $r_o = 1.48 \times 10^{-13}$ cm is closer to the correct one.

(b) ANOMALOUS SCATTERING OF CHARGED PARTICLES: Weisskopf and Ewing[42] considered the reverse penetration of the potential barrier using a sharp cut-off potential to interpret the experimental results. This potential is expected to give smaller radii. They ascertained that $r_o = 1.3 \times 10^{-13}$ cm fits best the experimental data on reactions initiated by charged particles in nuclei of medium atomic weight. This smaller value of r_o is in agreement with the theoretical ideas of Present[43] and Bethe,[44] which indicate that the nuclear density should decrease about 30 percent in going from the medium weight nuclei to the heaviest nuclei. This decrease is related to the rise in the packing fraction curve beyond the atomic weight 50.

(c) SCATTERING OF FAST NEUTRONS: Fast neutron scattering data are not easy to interpret because the effective wavelengths, $\lambda\!\!\!^{-}(=\lambda/2\pi)$, of the neutrons are comparable with nuclear radii. For example, $\lambda\!\!\!^{-} = 3 \times 10^{-13}$ cm for 2.5 Mev neutrons. Bethe[45] has shown that the effective radius for elastic scattering extends about 10^{-13} cm beyond the nuclear boundary, which when added to the 10^{-13} cm due to the range of the nuclear forces gives an effective radius 2×10^{-13} cm greater than the "constant density" radius. Using the data of Kikuchi, Aoki and Wakatuki,[46] Present[47] has calculated the constant density r_o:

$$Fe^{56}\ ;\ R = 6.5 \times 10^{-13}\text{ cm},\ r_o = 1.30 \times 10^{-13}\text{ cm.}$$
$$Pb^{207};\ R = 10.2 \times 10^{-13}\text{ cm},\ r_o = 1.48 \times 10^{-13}\text{ cm.}$$

[41] Bohr and Kalckar, *Danske Vid. Selsk.* 14, 9 (1937).
[42] Weisskopf and Ewing, *Phys. Rev.* 57, 472 (1940).
[43] R. D. Present, *Phys. Rev.* **60**, 28 (1941).
[44] H. A. Bethe, *Rev. Mod. Phys.* **9**, 69 (1937).
[45] H. A. Bethe, *Phys. Rev.* 57, 1125 (1940).
[46] Kikuchi, Aoki and Wakatuki, *Proc. Phys. Math. Soc. Japan* **21, 420** (1939).
[47] R. D. Present, *Phys. Rev.* **60**, 28 (1941).

This 13 percent increase in r_o is in fair agreement with theory. The values given here are the constant density values. The radii have been corrected from the scattering cross sections by the 2×10^{-13} cm mentioned above. Dunning et al[48] and Barschall and Kanner[49] have measured the total "scattering" cross section for fast neutrons on nuclei by *transmission* experiments. Although the interpretation of such results is not direct,[50] the cross sections followed an $A^{2/3}$ law over a wide range of atomic weights from aluminum to lead. Recently Sherr[51] has published results of the scattering of very fast neutrons, $E \sim 25$ Mev. There are no quantum effects due to long wavelengths at this energy. The results can be plotted in the form

$$R_{sc} = b + r_o A^{1/3},$$

giving a fairly good straight line from light to heavy elements with $b = 2.3 \times 10^{-13}$ cm and $r_o = 1.25 \times 10^{-13}$ cm.

(d) FITTING THEORETICAL FORMULAE TO MASS DEFECTS: This method will be discussed in the following section, which considers the general variation of atomic mass over the periodic system.

Nuclear masses and the semiempirical formula.—It was shown earlier that the short-range exchange forces, together with the Pauli exclusion principle and spins of $\frac{1}{2}\hbar$, produce a saturated arrangement when two neutrons and two protons are strongly interacting. This is, of course, the case of the alpha particle and also of those nuclei in which the alpha particle can move as a sub-unit. This is in agreement with the large binding energy of $_2He^4$ (27 Mev), compared with $_1D^2$ whose ground state is only 2.18 Mev below dissociation. Likewise, the low atomic weight nuclei which are multiples of $_2He^4$, such as $_6C^{12}$ and $_8O^{16}$, have lower packing fractions than the nuclei near them. The concept of alpha particles as sub-units of nuclei is not a precise one and meets with several objections, especially in the heavier nuclei. Bethe[52] has given a detailed discussion of the good and bad features of such a hypothesis. Nevertheless, we may conclude that two particles must be in similar quantum states in order to interact strongly. It is admittedly not a good approximation to speak of single particle quantum states, but the qualitative results are most easily obtained by doing so.

As long as the Coulomb forces are not very large, protons and neutrons move in very similar potential fields and can have

[48] Dunning, Pegram, Fink and Mitchell, *Phys. Rev.* **48**, 265 (1935).
[49] Barschall and Kanner, *Phys. Rev.* **61**, 129 (1942).
[50] E.g., I. Rabi, *Phys. Rev.* **43**, 838 (1933).
[51] R. Sherr, *Phys. Rev.* **68**, 245 (1945).
[52] H. A. Bethe, *Rev. Mod. Phys.* **8**, 169 (1936).

similar quantum states. This is another way of saying that the four particle arrangement can be realized in light nuclei. As the nuclear charge increases, the repulsive interaction between protons begins to raise the proton levels above the corresponding neutron levels, so that more neutron states than proton states lie below the dissociation energy. This condition is responsible for the shift to larger percentages of neutrons in the heavier nuclei. Another important consequence is that heavier nuclei tend to have an even number of both neutrons and protons if possible, for then the neutrons and protons themselves can be paired in quantum states of motion to give the strongest bonds. The effect of *n-n* and *p-p* pairing is noticeable above ${}_7N^{14}$ where there is no stable isotope of even atomic weight and odd charge number. If, then, two nuclear particles are added to an existing nucleus, say, ${}_{2Z}X^{2M}$, the only stable isotopes that can be produced are

$$ {}_{2(Z+1)}Y^{2(M+1)} \text{ and } {}_{2Z}X^{2(M+1)} $$

Not only does the Coulomb repulsion supply directly a positive electrostatic energy, but it can also cause enough difference in the wave functions of neutrons and protons to desaturate the neutron-proton bonds which contribute an important part of the binding energy of a nucleus. A third effect that works in the same direction is the net increase in the kinetic energy of the system associated with a shift from equal numbers of protons and neutrons. This can be seen most simply by considering a mixture of two Fermi gases in a container. If the total number of particles is fixed, the zero-point kinetic energy is a minimum when the numbers of both types are equal. In nuclei, this increase in kinetic energy is probably only about one-third as big as the increase in potential energy due to the decrease in the number of *n-p* bonds.

There has not yet been any successful theory that describes quantitatively the behavior of nuclear masses. However, statistical descriptions[53] do give formulae that have all the qualitative features described in the preceding paragraphs. The theory is unsuccessful in the respect that any attempt to ascribe an independent meaning to the constants appearing in it, and to determine them in a rational way as something other than adjustable parameters, results in poor agreement with experiment. However, the form of these expressions can be borrowed (and simplified) to give a "semiempirical" formula whose constants can then be adjusted to fit the packing fraction curve. The total binding energy

[53] H. A. Bethe, *Rev. Mod. Phys.* **9**, 149 (1937).

of a nucleus of atomic number Z and atomic weight A (Z protons and N neutrons) is given by:

$$E = NM_n + ZM_p - M(Z,A)$$

$$= \alpha A - \frac{\beta(N\text{-}Z)^2}{A} - 4\pi r_0^2 O A^{2/3} - \frac{3/5(Ze)^2}{r_0 A^{1/3}} \quad (3)$$

where $M(Z, A)$ is the mass of the nucleus and M_n and M_p the neutron and proton mass, respectively. The first term, αA, is the largest term and expresses the approximate constancy of the packing fraction of nuclei. The second term takes care of the effect of the n-p desaturation described above. The last term is the Coulomb energy, where $Z(Z\text{-}1)$ has been replaced by Z^2. The third term on the right results from the desaturation of the nuclear forces of the surface of the nucleus and is expressed in terms of a "surface tension" O. The surface energy is a much larger fraction of the total energy here than in a small liquid droplet because of the large fraction of the total number of particles that are surface particles. The rapid drop in the ratio of surface area to volume at the lower atomic weights is the cause of the steep slope of the packing fraction curve there. The desaturation is unimportant for the shape of the curve beyond, say, $A=50$, but even for the heaviest nuclei it represents a sizable part of the total energy. It is instructive to see the relative contribution of all the terms of equation 3 in the heavy nuclei. For U^{238}:

$$\alpha A \sim 3300 \text{ Mev}$$
$$-\beta(N\text{-}Z)^2/A \sim -250 \text{ Mev}$$
$$-4\pi R^2 O \sim -540 \text{ Mev}$$
$$-{}^3/_5 Z^2 e^2/R \sim -800 \text{ Mev}$$

There have been several determinations of the constants in equation 3 from the masses of the stable isotopes. The most reliable seems to be that of Feenberg,[54] whose results are used by Bohr and Wheeler.[55] He adjusted the constants of the semiempirical formula to fit Dempster's[56] packing fraction curve and found the values:

$$4\pi r_0^2 O = 13.3 \text{ Mev}$$

and $\quad R = 1.39 \times 10^{-13} A^{1/3}$ cm,

for $\quad 100 \leqq A \leqq 238$.

Feenberg's procedure seems to be less arbitrary than the others, which require the added assumption that certain nuclei are the "most stable" ones.

[54] E. Feenberg, *Phys. Rev.* **55**, 504 (1939).
[55] N. Bohr and J. A. Wheeler, *Phys. Rev.* **56**, 426 (1939).
[56] A. J. Dempster, *Phys. Rev.* **53**, 869 (1938).

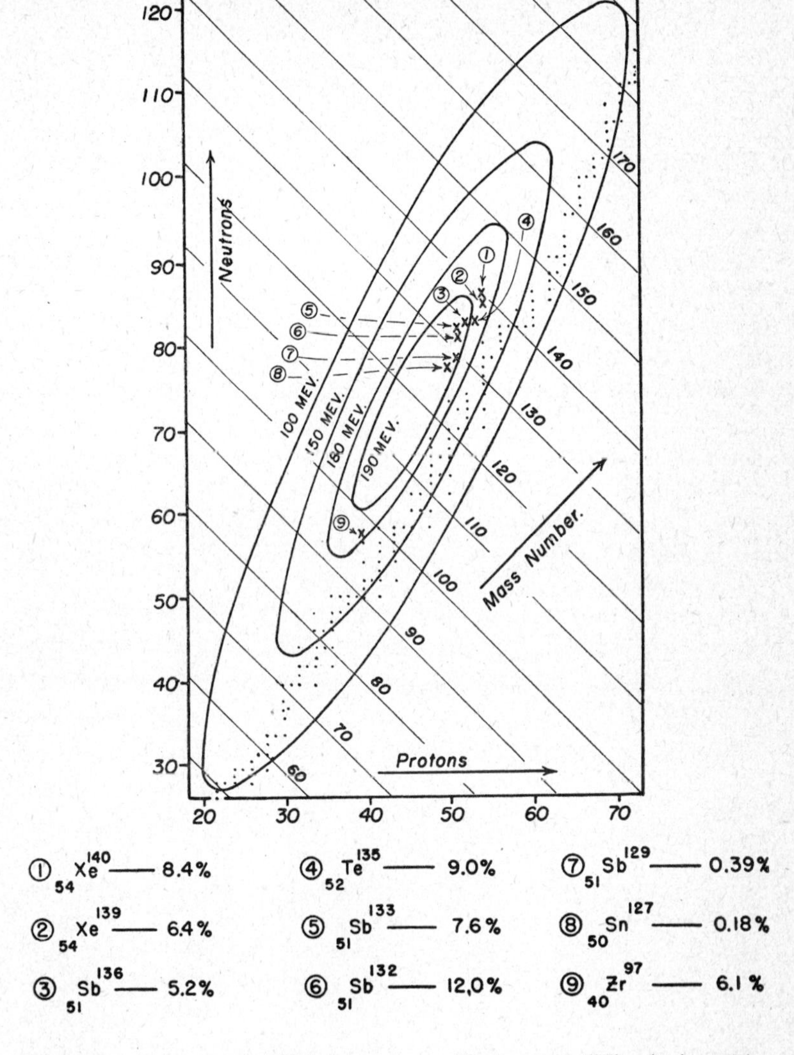

Fig. 23. Energy of U^{239} fragments (Bohr and Wheeler). The stable isotopes lie somewhat below the region of highest energy release. The slope of the axis of the ellipsoid corresponds to the neutron-proton ratio of U^{239}. Also shown are the fragments whose branching ratios have been measured (*cf.* chapter 4). Their position agrees with the estimated average energy released in fission if the assigned energies are lowered by somewhat over 10 mev; and their neutron-proton ratio is almost that of U^{239} except for $_{40}Zr^{96}$ (see text).

Equation 3 describes the average variation of the mass defect with Z and A, but it does not give any information about the fluctuations in the masses of neighboring isotopes which are evident from the rather irregular distribution of the vacancies in the table of stable isotopes (see figure 23). In general, an isotope will not be stable if it is heavier than either of its neighboring isobars. If it is heavier than its neighbor of higher charge, it can transform to it by emitting an electron; if its lower charge isobar is lighter, the nucleus can transform sometimes by positron emission and always by capture of a K electron. Most of the mass fluctuation is due to the odd-even effect resulting from the large n-n and p-p forces. If a particle, say, a neutron, is added to the nucleus ${}_{2Z}X^{2M}$, it will fall into an unoccupied quantum state. If still another neutron is added, it will occupy the same state with opposite spin. However, not only will it interact just like the first neutron with the rest of the system, but also the level will be depressed by the strong interaction between the two neutrons. That this depression is significant is shown by the fact that often the addition of either two neutrons or two protons will produce a stable isotope, whereas most of the time, if a single particle is to be added, it can be only one or the other, depending on which quantum state is the lower. It is seen that the binding energy of a particle in an even-even isotope is larger in general than that of the odd particle in an odd isotope (one with an odd total number of particles). Therefore, the same semiempirical formula cannot describe both the odd and even mass number isotopes. If it is adjusted to represent the odd nuclei, it will give values for the binding energy that are too small in the even-even isotopes and too large in the unstable odd-odd isotopes. It is reasonable to assume that the formula can be corrected up or down by an amount δ_A to give the corrected values for the even-even or odd-odd isotopes respectively. δ_A is expected to vary only slowly with A and is a measure of how much the n-n or p-p interaction depresses the energy levels. In the next section it will be shown how δ_A can be estimated just from the stability of isobars of even nuclei.

Binding energies and the energies released in fission.—The semiempirical formula can be applied to estimate the energies involved in fission, including the energy released in the fission process, the energy available in the fission fragments for subsequent beta decay or neutron emission and the neutron binding energy of the fissioning nucleus. This last gives the excitation energy of the compound nucleus after capture of a slow neutron. For different isotopes we shall see that the neutron binding energy may be more or less than the minimum energy needed to produce fission. This is the reason for the very different behavior of U^{235} and U^{238} when they capture slow neutrons. This treatment, and also many of the numbers quoted are due to Bohr and Wheeler,[57] whom we shall follow now that the preparatory discussion of nuclei is finished.

The energy released in the formation of the fragments M_1 from the fission of the nucleus M_0 is given by the Einstein relation.

$$E = (M_0 - \Sigma M_1)c^2, \qquad (4)$$

where M_0 and M_1 are, respectively, the rest masses of the unexcited original nucleus and the product nuclei. The product nuclei have in general an abnormally high neutron-proton ratio characteristic of the original heavy nucleus, so that to obtain their mass for use in equation 4 we must extrapolate from the masses of the stable isotopes of the same mass number. The extrapolation will be described here. If there were a universal expression for the mass $M(Z,A)$ of nuclei in terms of Z and A, then for a given value, A, there would be a value Z_A corresponding to the most stable isobar. Z_A is not necessarily an integer, but it will be near a stable isobar. Consider the odd nuclei and assume that the semiempirical formula holds. Then Z_A can be found by setting the derivative of equation 3 equal to zero:

$$d(M(Z,A))/dZ = 0 = -\frac{2\beta}{A}(A-2Z_A) + 6\, Z_A e^2/5R - (M_n - M_p):$$

$$Z_A = \frac{A}{2}\,\frac{[\beta + \tfrac{1}{2}(M_n - M_p)]}{[\beta + {}^3/_{10}\, e^2 A^{2/3}/r_0]} \qquad (5)$$

$M(Z,A)$ will vary parabolically with Z in the neighborhood of Z_A.

$$M(Z,A) - M(Z_A,A) = \tfrac{1}{2}\, B_A(Z-Z_A)^2. \qquad (6)$$

B_A comes from equation 3 by a second differentiation:

$$B_A = \left\{ \frac{d^2}{dZ^2}\quad M(Z,A) \right\}_{Z_A} = 4\beta/A + 6e^2/5R \qquad (7)$$

Eliminating β from equations 5 and 7 gives

$$B_A = \tfrac{1}{2}(M_p - M_n + \frac{3}{5}\,\frac{A^{2/3}\, e^2}{r_0})\;(A-2Z_A). \qquad (8)$$

Now Z_A lies between two values, Z and $Z+1$, one or both of which is the stable nucleus of number A, so that Z_A is at most ½ unit from a stable isotope. Instead of using equation 5 Bohr and

[57] N. Bohr and J. A. Wheeler, *Phys. Rev.* **56**, **426** (**1939**).

Wheeler use the above property of Z_A to estimate it directly. They draw a smooth line through the table of stable, odd isotopes (Z plotted against A) so as to pass within ½ unit of each stable odd isotope and use this line to define Z_A. It turns out that the range of possible lines is very small and, as might be expected, the Z_A line is above the integral value of Z as often as it is below it. Thus the parameter β has been replaced by Z_A and equations 6 and 8 give the desired extrapolation. The value of r_o used is 1.48×10^{-13} cm, obtained from alpha particle radioactivity.

Similarly, the "most stable" mass, $M(Z_A,A)$, may be obtained directly from the average value, f_A, of the packing fraction over a small region of atomic weights.

$$M(Z_A,A) = A(1+f_A). \tag{9}$$

In averaging f_A, both the even and odd isotopes are included, because then the positive term $\frac{1}{2}B_A(Z-Z_A)^2$ is largely canceled by the negative $-\delta_A$ of the even isotopes. From equations 6 and 9 and our previous discussion we may write

$$M(Z,A)=A(1+f_A)+\tfrac{1}{2}B_A(Z-Z_A)^2 \begin{matrix} +0 \\ -\delta_A \\ +\delta_A \end{matrix} \left\{ \begin{matrix} A \text{ odd} \\ A \text{ even, } Z \text{ even} \\ A \text{ even, } Z \text{ odd} \end{matrix} \right\} \tag{10}$$

$M(Z,A) - M(Z+1,A)$ is the maximum energy release, E_β, in a beta disintegration.

$$E_\beta = B_A\left[(Z_A-Z) - \tfrac{1}{2}\right] \begin{matrix} +0 \\ -2\delta_A \\ +2\delta_A \end{matrix} \left\{ \begin{matrix} A \text{ odd} \\ A \text{ even, } Z \text{ even} \\ A \text{ even, } Z \text{ odd} \end{matrix} \right\} \tag{11}$$

Similarly, for the K capture process the energy release is $M(Z,A) - M(Z-1,A)$, which is the same as equation 11 with $(Z-Z_A)$ instead of (Z_A-Z). The energy of either process is then

$$E_{\beta\pm} = B_A\left[|Z_A-Z| - \tfrac{1}{2}\right] + \text{odd-even terms.} \tag{12}$$

δ_A can be bracketed closely by examining the stability of the even-even isobars. If an isobar is stable, $E_{\beta\pm}<0$; if it is unstable, $E_{\beta\pm}>0$. For the heavy nuclei δ_A can be determined from the energies of successive beta ray disintegrations. For example, the uranium series:

$U^{238} \longrightarrow UX_1 + \alpha + 4.05$ Mev

$\quad \lfloor\longrightarrow UX_2 + \beta_- + 0.13$ Mev

$\quad\quad \lfloor\longrightarrow U\,II + \beta_- + 2.32$ Mev

$\quad\quad\quad \lfloor\longrightarrow$ etc.

(We have shown only the lower isomer of UX_2.) From equation 10 it follows that the difference between the two beta ray energies is $4\delta_A$, so that $\delta_A = 0.5$ Mev.

Bohr and Wheeler have carried out these quantitative estimates of the fragment masses as well as the beta ray energies released in the subsequent disintegration chains. Table 8 and figure 23, which have been reproduced from their paper,[58] show the energies that can be released by fission into two fragment nuclei. Figure 23 shows the energy released by division of the compound nucleus U^{239} into two fragments. It is seen that there is a large range of masses of the fragments (a range of about 20 for the *heavier* fragments) for which nearly the maximum possible energy release is attained. On the other hand, there is only a narrow range of charge numbers in this region, which is separated from the stable isotopes by about three to five beta emissions. According to the distribution of the fragments the average energy released by the division alone (neglecting subsequent beta emission) should be somewhat over 190 Mev. This amount is not in agreement with experimental results. Measurements of the ionization produced

Table 8

(From Bohr and Wheeler)

THE ENERGY RELEASED ON DIVISION INTO TWO EQUAL FRAGMENTS

(Fission is exothermic down to atomic weights of about 100, where the relatively large surface effect predominates (see page 74), i.e., for fragments of atomic weight below about 50).

Original	Two products	Energy release on division	Energy released in subsequent beta decay
$_{28}Ni^{61}$	$_{14}Si^{30,31}$	−11 Mev	2 Mev
$_{50}Sn^{117}$	$_{25}Mn^{58,59}$	10	12
$_{68}Er^{167}$	$_{34}Se^{83,84}$	94	13
$_{82}Pb^{206}$	$_{41}Nb^{103,104}$	120	32
$_{92}U^{239}$	$_{46}Pd^{119,120}$	200	31

by the recoils (page 18) give 163 Mev. This is just the translational kinetic energy of the fragments. The distortion of the nucleus immediately before splitting is so large that the product nuclei themselves have a large excitation energy aside from that due to their high neutron-proton ratios. This is probably of the order of 10 or 15 Mev and it is not expected that it would be detected in

[58] N. Bohr and J. A. Wheeler, *Phys. Rev.* **56**, 426 (1939).

an ionization chamber. On the other hand, calorimetric measurements give 177 ± 2 Mev for the average total energy released per fission (page 18), not counting that carried off by hard gamma rays, neutrons and neutrinos, as well as by beta activity with half-lives longer than several minutes. The neutrinos and neutrons carry away over two-thirds of the total energy available for the beta decay from the direct fission products to stable isotopes. Table 8 shows that this total energy is about 30 Mev. Thus it seems that there is a slight discrepancy between the actual and the expected difference between the ionization and calorimetric measurements. Furthermore, the kinetic energy of the fragments plus their excitation energy is only about 180 Mev, which is at least 10 Mev below the values assigned in figure 23. It is not improbable that the estimates in figure 23 are in error by that amount.

The paired fragments expected to accompany a fission should lie opposite each other along a line through the center in figure 23 and on the same energy contour. Except for Zr^{97}, no branching ratio measurements for the light fragments have been published. The fragments shown in figure 23 are the first identified members of rather long disintegration chains and it is probable that they are close to the direct fission products. Their neutron-proton ratios are approximately the same as that of U^{239}, which is to be expected from the liquid drop model. The exception, ${}_{40}Zr^{97}$, is possibly only an apparent one since only two members of its chain of disintegrations have been detected.

The possibility of producing fission by neutron bombardment depends on the degree of excitation of the compound nucleus resulting from the capture of a neutron. This is determined by the binding energy of the neutron in the compound state. In principle, equation 10 can be used to calculate neutron binding energies. However, it is more reliable to use the mass differences of neighboring isotopes if these are available. For the heaviest nuclei it is necessary to use the energy released in the various radioactive transformations, since the isotope masses themselves have been calculated from them.[59] As an example let us consider the U^{238} sequence. The steps from U^{238} to U^{234} correspond to the removal of either four neutrons or one alpha particle and two beta particles. If *B.E.* is the average binding energy of the 4 neutrons for U^{234} to U^{238}, the mass equation can be written

$$U^{238} = U^{234} + 4\ {}_0n^1 - 4\ B.E. = U^{234} + He^4 + 6.5 \text{ Mev}.$$

6.5 Mev is the total energy released in the transformation.

$$B.E. = \tfrac{1}{4}\{(4 \times 1.00893 - 4.00336)\ 931 - 6.5\} \text{Mev} = 5.9 \text{ Mev}.$$

[59] A. J. Dempster, *Phys. Rev.* **53**, 869 (1938).

The odd isotopes U^{235} and U^{237} have binding energies smaller by $\delta_0=0.5$ Mev; the even isotope binding energies are larger by δ_0. In a similar manner Bohr and Wheeler calculated the neutron binding energy for some of the interesting compound nuclei. These are included in table 9. The values for plutonium and neptunium were estimated from the behavior of the second term of equation 10,

$$\tfrac{1}{2} B_A (Z-Z_A)^2.$$

In terms of equation 10, the neutron binding energy of the nucleus (Z,A) can be written

$$B.E.=M(Z,A-1)+M_N-M(Z,A)$$

$$=M_N-1-f_A-\tfrac{1}{2}\frac{d}{dA}\{B_A(Z-Z_A)^2\}+ \text{terms due to odd-even effect.} \quad (13)$$

Table 9

ESTIMATED VALUES OF THE NEUTRON BINDING ENERGY OF THE DIVIDING NUCLEUS

(The estimates were made by the method indicated by Bohr and Wheeler. The odd-even fluctuation is clearly shown.)

Compound nucleus	Neutron binding energy
$_{90}Th^{232}$	6.2 Mev
$_{90}Th^{233}$	5.2
$_{91}Pa^{231}$	6.4
$_{91}Pa^{232}$	5.4
$_{92}U^{234}$	6.5
$_{92}U^{235}$	5.4
$_{92}U^{236}$	6.4
$_{92}U^{237}$	5.2
$_{92}U^{238}$	6.1
$_{92}U^{239}$	5.1
$_{93}Np^{239}$	$\sim$ 6.3
$_{93}Np^{240}$	$\sim$ 5.3
$_{94}Pu^{239}$	$\sim$ 5.4
$_{94}Pu^{240}$	$\sim$ 6.4

From the values of B_A and Z_A tabulated by Bohr and Wheeler the derivative term can be estimated for the nuclei in table 9. For example, the slight difference between U^{239} and U^{235} can be attributed to this term. For both U^{235} and Pu^{239} that term is negligible and so the value 5.4 Mev was assigned to Pu^{239}. The larger neutron binding energy of U^{236} compared to U^{239} shows that capture of neutrons by U^{235} will produce a more highly excited compound nucleus than capture by U^{238}. This is a result of the odd-even fluctuations. On the other hand, the critical energy for fission is probably a smoothly varying function of the atomic weight and number.

Equation 13 can be used to estimate the neutron binding energies of the nuclei in the beta radioactive sequences that result from most direct fission products. Likewise equation 11 will give the maximum beta ray energies available. For reasons similar to those of the preceding paragraph the values obtained can fluctuate over several Mev, and there occur cases where the available beta ray energy in going from nucleus A to nucleus B is appreciably larger than the neutron binding energy of the resultant nucleus. This fact, combined with the fair probability that the beta transition will use only a small amount of the available energy, may explain the delayed neutron emission (cf. chapter 5). The nucleus B may be left with sufficient energy to emit a neutron with appreciable energy; if so, the neutron emission will have *no* competition from the beta decay, although there might be some from gamma ray emission. If more complete data on fission branching ratios were available, it might be possible to show that this mechanism accounts for the number of delayed neutrons per fission.

The fission threshold energy.—In this section only the energies required to produce fission will be discussed. We shall not consider the manner in which the excitation energy of the fissioning nucleus is interchanged among the nuclear particles until finally a configuration is reached which leads to fission. This energy interchange belongs strictly to the dynamics of the fission process and will be taken up in the next chapter. For our purpose it is sufficient to assume that if enough energy is available to produce fission (and no other processes compete in carrying off this energy) the configuration leading to fission will eventually be reached. It follows from this that we must consider every possible configuration of the nucleus consistent with energy conservation.

The problem is simplified greatly by the statistical treatment of the nucleus. The nucleus will be considered as a charged liquid droplet with a surface tension calculated from the semiempirical formula, equation 3. The configurations to be considered are the possible distortions of this droplet. Because of the large energies needed to compress nuclear matter, it will be sufficient to consider only those distortions that leave the total volume unchanged. The charge distribution is assumed to be uniform because of the large n-p forces compared to the weaker Coulomb forces. Feenberg[60]

[60] E. Feenberg, *Phys. Rev.* 59, 593 (1941).

has made calculations indicating that the protons do tend to concentrate on the outside of the nucleus. This effect is associated with the progressive change in the density of nuclei with increasing atomic weight mentioned on page 71. However, this tendency is not great and produces only a very small change in the energy of the nucleus and so we shall use the simpler assumption of constant charge density.

Meitner and Frisch[61] were the first to recognize the essential aspects of fission: that the problem is nearly a classical one because of the large masses ($\sim$ one-half the uranium nucleus) that are in motion; and that for the heaviest nuclei the effect of the surface tension in resisting distortions is almost neutralized by the electrical forces. Bohr[62] discussed these ideas in more detail and fitted them into his general theory of nuclear reactions.

Before the general problem is discussed, it is interesting to find the limit of stability of nuclei against fission. This was derived at almost the same time by several persons.[63] We wish to know the values of Z and A for which the decrease of the electrostatic energy associated with an infinitesimal distortion exactly cancels the increase caused by the larger surface area. Feenberg, Frenkel and Weizsäcker considered spheroidal distortions. The result of their calculations is that the electrostatic energy must equal *twice* the surface energy, or that

$$(Z^2/A)_{\textit{limiting}} = 10(4\pi/3)r_o^3\, O/e^2 \tag{14}$$

From the values of the surface and electrostatic energy given on page 74 it is seen that this ratio for normal U^{238} is about 20 percent below this limit.

The critical deformation energy is defined in the following manner: any distortion of the nucleus requires a potential energy, say Ω. This energy can be plotted as a function of the coordinates used to specify the deformation. The function Ω has many minima; one occurs, of course, when the nucleus is undeformed and the others occur when it is broken into several separated spherical fragment nuclei. Of all the ways of going from the undistorted nucleus to two separate fragments, one will require the least energy. The value of this energy, E_c, is called the critical energy for fission. According to our assumption at the beginning of this section, it is also the minimum excitation necessary for fission (neglecting quantum effects). An exact evaluation of E_c for all values

[61] Meitner and Frisch, *Nature* **143**, 239 (1939).
[62] N. Bohr, *Nature* **143**, 330 (1939).
[63] E. Feenberg, *Phys. Rev.* **55**, 504 (1939); F. Weizsäcker, *Naturwiss.* **27**, 133 (1939); J. Frenkel, *Phys. Rev.* **55**, 487 (1939); Bohr and Wheeler, *Phys. Rev.* **56**, 426 (1939).

of Z and A (or, as it turns out, for all values of the ratio Z^2/A) is a very difficult mathematical problem. Fortunately, it is possible to use some approximations because uranium and other nuclei observed to fission are near the limiting value of (Z^2/A) and because at the limit the critical deformations are only slightly different from simple shapes. For $(Z^2/A) \sim (Z^2/A)_{limiting}$, the critical shape is a slightly prolate spheroid; in the other limit of (Z^2/A) very small, the critical shape is two spheres connected by a small neck to balance the weak electrostatic repulsion. Bohr and Wheeler carried out the approximate calculations and obtained a function, $f(x)$, where $4\pi r_o^2 A^{2/3} f(x)$ gives the value of E_c, and x is defined by

$$x = (Z^2/A) / (Z^2/A)_{limiting} .$$

They considered only axially symmetric deformations and specified the deformation by the coordinates $\alpha_o, \alpha_1, \alpha_2 \ldots .$ where α's are the coefficients in the expansion

$$r(\Theta) = R\,[1 + \alpha_o + \alpha_1\, P_1\,(\cos \Theta) + \ldots .]; \qquad (15)$$

$r\,(\Theta)$ is the radius corresponding to the colatitude Θ; P_n is the nth spherical harmonic. For small E_c, i.e., nuclei with x close to unity the value E_c is attained by deformations which are symmetric about some center point. The calculations seemed to indicate that this was also true for the critical deformation of uranium and thorium and the function $f(x)$ plotted in figure 24 was derived using this assumption. Actually a smooth curve is drawn connecting the pieces near $x = 0$ and $x = 1$, which can be calculated by suitable approximations.[64] By using the value of $A^{2/3}$ for U^{235} and the measured fission threshold energies[65], E_c can be assigned to the curve and the values of x for the nuclei of interest can be determined. These values of x lie between about 0.70 and 0.80. Figure 24 shows the region of interest with the assigned values of E_c in Mev. This is a revision of the curve given by Bohr and Wheeler to agree with the more recent threshold measurements described in chapter 2. The vertical lines through the points Th^{232} and U^{238} give the probable errors of the most precise measurements published to date.

Actually the value of x for a particular nucleus can be calculated from the value $(Z^2/A)_{limiting}$ given by equation 14; however, this requires knowing the value of r_o and so it is better to get x from the fission phenomenon itself. Incidentally, as was pointed out by Bohr and Wheeler, this provides an independent determination of the value of r_o, namely, $r_o = 1.46 \times 10^{-13}$ cm.

[64] N. Bohr and J. A. Wheeler, *Phys. Rev.* **56**, 426 (1939).
[65] Cf. chapter 2. The thresholds for photofission are the most reliable.

The curve in figure 24 was drawn through Th^{232} because that seems to give the best agreement with the results that have been published (see chapter 2). The value of E_c for U^{238} is about 0.5 Mev high. Both this and the value for Th^{232} were taken from the photofission threshold measurements by Koch,[66] which give E_c directly. The thorium threshold agrees with the photofission results of Haxby, Shoupp, Stephens and Wells (see chapter 2), who estimated $E_c \sim 6.1$ Mev. This, however, was just an estimate that seemed consistent with their cross section measurements and so there are only the two precise direct measurements of E_c reported to date. However, figure 24 is in good agreement with

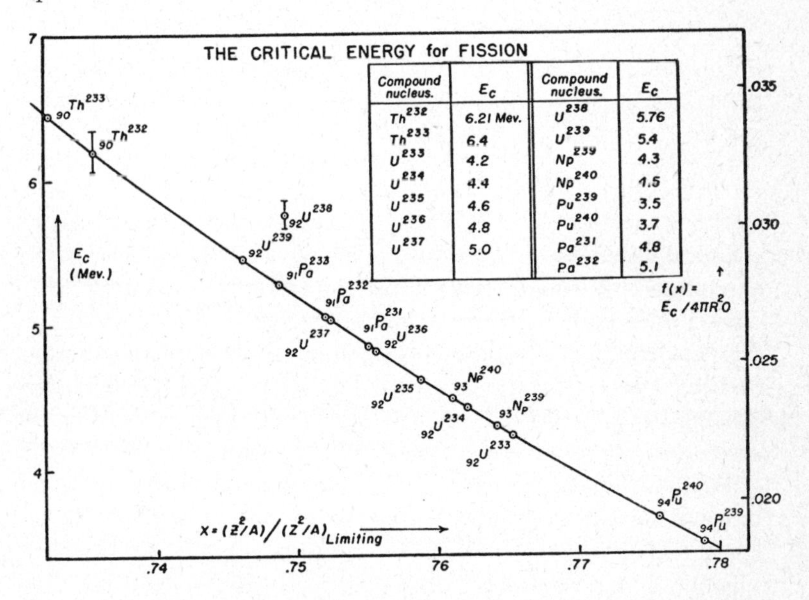

Compound nucleus.	E_c	Compound nucleus.	E_c
Th^{232}	6.21 Mev.	U^{238}	5.76
Th^{233}	6.4	U^{239}	5.4
U^{233}	4.2	Np^{239}	4.3
U^{234}	4.4	Np^{240}	4.5
U^{235}	4.6	Pu^{239}	3.5
U^{236}	4.8	Pu^{240}	3.7
U^{237}	5.0	Pa^{231}	4.8
		Pa^{232}	5.1

Fig. 24. The critical energy for fission, showing the values of the function $f(x) = E_c/4\pi r_o^2$ and the assigned values of the critical energy E_c for the region of interest. The assignment was made to fit best the recent determinations of the thresholds for fission (see text).

the threshold for neutron induced fission. From table 10 the neutron binding energies for U^{239} and Th^{233} are 5.1 and 5.2 Mev, respectively. The measured fission threshold energies are, in order, 0.35 ± 0.1 and 1.1 ± 0.1 Mev. These add up to give E_c (U^{239}) $= 5.45 \pm 0.1$ and E_c Th (233) $= 6.3 \pm 0.1$ Mev, which agree with figure 24.

It is unlikely that the photofission threshold of U^{238} was obscured by the strong neutron or gamma ray competition (which

[66] H. W. Koch, *University of Illinois Thesis* (1944).

we shall discuss in the next chapter).[67] The neutron binding energy is 6.1 Mev according to table 9. Even if this estimate were 0.3 Mev or more too high, which does not seem probable, the binding energy would still be larger than the fission critical energy predicted by the curve in figure 24. Thus no neutrons could be emitted and only the emission of gamma rays could compete with fission. We should then be forced to the *ad hoc* assumption that radiation is much more probable in the compound nucleus U^{238} than in the corresponding nuclei Th^{232}, Th^{233} and U^{239}, where the observed thresholds are self-consistent. It is more reasonable to attribute the higher threshold in part to some specific quantum effect such as will be described in the next chapter. Figure 24 was calculated classically and cannot show such fluctuations. Some of the 0.5 Mev. discrepancy is still unaccountable. According to table 9 and figure 24, protactinium should fission with slow neutrons, since the neutron binding energy of Pa^{232} is 5.4 Mev and $E_c = 5.1$ Mev. However, this phenomenon has not been observed. It is possible that the estimated neutron binding energy of Pa^{232} is in error by enough to explain this fact since there is no convenient radioactive sequence such as there is for uranium from which mass defects can be calculated accurately.

The important prediction of this theory is, of course, that the rarer isotope U^{235} will fission with slow neutrons. Bohr[68] was the first to point out that the rare isotope was almost solely responsible for the slow neutron fission cross section of natural uranium and that this was a consequence of the odd-even fluctuations in neutron binding energy described above. Plutonium would also be expected to fission like U^{235} for it is still more highly excited above the fission threshold by capture of a thermal energy neutron:

$$E_B - E_c \sim 6.4 - 3.7 = 2.7 \text{ Mev extra energy.}$$

Similarly for U^{233} and Np^{239}.

The possibility of symmetric and asymmetric fission.—A look at the table of identified fragment nuclei in chapter 4 shows that most of the splitting is into two unequal parts. To understand this we must consider the distortions leading to fission in somewhat more detail than in the preceding section.

The distortion potential energy can be plotted as a function of the distortion coordinates, say, $(\alpha_0, \alpha_1, \alpha_2, \ldots)$ of equation 15. This will give a "surface" in the hyperspace $(\alpha_0, \alpha_1, \alpha_2 \ldots,$

[67] If this competition were large, the fission yield would be very small for energies near the threshold; and, in the nonlinear extrapolation back to where the fissions first seem to appear, a higher value of E_c might be obtained.

[68] N. Bohr, *Phys. Rev.* 55, 418 (1939).

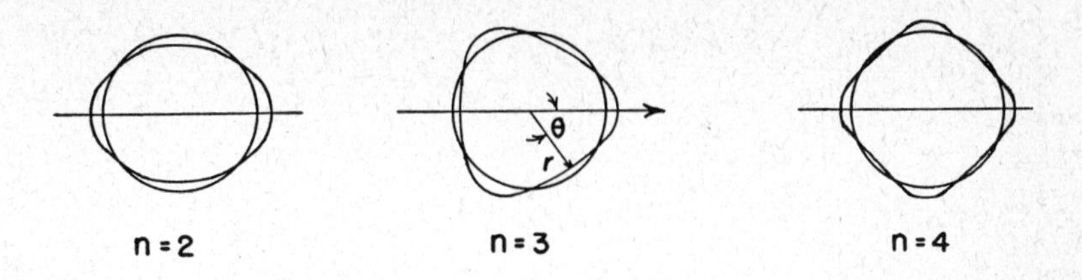

Fig. 25. Deformations of the nucleus corresponding to α_2, α_3, and α_4 of Equation 15. The cases shown are for small values of the α's. For larger α_2 the nucleus will develop a neck connecting two equal portions of the "liquid." For a given increase in surface area the deformation $n=2$ tends to decrease the electrostatic energy faster than any other deformation.

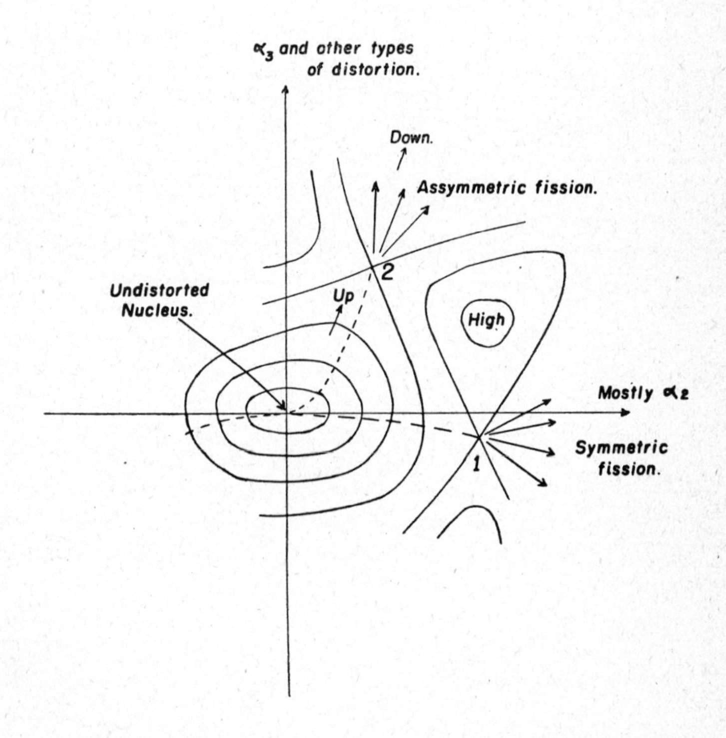

Fig. 26. Potential energy diagram for nuclear deformations. This shows the possibility of two different critical fission energies E_{c1} and E_{c2}, the first corresponding to a symmetric distortion and the second to division into unequal fragments. In the case shown here $E_{c1} > E_{c2}$ which seems to be indicated by experiment.

Ω). This is drawn schematically in contour map fashion in figure 26, where α_2 is one coordinate and all the other coordinates are drawn "perpendicular" to it. The contours of constant potential energy Ω give the limits of the possible distortion of the nucleus which has the excitation energy Ω. The coordinate α_2 has been assigned the special role because it represents that distortion that elongates the nucleus in such a way as to decrease the electrostatic potential energy compared to the increase in surface energy faster than any other. Figure 25 shows the type of deformation corresponding to α_2, α_3 and α_4.

The potential surface will have a pass, or saddle point, at the place corresponding to the critical energy E_c, because E_c is the maximum energy of distortion along the easiest path of deformation passing through that point (see figure 26) and is also by its definition lower than the maximum for any neighboring path. At the critical deformation the nucleus will be in unstable equilibrium and may have developed a neck that weakens the restoring force of the surface tension compared to the disrupting force of electrostatic repulsion. If the nuclear charge is not large, (Z^2/A) not close to $(Z^2/A)_{limiting}$, the neck effect will be pronounced and the mathematics of the distortion becomes complicated, and so there is the possibility of more than one saddle point in some other "direction" of fission. This is shown schematically in figure 26, where the one pass, approximately in the α_2 direction, corresponds to the symmetric distortion considered by Bohr and Wheeler and the other pass corresponds to a strong additional contribution from the α_3 type of deformation (figure 25), which will evidently lead to a fission into unequal fragments. When there is more energy than E_c available there are many paths through the saddle and a variety of fragments.

The possibility of a lower asymmetric fission threshold was suggested by Present and Knipp[69], who investigated larger deformations of the nucleus than did Bohr and Wheeler. They found two interesting saddle points, the first a symmetric distortion ($\alpha_3=0$) and the second with $\alpha_3\neq0$. They did not determine which distortion has the lower value of E_c. The estimated ratio of fragment masses for the asymmetric case is about 2/3, a value which is consistent with the ionization measurements described on page 18, as well as with the fission product results of chapter 4.

In line with this there is also further experimental evidence. Lark-Horovitz and Schreiber[70] observed a change in the type of ionization pulses produced by the fragments in going from slow

[69] Present and Knipp, *Phys. Rev.* 57, 751, 1188 (1940).
[70] Lark-Horovitz and Schreiber, *Phys. Rev.* **60**, 156 (1941).

neutron to very fast neutron induced fission. For slow neutrons they observed two different peaks in the number versus ionization curve. For fast neutrons an additional group corresponding to approximately equal fragments was observed. Similarly, Segrè and Seaborg[71] and others detected fission products from symmetric fission for neutron energies above 10 Mev. The increase in the fission cross section of uranium for neutrons above 10 Mev observed by Ageno et al[72] is possibly connected with the appearance of symmetric fission.

One possible explanation is that there are *two* fission thresholds, as suggested by Present and Knipp, the lower one leading to asymmetric splitting and the other to approximately symmetric splitting. The experiments just mentioned seem to indicate that the latter threshold is at least many million volts above the asymmetric fission threshold, although theoretical calculations[73] have not yet established which is higher! However, it is probable that no symmetric fission will be observed for energies near the threshold for symmetric fission because of the strong competition from other processes. As will be discussed in the next chapter, when a fissionable nucleus is excited there are several competing processes, neutron emission, fission and radiation, which try to carry away the energy; the outcome of this competition determines the fission yield and, therefore, the fission cross section. Thus symmetric splitting may not be detected until the excitation energy is much higher than the threshold value. At these energies, because there are more ways available for the nucleus to fission, one might expect some increase in the relative yield of fissions compared with the neutron emission and so an increase in cross section, as is observed.

Most of the observed increase in the neutron-induced fission cross section of U^{238} is probably due to another process suggested by Bohr.[74] For sufficiently high excitation energy the U^{239} nucleus can emit a neutron and *still* have enough residual energy to fission. Bohr has shown that this double process gives the proper magnitude for the increase in the fission cross section. He has shown also that a much larger relative increase in cross section should result from bombarding thorium with high-energy neutrons. In both cases the fission following the emission of a neutron is favored because the residual nuclei U^{238} and Th^{232} are even and have

[71] Segrè and Seaborg, *Phys. Rev.* **59**, 212 (1941); Nishina, Yasaki and Ikawa, *Phys. Rev.* **58**, 660 (1940); **59**, 323 (1941).
[72] Ageno, Amaldi, Bocciarelli, Cacciapuoti and Trabacchi, *Phys. Rev.* **60**, 67 (1941).
[73] Present and Knipp, *Phys. Rev.* **57**, 1188 (1940).
[74] N. Bohr, *Phys. Rev.* **58**, 864 (1940).

a larger neutron binding energy than the parents. This greatly decreases the neutron competition.

This double process might also be an alternative (or supplementary) mechanism for symmetric fission. First let us assume that the symmetric distortion threshold is the lowest and the only one of interest. For a nucleus to break into equal parts it must cross over the fission barrier in the vicinity of the saddle point, i.e., with energy not much higher than the critical energy for fission. This is not improbable according to the theory of nuclear reactions. If a nucleus has an excitation energy E, then it can be characterized by a "temperature" T (in energy units), which is much lower than E (see page 95). The probability that the nucleus will emit a neutron of energy ϵ is given by the Boltzmann law.

$$W(\epsilon)d\epsilon = \frac{\epsilon}{T^2} e^{-\epsilon/T} d\epsilon. \tag{16}$$

Consider, for example, an incident neutron of 10 Mev energy. It will be captured, leaving the compound nucleus with about 16 Mev excitation energy (6 Mev binding energy). The temperature corresponding to this excitation energy is about 2 Mev. Now let us assume a reasonable value of 0.10 Mev for the necessary proximity of the excitation energy of the nucleus to the critical energy, E_c, to produce nearly equal fragments. (This energy comes from the quantum mechanical description to be discussed in chapter 8. There are only a few states of motion of the nucleus with energy near the critical energy that can produce fission. The lowest states are separated by about the 0.10 Mev used here. We are assuming then that at least the lowest state leads to symmetric fission.) To leave the residual nucleus with just the energy E_c, a neutron must escape with energy $(10 - E_c)$ Mev. Substituting in equation 16 and using $E_c \sim 6$ Mev we get

$$W(\epsilon)d\epsilon \sim 2 \text{ percent},$$

which agrees fairly well with the observed fraction of fissions that are symmetrical, as reported by Lark-Horovitz and Schreiber. It would be interesting to do photofission experiments in which symmetrical fission is looked for as the gamma ray energy is increased in the neighborhood of the fission threshold. A simpler experiment is to examine the fission products from the 6.3 Mev F(p,γ) gamma rays on thorium, whose threshold is 6.2 Mev.

The path of fission (as represented by the dotted lines in figure 26 showing successive distortions) may have a wide range of directions after passing over the saddle point. This is especially important when the nucleus passes over the fission barrier with energy in excess of E_c. Therefore, the saddle point 1 does not necessarily lead to symmetric fission or even to binary fission. Present [75] has shown that it is possible to cross through the saddle region at such an angle as to lead to division into three fragments and the release of about 20 Mev more energy than in symmetric fission. Apparently this is only probable at higher excitation energies,[75] although the threshold is the same as for binary fission.

[75] R. D. Present, *Phys. Rev.* **59**, 466 (1941).

CHAPTER 8

DYNAMICS OF FISSION

In chapter 7 we discussed the possibility of the occurrence of fission; here we shall consider the probability of its occurring once it is possible. Fission is like other nuclear processes in that it can be described in terms of the initial formation of an excited compound nucleus and the subsequent competition among the possible modes of releasing the excitation energy. This description was initiated by Bohr[76] and developed by many others.[77]

Since the nucleus is a "dense" system, an impinging particle quickly loses its energy, which is rapidly distributed among the nuclear particles. The energy of excitation is then dispersed over all the nucleus, each nuclear particle having on the average a very small fraction. The compound nucleus remains in this state of excitation until, by chance, enough energy is concentrated on an individual particle to permit its escape. If a particle of the incident type escapes, the process is inelastic scattering. If only a gamma ray is emitted, the incident particle is "captured." Other transmutations involve the ejection of a neutron, proton or alpha particle from the compound nucleus with or without a subsequent emission of radiation. In fission a mode of "surface tension" oscillation is excited that leads to rupture of the heavy compound nucleus.

For a quantitative discussion of the dynamics of the fission process we must consider more closely the levels of the compound nucleus. Consequently, we shall first discuss the many particle model and nuclear energy levels in general.[78] Level widths and competition among disintegration processes, the dispersion formula and the statistical method of considering closely spaced levels will be described. In later sections we shall apply these ideas to the calculation of reaction rates for fission and compare them with experiment following Bohr and Wheeler.[79]

Many-particle model and nuclear energy levels.—When a heavy nucleus is bombarded by neutrons of, say, 1 Mev energy,

[76] N. Bohr, *Nature* **137, 344, 351 (1936).**
[77] E.g., H. A. Bethe, *Rev. Mod. Phys.* 9, pt. B (1937).
[78] Bohr and Kalckar, *Danske Vid. Selsk.*, **14,** 9 (1937); H. A. Bethe, *Rev. Mod. Phys.* **9,** pt. B (1937).
[79] Bohr and Wheeler, *Phys. Rev.* **56,** 426 (1939).

the most common process observed is scattering, both elastic and inelastic.[80] The inelastic scattering is accompanied by the emission of gamma rays whose line breadths are fairly small, about 0.10 volt for 10^5 volt energy quanta. Since the neutron could traverse the nucleus in a time $\tau \sim 10^{-21}$ seconds, it is difficult to see how in this short time the nucleus could radiate a *sharp* line of frequency $\sim 10^{20}$ per second. In fact, the neutron must be in the nucleus about one million times larger than τ. Similar considerations of the capture of charged particles in lighter nuclei and the resonances observed in their transmutation[81] led Bohr[82] to discard the former potential well description of the nucleus and to propose the intermediate-state process described in the introduction. The formation of a compound nucleus for high-energy incident particles may be simply described classically. The particle reaching the surface of the nucleus suffers an inelastic impact with the surface particles in its vicinity and may continue losing energy by impact until it is no longer distinguishable from any other nuclear particle. This process is unlike ordinary atomic collisions, which are mostly elastic and which are characterized by the collective action of all the orbital electrons on the incident particle. High-energy particles incident on a nucleus will almost always form a compound nucleus. Low-energy particles need a wave description, and a compound nucleus can be formed only if the energy of the incident particle plus target nucleus coincides with an energy level of the compound system. Such capture may then be a highly selective process. This change in character of reactions in going from low to high incident particle energies results from the smearing out of the energy levels of the compound nucleus into a continuum of closely spaced overlapping levels. One consequence of this difference between a continuum and a discrete level system has been pointed out by Kalckar, Oppenheimer and Serber.[83] In a discrete level, the compound nucleus has a well-defined probability for disintegration into its possible products, which is independent of the way it was formed. In a continuum the compound state consists of a combination of many neighboring states, and their phase relationship can be characteristic of the mode of formation; if so, the disintegration probability may also depend

[80] Dunning, Pegram, Fink and Mitchell, *Phys. Rev.* **48**, 265 (1935). See also a recent paper by Sherr, which summarizes the results of fast neutron scattering: R. Sherr, *Phys. Rev.* **68**, 240 (1945).
[81] E.g., Hafstad, Heydenberg and Tuve, *Phys. Rev.* **50**, 504 (1936).
[82] N. Bohr, *Nature* **137**, 344, 351 (1936).
[83] Kalckar, Oppenheimer and Serber, *Phys. Rev.* **52**, 273 (1937).

on the method of formation. In particular, the probability of re-emission of the incident particle may be abnormally high.[84]

The energy levels of a nucleus are states of collective motion of all the particles. The analogy to the oscillations of a liquid droplet or to the vibrations of a solid lattice has been mentioned before (chapter 7). This analogy can be extended to give a semi-quantitative description of the distribution of energy levels. In a heavy nucleus there is such an enormous number of ways in which energy can be distributed over the nuclear particles that the energy level spacing decreases very rapidly with increasing excitation energy. This spacing variation can be estimated on a purely statistical basis with the droplet model.[85] The collective type of motion of the nuclear particles is assumed to have fundamental frequencies and "harmonics." If the energy of excitation E is expressed as a multiple of the quantum energy $h\nu$, then $nh\nu$ can be divided among the nuclear particles in $p(n)$ ways. "$p(n)$" is the "partitio numerorum," or number of ways in which the integer n can be written as the sum of integers smaller than itself. For large n the asymptotic form of $p(n)$ is

$$p(n) = \frac{1}{4\sqrt{3}n} e^{\pi\sqrt{2n/3}} \tag{17}$$

In our case, n is the ratio $E/h\nu$, where E is the excitation energy. $h\nu$ can be estimated for heavy nuclei from gamma ray evidence and the fine structure of alpha particle radioactivity;[86] it is about 1—2x 10^5 ev. ν can also be calculated from the frequency of surface oscillations in the liquid drop model. For heavy nuclei ($A > 100$) this model gives a value of $h\nu$ somewhat larger than 2×10^5 ev.[87] If we use the value $h\nu = 2 \times 10^5$ ev, then for an excitation energy $E = 8$ Mev, n is 40 and $p(40)$ is about 2×10^4. This represents the number of levels in the unit interval $h\nu$. Therefore, the level spacing is about 10 volts. 8 Mev corresponds approximately to the excitation of a heavy nucleus by capture of a slow neutron. The distance between levels for incident neutrons of 2.5 Mev energy is about 0.5 volts.

Although equation 17 is based on crude considerations, it agrees with more elaborate derivations and with experimental observations on the capture of slow neutrons by heavy nuclei. These experiments indicate a level spacing of about 10 or 20 volts for neutron capture.

[84] Bohr, Peierls and Placzek, *Nature* **144**, 200 (1939).
[85] Bohr and Kalckar, *Danske Vid. Selsk.* **14**, 9 (1937).
[86] H. A. Bethe, *Rev. Mod. Phys.* **9**, sec. 69 (1937).
[87] H. A. Bethe, *Rev. Mod. Phys.* **9**, sec 69 (1937).

If we make the basic assumption that all possible states of distribution of the energy E are equally probable, we have an analogue of the different configurations of gas molecules in a container. Accordingly, we should be able to assign some temperature to a nucleus of energy E just as we can to a quantity of gas of known energy. The nuclear "temperature" so defined is a very useful concept for describing nuclear reactions. Moreover, this assumption of thermal equilibrium among the particles is necessary for any more refined calculation of the energy level distribution. Several such calculations have been made.[88] The particular model used determines the nuclear temperature T as a function of the excitation energy. For example, if the nucleus is considered as a Fermi gas, then

$$E = \alpha T^2 \qquad \text{or} \qquad T = (E/\alpha)^{1/2}$$

just as for the free electrons in a metal. Thus, the level density $\rho(E \text{ or } T)$ can be obtained directly by first finding the entropy $S(T)$ and using the Boltzmann relation $\rho \sim e^{S(T)}$. As might be expected, the liquid droplet model gives fair results for the heavier nuclei but not for the light nuclei.

Our knowledge (both theoretical and experimental) of the spacing of nuclear levels can be summarized briefly. For the light nuclei, i.e., those of atomic weight about 15, both theory and experiment give average level spacings of about 1 Mev for 10 Mev excitation energy. The experimental evidence comes from the distribution of the resonances of alpha particle and proton reactions[89] as well as from the gamma rays emitted.

For somewhat heavier nuclei ($A \sim 30$) the spacing is about 0.5 Mev for the same excitation energy and decreases to the order of 10^5 volts at about 15 Mev energy. This contrasts with the heavy nuclei whose lowest excited states are some 10^5 volts above the ground state and are spaced about 10 volts apart at the neutron dissociation energy (between 6 and 7 Mev). The difference is due to the much larger number (~ 200) of particles, which greatly increases the number of ways of sharing the energy E. Slow neutrons can be captured only into compound levels whose angular momentum J differs by no more than one-half from that of the original nucleus. Thus the level spacing observed for neutron capture in the heavy nuclei may be larger than the actual spacing.

[88] V. Weisskopf, *Phys. Rev.* **52**, 245 (1937); L. Landau, *Zeit. Sowjetunion*, **11**, 556 (1937); H. A. Bethe, *Rev. Mod. Phys.* **9**, pt. B (1937); Oppenheimer and Serber, *Phys. Rev.* **50**, 391 (1936); Bardeen, *Phys. Rev.* **51**, 799 (1937).

[89] H. A. Bethe, *Rev. Mod. Phys.* **9**, 208 (1937).

Level widths and competition among disintegration processes.—Like the excited states of atoms, the higher-energy states of a nucleus are not stationary but decay with time by making transitions to other levels with the emission of radiation or material particles. Because our assumed compound state is not an exact solution of the Schrödinger equation, it cannot correspond to a single energy eigenvalue, and the state is taken to have an uncertainty in its energy or a level width $\triangle E$. The uncertainty will be larger the less correct the description of the compound state is, that is, the more important the asymptotic behavior of the wave function which corresponds to the emission of particles becomes. This fact demonstrates the relationship between the level width and the probability of disintegration of the compound state. The lifetime of the state, λ, is simply connected to $\triangle E$. If the initial amplitude of the compound state is A, the subsequent behavior is $Ae^{-t/2\lambda}$. Then

$$\triangle E = \hbar/\lambda = \hbar\,\omega \tag{18}$$

ω is the transition probability per unit time. $\triangle E$ denotes the half width of the distribution in energy of the emitted particles. Equation 18 is just a statement of the Uncertainty Principle and is exactly analogous to the case of optical transitions in atoms. In fact, much of the formalism of nuclear theory is similar to the description of the optical phenomena of line breadth,[90] resonance and dispersion, the sole difference being the replacement of particles for light quanta. There is a larger variety of final states in nuclear transitions corresponding to the different possible product particles as well as the energy levels of the final state of the nucleus.

It is customary to use the symbol Γ instead of $\triangle E$ in equation 18; i.e.,

$$\omega = \Gamma/\hbar. \tag{19}$$

$\Gamma/\hbar$ is the total transition probability of the compound state; it can be separated into contributions from all the possible final states (all the ways in which compound nucleus can disintegrate).

$$\Gamma = \sum_{qP} \Gamma_{qP}. \tag{20}$$

Γ_{qP} refers to the emission of the particle q, leaving the residual nucleus in the state P. The partial transition probabilities are proportional to the squared matrix elements of the Hamiltonian taken with respect to the initial and final systems. The expression for Γ_{qP} is

$$\Gamma_{qP} = 2\,\pi\,\rho(E) \quad |\,H_{qP}{}^{C}\,|^{\,2}. \tag{21}$$

[90] See for example: Weisskopf and Wigner, *Zeit. für Phys.* **65**, 18 (1930).

H_{qP}^C is the matrix element referred to above, the superscript C denoting the compound state. The initial and final states are normalized to unity. $\rho(E)$ is the density per unit energy of the final states, which may be either free-particle states or states of the radiation field.

The theoretical definition of Γ is of little help in calculating it because of our incomplete knowledge of nuclear forces and our still more meager knowledge of the states of motion under such forces. Most of our information about level widths is deduced from direct or indirect measurements of nuclear resonances and the relative yield of the different reactions. We shall discuss the pertinent results in this section and make some estimates of the neutron width Γ_n. The neutron width is rather important, since it is the determining factor in most fission reactions.

Equation 21 is very useful in correlating experimental observations and in extending experimental results into regions of energy not easily investigated. For example, equation 21 tells how the particle widths vary with the energy of the particle emitted and with its angular momentum. These considerations are also applicable to the capture process as well as to emission, and $\Gamma_{qP}^C/\hbar$ plays the converse, though not quite so simple, role in the formation of the compound state C from the free particle q and initial nucleus P. The variation of Γ with energy is responsible for the well known "$1/v$ law" for slow neutron capture. The first term $\rho(E)$, for material particles of mass m, energy E and wave vector k (where $E = \hbar^2k^2/2m$), contained in a volume is

$$\rho(E) = \frac{\Omega}{4\pi^2}\left(\frac{2m}{\hbar^2}\right)^{3/2}\sqrt{E}$$

$$= \frac{\Omega\, m}{2\pi^2\,\hbar^2}\,k$$

$\rho(E)$ is proportional to the particle velocity. For gamma rays $\rho(E)$ is somewhat different, being

$$\rho(E) = \frac{\Omega\; E^2}{2\,\pi^2\,\hbar^3\,c^3},$$

$$= \frac{\Omega\; k^2}{2\pi^2\;\hbar c}.$$

We therefore expect an essential difference between the variation with energy of Γ_γ, the gamma ray width, and Γ_x, the width for emission of particle x. This is only an illustration; there are more profound differences between particle and radiation widths. For the present, in preparation for the "$1/v$ law", we shall consider only the particle width, Γ_n. The terms $H_{qP}{}^C$ in equation 21 is a matrix element of the form

$$H_{qP}{}^C \sim \int \psi_C{}^* H x_P \Phi_q d\tau \quad ,$$

where ψ_C is the wave function of the compound state, x_P that of the initial nucleus and Φ_q the incident particle wave function, which for simplicity we will take to be a plane wave. Thus,

$$\Phi_q = \frac{1}{\sqrt{\Omega}} e^{i\vec{k}\cdot\vec{r}}$$

A plane wave corresponds to a uniform stream of incident particles which can have all values of the angular momentum $l\hbar$ with respect to the nucleus P. In fact Φ_q, and therefore $H_{qP}{}^C$, can be split into the sum of parts corresponding to particles with different values of l. It is easy to show that the matrix element corresponding to $l = 0$ does not contain k, the matrix element corresponding to $l = 1$ contains k to the first power, etc. For slow neutrons the wavelength is so large that only terms of $l = 0$ are important. Thus, the partial width Γ corresponding to emission of a slow particle will vary directly as the velocity of the particle.

An important consequence of this variation with velocity is the large probability of capture compared to scattering of slow neutrons. If a heavy nucleus captures a slow neutron, the compound nucleus has only two alternatives of appreciable probability. It can either radiate a gamma ray and fall into state of lower energy, or it can re-emit the slow neutron. The latter probability is small for small velocities, so that radiative capture is the most probable process. In general, the relative probability of the occurrence of the process x in a compound state is

$$\Gamma_x / \sum_i \Gamma_i = \Gamma_x / \Gamma,$$

where the total width Γ is also written as the sum of the partial widths Γ_i.

To summarize the information available on level widths:

(a) GAMMA RAY WIDTHS: The total gamma ray widths are remarkably constant both with respect to different nuclei and to energy of the compound state. From the resonance widths of slow neutron capture in heavy nuclei the values of Γ_γ are

$$\Gamma_\gamma \sim 0.1 - 1 \text{ volt.}$$

In this case, because of the small value of Γ_n as mentioned above, the total width of the capturing level is $\Gamma = \Gamma_\gamma$. For the lighter nuclei at somewhat higher excitation energies Γ_γ is about 1 to 10 volts. The very small radiation probability (0.1 volt corresponds to 10^{14} sec^{-1}) peculiar to nuclei results from the approximately uniform distribution of charge, which makes dipole radiation very improbable. In fact, the dipole moments are of the order of 10^{-3} times the nuclear radius. Most nuclear radiation is quadrupole radiation, i.e., radiation emitted by a uniformly charged sphere which is oscillating in ellipsoidal deformations. The total gamma width Γ_γ includes the possibility of radiating many different frequencies ($\sim 10^5$ possible final levels in heavy nuclei). The partial width for a single transition, Γ'_γ, probably increases with about the fifth power of the frequency radiated;[91] similarly for the reverse probability that the nucleus will absorb a quantum.

(b) PARTICLE WIDTHS.—For neutron emission the width Γ_n depends only on the probability that a fluctuation will occur in the distribution of the excitation energy so as to concentrate enough energy on the neutron to separate it from the rest of the nuclear matter. In the case of charged particles this must also be multiplied by a penetration factor because of the coulomb barrier. This makes charged particle emission very improbable for nuclei of medium atomic weight. This effect is small for light nuclei, $A \sim 15$, because the barrier for low Z is rather low.

Other things being equal, the probability of concentrating the energy on a neutron should be about the same on the average as for protons and alpha particles. Which will be the most probable in a particular compound nucleus depends on the energy evolution. For example, if the compound nucleus is a light one and contains an integral number of alpha particles, then because of the low internal energy of the alpha particle its creation will leave the residual nucleus with a high excitation. This makes many residual states available, and alpha emission should be predominant. This observation is verified in the light nuclei transmutations

$$Li^7 + H^1 \rightarrow (Be^8) \rightarrow 2He^4,$$
$$F^{19} + H^1 \rightarrow (Ne^{20}) \rightarrow O^{16} + He^4, \text{ etc.}$$

In fact, since the energy of charged particles is easier to measure, most of our information about particle widths in light nuclei ($A < 30$) is obtained from emitted alpha particles and protons. The neutron widths should be comparable. The particle widths for light nuclei are much larger than for heavy nuclei. For very light elements Γ_{particle} is about 10^5 volts for 10 Mev excitation,

[91] See for example, H. A. Bethe, *Rev. Mod. Phys.* **9**, 227 (1937).

if the transition is not forbidden by any selection rule. This is the case in the B^{10} reaction, which is a sensitive detector of slow neutrons:

$$B^{10} + n^1 \rightarrow (B^{11}) \rightarrow Li^7 + He^4.$$

In the next section it will be shown how the large value of Γ_α makes the boron reaction useful for measuring neutron energies. It should be emphasized that the partial width Γ_α for light nuclei refers to the emission of particles with only a small number of different energies (~ 2). In heavy nuclei such a Γ would be extremely small and the large neutron widths that are observed result from the increased number of possible final states. Primes will be used to denote that the partial width refers to a single final state; no prime indicates the total probability of emission of the particle in question. Thus for gamma rays $\Gamma_\gamma \sim 1$ volt, but the value obtained for a single transition is $\Gamma'_\gamma \sim 0.01$ to 10^{-4} volts.[92]

The partial width Γ_n' for neutron emission has been estimated from the resonance absorption of slow neutrons. We have already seen that $\Gamma_n' \propto E_n^{1/2}$. The constant of proportionality or reduced partial width j is remarkably constant from element to element.

$$\Gamma_n' = j \ (E \text{ in Mev})^{1/2} \text{ volts,}$$

where j is between 0.1 and 1 volt. This is in agreement with the large capture probability of slow neutrons, since for neutrons of about ten volts energy

$$\Gamma_n'/\Gamma_\gamma \sim 10^{-3}.$$

Γ_n' corresponds to a transition that leaves the residual nucleus in its ground state. Since the next highest level is some 200 kv above the ground state, for neutron energies less than about 200 kv, $\Gamma_n' = \Gamma_n$. Above this energy more final levels are available and Γ_n increases more rapidly than $E^{1/2}$. For excitation energies of the compound nucleus about 5 Mev above the neutron threshold $\Gamma_n \sim 10^4 \ \Gamma_\gamma$ and radiation is negligible as a competing process Neutron emission is already predominant at about ½ Mev above the threshold. This means that the most common process in heavy nuclei bombarded by fast neutrons is inelastic scattering. The probability of elastic scattering is Γ_n'/Γ_n where Γ_n' refers to the emission of a neutron with the same energy as the incident neutron. This ratio decreases approximately inversely as the number of final states available and is negligible for a bombarding energy of several Mev. Γ_n' also measures the probability of formation of the compound state by capture of the incident neutron.

[92] From the photo disintegration of light nuclei. See Kalckar, Oppenheimer and Serber, *Phys. Rev.* **52**, 273 (1937).

In the next section we shall make some estimate of the total neutron width Γ_n and also of the energy spectrum of the neutrons emitted. This latter is obviously of importance in any fast neutron chain reaction. The total width Γ_n will determine the effectiveness of the neutron emission in limiting the fission yield. The relatively low degree of concentration of energy associated with fission will make the fission width Γ_f large.

The dispersion formula and the statistical method.—The level system of a compound nucleus does not change in character when the energy exceeds the binding energy of a neutron. Suppose the distance between levels is still large compared to their widths in this energy region. Then the compound states can be considered as discrete, approximately stationary states of well-defined energy. Because of the principle of energy conservation, an incident particle cannot be captured by the original nucleus unless the total energy (kinetic + internal) of the initial system coincides with an energy level of the compound system. The capture of a neutron is therefore a very selective process at these energies. The process is exactly analogous to the optical phenomenon of resonance absorption.

Most nuclear reactions are so-called double processes, i.e., they proceed via an intermediate state. However, there can be some direct transitions from the initial to the final state. This is the case of potential scattering where the incident particle never enters the nucleus but may interact with some sort of average field close to the surface of the nucleus. For charged particles this average field extends to great distances as a pure coulomb field and is responsible for most of the scattering at moderate energies. For neutrons, on the other hand, the potential field is only effective at the nuclear surface, and Bethe[93] has shown that its effect is similar to that of a rigid sphere of the same radius. At sufficiently high neutron energies the effective wavelength $\lambda\!\!\!^-$ is smaller than the nuclear radius, and the total cross section presented by the nucleus can be written as the sum of the potential scattering cross section πR^2 and the cross section for formation of a compound state. In any case the distinction between the two types of processes is purely formal and becomes untenable at lower neutron energies, where the two processes can interfere.

Let us consider reactions accompanied by the formation of a compound state. For simplicity, only a single compound level of angular momentum J will be considered important. Then the yield

[93] H. A. Bethe, *Rev. Mod. Phys.* **9**, 91 (1937).

of the reaction produced by bombardment with neutrons of kinetic energy E is described by the dispersion formula:

$$\sigma_x = \frac{\pi \lambda\!\!\!^{-2} (2J+1)}{(2i+1)\ (2S+1)} \quad \frac{\Gamma_n' \Gamma_x}{(E-E_o)^2+(\Gamma/2)^2}; \qquad (22)$$

σ_x is the cross section for production of particle or process x, Γ_x is the corresponding partial width and Γ_n' is the partial neutron width discussed above, which measures the capture probability. Γ is the total width for all ways of disintegration of the compound state. E_o is the kinetic energy of the incident particle at "resonance," $\lambda\!\!\!^{-}$ is the effective wavelength of the incident particle in the reduced mass system and S is its spin, equal to ½. i is the angular momentum of the initial nucleus and is probably zero if the nucleus contains an even number of particles. A dispersion formula of this type was first applied to nuclear reactions by Breit and Wigner.[94] Bethe and Placzek[95] extended the formula to the case where many compound levels are important at the same time; they also took into account the angular momentum of the system.

The competition between different processes is evident in equation 22; the ratio of the yields of particle x and y is just Γ_x/Γ_y. If the width of the level Γ is small, the resonance is sharp and the yields smaller for values of E off resonance. For low energies the quantity Γ_n' depends on the energy of the captured neutron according to equation 21, i.e., it is proportional to neutron velocity. In general, for slow neutron capture $\Gamma \sim \Gamma_y$ or Γ_f, depending on which is larger where $\Gamma_f/\hbar$ is the probability per unit time of fission. Neither of these last two quantities will change with a small variation of the energy of the incident particle in the neighborhood of zero energy. Thus for sufficiently small energies (for which $E_o - E \sim E_o$), equation 22 becomes

$$\sigma_x = \text{const.} \times \lambda\!\!\!^{-} = \frac{\text{const.}}{v},$$

which is the $1/v$ law. In several cases this behavior extends to neutron energies of many kilovolts. Substances that absorb neutrons by the $1/v$ law can be used to determine the energy of neutrons that are responsible for some reaction when monochromatic neutrons are not available. The absorption coefficient $K(E)$ of these reaction producing neutrons in the $1/v$ absorber is com-

[94] Breit and Wigner, *Phys. Rev.* **49**, 519 (1936).
[95] Bethe and Placzek, *Phys. Rev.* **51**, 450 (1937).

pared with that for neutrons of known energy E_1 and the energy calculated from

$$\frac{E}{E_1} = \frac{K^2(E_1)}{K^2(E)},$$

For example, the resonance energy for capture of neutrons by U^{238} was measured in this way by Anderson,[96] who found it to be about 5 volts. Boron is commonly used for this purpose.

The $1/v$ law always holds for sufficiently low energies. The extent of its validity depends on how rapidly the resonance term varies compared to the $1/v$ term. Bethe[97] has shown the conditions for the validity of the $1/v$ law to be

$$E \ll |E_o|,$$

or

$$E \ll \Gamma,$$

whichever is larger, where E_o and Γ are the resonance energy and width, respectively, of the level nearest to $E = 0$. For the compound nucleus B^{11}, both E_o and Γ are about 10^5 volts. On the other hand, the peculiar behavior of cadmium in the thermal energy region results from a resonance level that lies at a fraction of a volt neutron energy. The $1/v$ law is not valid in cadmium until the neutron energy is even lower than the average thermal energy at room temperature.

For heavy nuclei, it is only to incident neutron energies near zero that the one-level formula may be applicable, if at all. Even for neutron energies of several hundred kilovolts the spacing of the levels of the compound nucleus is of the order of 5 volts, which is so small compared to the energy spread of the neutron energy that any determination of cross sections, etc., measures the average of that quantity taken over many neighboring states of the compound system. If the energy levels are still discrete in the sense that the one-level formula, equation 22, can be applied in the neighborhood of each resonance, we may simply average many such terms over an energy interval ΔE, large compared to the distance between levels. All characteristic resonance fluctuations in yield disappear and are replaced by a smoothed out mean variation. The cross section σ_x becomes

$$\sigma_x = \frac{\pi \lambda\!\!\!^{-2}}{(2i+1)(2S+1)\Delta E} \Sigma(2J+1)\Sigma \frac{\Gamma_n^{Ji'}\, \Gamma_x^{Ji}}{(E-E_i^J)^2+(\Gamma^{Ji}/2)^2} \qquad (23)$$

The first sum is with respect to J and the second sum is over all levels in ΔE.

[96] H. L. Anderson, *Phys. Rev.* **57**, 556 (1940).
[97] H. A. Bethe, *Rev. Mod. Phys.* **9**, 117 (1937).

The last summation is simplified by writing it as an integral introducing the level spacing $d(E)$ and an average value of $\Gamma_n^{Ji'}\,\Gamma_x^{Ji}$. It becomes

$$\Gamma_n^{J'}\Gamma_x^{J}\int_{\Delta E}\frac{dE'}{d(E')\{(E-E')^2+(\Gamma/2)^2\}}=\frac{2\pi\Gamma_n^{J'}\Gamma_x^{J}}{\Gamma^J}\,\frac{\Delta E}{d(E)}\ .$$

Substituting this back in equation 23 and again averaging Γ_x^J/Γ^J over J, we get

$$\sigma_x=\frac{\pi\lambda\!\!\!^{-2}}{(2i+1)\,(2S+1)}\left(\frac{2\pi}{d}\right)\frac{\Gamma_x}{\Gamma}\sum_J(2J+1)\Gamma_n^{J'}\ . \qquad (24)$$

When the incident neutron energy is several Mev, the level density and level widths have so increased that there is now a quasi continuum of overlapping states. The one-level formula, equation 22, is no longer applicable at all, and it is not obvious what kind of an average must be taken. For our purpose it will be permissible to apply equation 24 [98] although the interference among the overlapping states comprising the compound state sometimes requires special consideration.[99]

We should expect equation 24 to be amenable to a classical description when the neutron energy is sufficiently high so that its wavelength $\lambda\!\!\!^{-}$ is small compared to the nuclear radius. For neutrons

$$\lambda\!\!\!^{-}\sim\frac{0.4\times10^{-12}}{\sqrt{E}\ (\text{Mev})}\ \text{cm}\ .$$

Therefore, neutrons with energy above 1 Mev should be capable of classical consideration. Equation 22 describes a specific quantum phenomenon, and it is really rather arbitrary to separate it into two factors, one for the rate of formation and the other for the rate of disintegration. However, in equation 24 we can lump everything but Γ_x/Γ together and call it the cross section for formation of a compound system. In the classical limit, Bohr's picture of nuclear reactions would lead us to expect this cross section is πR^2. In other words, we expect the relationship

$$\frac{R^2}{\lambda\!\!\!^{-2}}\left(\frac{d}{2\pi}\right)(2S+1)\,(2i+1)=\sum_J(2J+1)\Gamma_n^{J'}\ . \qquad (25)$$

Although it will not be shown here, equation 25 can be proved by using statistical arguments pertaining to the rates of formation

[98] Bohr and Wheeler, *Phys. Rev.* **56**, 438 (1939).
[99] Kalckar, Oppenheimer and Serber, *Phys. Rev.* **52**, 275 (1937).

and disintegration of the compound nucleus.[1] Putting πR^2 in equation 24 gives the very simple result

$$\sigma_x = \pi R^2 \Gamma_x / \Gamma \quad , \tag{26}$$

which needs no further interpretation.

The three formulae 22, 24, and 26 each have a well-defined range of applicability. Equation 26 will be used to discuss the fast neutron fission cross sections; in the slow neutron region it often cannot be said beforehand whether the conditions require the *one* level formula 22 or the *many* level formula 24. Both must be tried to see which gives the more consistent interpretation of the observed fission cross sections.

By returning briefly to equation 25 we obtain some very useful further information. The terms on the right represent the rate of disintegration of the compound nucleus into a specified residual nucleus (in this case the ground state). The left side really tells the rate of the converse process, i.e., that a neutron of velocity $v = \hbar / m\lambda\!\!\!^{-}$ will strike the surface of the initial nucleus, area $4\pi R^2$. The equilibrium between the two rates is exactly the same as in the theory of the rate of evaporation from a solid or a liquid surface. In fact, Frenkel[2] proposed to calculate the probability of neutron emission from a nucleus of temperature T (see page 95) by the simple evaporation formula,

$$\Gamma_n / \hbar \sim N^{2/3} t^{-1} e^{-E_B/kT}, \tag{27}$$

where $N^{2/3}$ is the number of surface neutrons, t is some characteristic time of the order of 10^{-22} sec and E_B is the neutron binding energy. Weisskopf [3] has refined these ideas somewhat, and we shall use his results to estimate the neutron width Γ_n. Γ_n includes all values of the energy of the emitted neutron, whereas equation 25 pertains to only one. Therefore,

$$\Gamma_n = \sum \Gamma_n'$$

$$\text{(All possible final states)},$$

$$= \int_0^{E_{max}} \rho(E)\, \Gamma_n' dE \quad . \tag{28}$$

$\rho(E)$ is the density of states in the residual nucleus when the neutron is emitted with energy E. Each of the $(2J+1)$ states belonging to a given value of J has been counted separately. If the com-

[1] See for example H. A. Bethe, *Rev. Mod. Phys.* **9**, 98 (1937). Bethe's result is not in quite the same form as equation 25.

[2] J. Frenkel, *Zeit. Sowjetunion* **9**, 533 (1936).

[3] V. Weisskopf, *Phys. Rev.* **52**, 295 (1937).

pound nucleus α has an excitation energy E_a its level density is $\rho_a(E_a) \sim e^{S_a(E_a)}$; similarly, the residual nucleus after emission of a neutron of energy E will have an excitation energy $E_\beta = E_a - E - E_B$ and a level density $\rho_\beta(E_\beta) \sim e^{S_\beta(E_\beta)}$. (We are using the ideas of nuclear temperature T and entropy S discussed on page 95). Incorporating with equation 25, equation 28 then becomes

$$\Gamma_n/\hbar = \frac{m\pi R^2(2S+1)(2i+1)}{\pi^2\hbar^3} e^{-S_a(E_a)}$$

$$\times \int_0^{E_{max}} e^{S_\beta(E_a - E_B - E)} E\,dE \quad .$$

This rather formidable expression can be simplified by expanding the exponential term inside the integral and using the thermodynamic relation

$$\frac{d\,S_\beta}{d\,E}(E_a - E_B) = \frac{1}{T_\beta\,(E_a - E_B)},$$

where $T_\beta\,(E_a - E_B)$ is the temperature of the residual nucleus if the neutron is emitted with zero energy.

$$\Gamma_n/\hbar \cong \text{const.}\; e^{-S_a(E_a) + S_\beta(E_a - E_B)} \int_0^{E_{max}} e^{-E/T_B} E\,dE \qquad (29)$$

$$\cong \text{const.}\; T_\beta^2 \exp{-S_a(E_a) + S_\beta(E_a - E_B)} \qquad (30)$$

The nuclear temperature is generally low compared to the maximum neutron energy, so that the upper limit can be replaced by infinity. It is seen by considering just the part of equation 29 from E to $E + \Delta E$ that the distribution in energy of the neutrons emitted is Maxwellian corresponding to the highest possible temperature of the *residual* nucleus, $T_\beta(E\alpha - E_B)$. This fact was used in chapter 7.

An estimate of the behavior of Γ_n can be made by using the two equations 25 and 30 and the fact that for low energies only one final state is available so that $\Gamma_n = \Gamma_n' = j\,E^{1/2}$. This last relationship holds up to about 100 kv with $j \sim 0.001$ volts $^{1/2}$. Actually this $E^{1/2}$ law is valid for any particular Γ_n' until the outgoing neutron has a wavelength $\lambdabar$ comparable with the nuclear radius, i.e., $E \sim 0.25$ Mev. Above this energy equation 25 shows that Γ_n' depends linearly on energy and therefore Γ_n varies roughly quadratically with E because of the increasing number of final states available. Above about 1 Mev, equation 30 can be applied. Figure 27 shows the estimated behavior for the heaviest nuclei of Γ_n as a function of the maximum neutron energy. For the entropy S, or rather for the level density, ρ, equation 17 was used. The constant energy difference $h\nu$ (see page 94) was chosen to give a level spacing of 20 volts at the neutron dissociation energy. Figure 27 shows also the estimated level spacing and nuclear temperature in Mev.

The fission width and spontaneous fission.[4]—It is more difficult to get as accurate an estimate for the fission width Γ_f as for the neutron width Γ_n, since there is no simple reverse process that can be conveniently used. In fact, the reuniting of the two fragments (to say nothing of the extra neutrons) is more complicated to consider than the splitting itself. Bohr and Wheeler have emphasized the irreversible nature of fission.

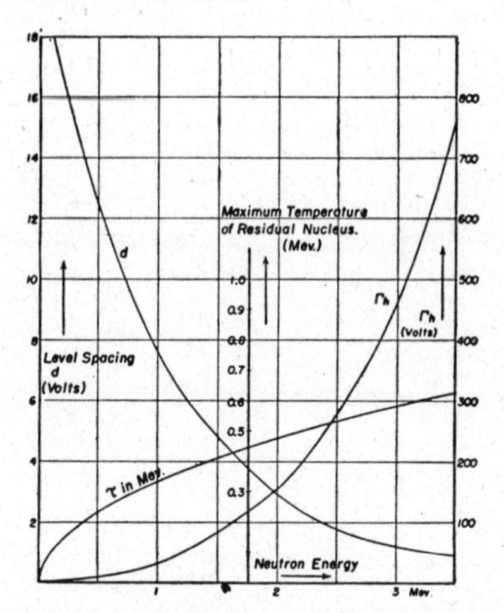

Fig. 27. The total neutron width in volts is shown as a function of the maximum kinetic energy of the escaping neutrons. An attempt was made to take account of the odd-even effect which makes the level spacing in an odd-even nuclei lower than an even-even nuclei for the same excitation energy. (Bohr and Wheeler, *Phys. Rev.* **56**, 442 (1939)).

The surface tension oscillations of the excited nucleus can be considered as a set of closely coupled oscillators which continually exchange energy with each other until finally a large amount of the energy accumulates in a mode of vibration that produces fission (see figure 25). This energy interchange is complicated, but it turns out to be unnecessary to discuss it in detail and a statistical method can be used. In essence, we examine a large number of identical compound nuclei whose excitation energy lies in the range

[4] Following Bohr and Wheeler, *Phys. Rev.* 56, 435 (1939).

E to $E+dE$, and we ask what fraction of them are in a state leading to fission at that instant. Referring to the potential energy diagram, figure 26, those nuclei which are about to fission will have their representative point in the vicinity of the fission barrier. Actually, it is necessary to consider the quantum character of the system because the zero point oscillations in the direction "perpendicular" to fission have large energies of the order of 0.5 Mev. This, of course, is a result of the Uncertainty Principle, which makes meaningless the statement that the nucleus crosses over the fission barrier at its lowest point.

We then ask how many quantum states of motion there are in the vicinity of the fission barrier. The saddle point corresponds to an unstable equilibrium deformation of the nucleus. A nucleus can execute oscillations about the potential minimum, perpendicular to the fission direction, but if its kinetic energy in the fission direction is sufficiently large, it moves across the barrier as a free particle. The spectrum of the nonfission oscillations should not be different from that of an ordinary heavy nucleus about equilibrium. Thus, the quantum states in the vicinity of the barrier will correspond to continuum states with momentum $\vec{p}$ in the direction of fission and to discrete levels in the perpendicular direction resembling the lowest states of a heavy nucleus. (The concept of quantum states in the vicinity of the barrier can be clarified somewhat. Because of the short wavelengths for even moderate kinetic energies in the direction of fission we can construct wave packets of similar states which define fairly well the normal distance of the representative state from the barrier. The discrete states of motion in the nonfission direction are spaced about 50 to 100 kv apart so that no packets can be set up.) A given nucleus with excitation energy E crosses the fission barrier with kinetic energy $K=E-E^*-E_c$, where E_c is the critical energy for fission and E^* is the energy of its nonfission quantum state. The number of states available in a nucleus of energy E is obviously just $N^*(E)$, where $N^*(E)$ is the number of the nonfission states with energy E^* less than the maximum available kinetic energy $E-E_c$. In our original ensemble of nuclei, it is reasonable to expect then that the fraction of nuclei in the energy range from E to $E+dE$ about to fission is proportional to $N^*(E)$. This fraction is also inversely proportional to the total number of possible levels $\rho(E)dE$ between E and $E+dE$, where $\rho(E)$ is the level density of the compound nucleus with excitation energy E.

In fact, it can easily be shown that [5]

$$\Gamma_f = \frac{N^*}{2\pi\rho(E)} = \left(\frac{d}{2\pi}\right) N^* . \quad (31)$$

It is obvious that we cannot make any precise estimate of Γ_f from this equation. Figure 28 shows the estimated value of the ratio Γ_f/d as a function of the excess energy $E—E_c$. Actually even for E less than E_c the fission yield is not zero because of the possibility of quantum mechanical tunneling through the fission barrier. This effect will be considered below. From equation 31, or from the considerations leading up to it, it is apparent that near the threshold energy the fission yield should show characteristic steps as the first few values of N^* are reached. It also seems that the actual observed threshold should correspond to the lowest non-fission energy level rather than to the critical energy E_c. The fluctuation in the position of this level with respect to E_c was offered as an explanation for part of the "inconsistency" in the observed photo-fission thresholds on page 86. From equation 31 it is seen that, when the fission threshold is reached, the fission width Γ_f is much larger than the radiation width $\Gamma_\gamma \sim 0.1$ volt, so that gamma emission is never a serious competitor.

The question arises as to the half-lives for spontaneous fission of the fissionable nuclei and whether these are comparable with the half-lives of other radioactive decay. A crude estimate can be made of the probability of fission by tunneling through the fission barrier. Figure 29 shows a diagrammatic cross section through the fission barrier. The energy level corresponds to, say, the ground state of the oscillation $n=2$, which is most favorable to fission. If we consider U^{235}, the barrier height is about 4.6 Mev. The zero-point energy has been shown[6] to be $\hbar\omega_f/2 \sim 0.4$ Mev where ω_f is the frequency of the $n=2$ mode of oscillation. Thus, we may approximate the penetration probability

$$\tau^{-1} \sim \omega_f \exp - \frac{2}{\hbar} \int_{P_1}^{P_2} \sqrt{2m(V—E)}\, dx$$

where $P_1$and P_2 are the entrance and exits to the "tunnel" as shown and τ is the half-life. Even if it were clear what to substitute for the mass m, the integral would require an accurate knowledge of the other quantities V and E, so that at best we can make only a

[5] Bohr and Wheeler, *Phys. Rev.* **56, 436 (1939).**
[6] Bohr and Wheeler, *Phys. Rev.* **56, 435 (1939).**

very crude estimate which will provide a means of comparison of the heavy nuclei.[7] Just to get some idea of the size of τ, m will be taken as the mass of U^{235}, and $V-E$ will be assumed constant and equal to 4.2 Mev for a distance P_1P_2 equal to say $\alpha/2$. α will be of the order of the nuclear radius and perhaps somewhat larger because the deformations are such as to try to separate the nascent nuclei as far as possible before the rupture occurs. Thus

$$\tau \sim 10^{-21} \times 10^{41} = 10^{20} \text{sec} \sim 10^{12} \text{ years.}$$

It is probable that the measured critical energy for fission plotted in figure 24 should be counted from the zero point energy so that 4.6 Mev should be used instead of the 4.2 Mev above. This changes

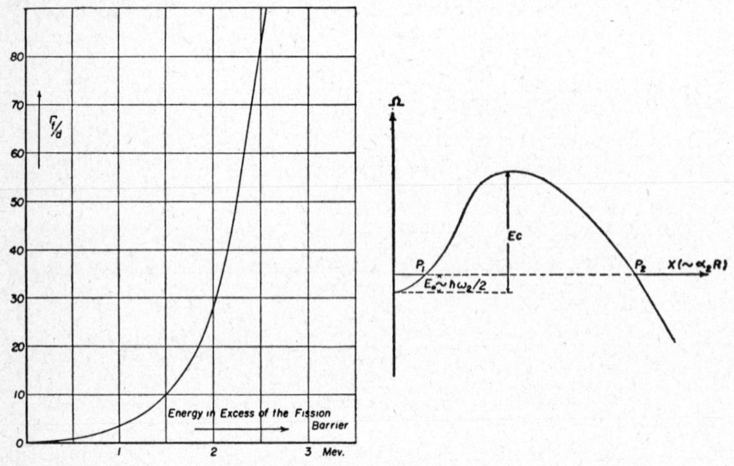

Fig. 28. (Bohr and Wheeler). The fission width divided by the level spacing of the compound nucleus is plotted against the energy in excess of the critical energy for fission. For example, with thermal neutrons on U^{235}, d is about 20 volts and the excess energy about 1.6 Mev so that Γ_f is about 200 volts, which is much larger than the radiation width.

Fig. 29. Diagrammatic view showing cross section through the fission barrier.

our estimate to about 10^{15} years. This is a very rough estimate and may be off by several orders of magnitude. However, the exponent $\sqrt{2m(E_c-E_0)}\ \alpha/\hbar$ is a convenient quantity to compare in different nuclei since the variations of all its terms are approximately known. Turner[8] has done this for the fissionable heavy nuclei and the transuranic nuclei to ascertain whether spontaneous fission can account for their nonoccurrence in nature (see chapter 6).

[7] Bohr and Wheeler give a more correct form of the penetration integral, which removes the uncertainty in the meaning of m but is no more amenable to accurate calculation.

[8] L. A. Turner, *Rev. Mod. Phys.* 17, 293 (1945).

Comparison with experiment and estimates of fission cross section.—It is well known now that the thermal neutron fission of uranium is due mainly to the isotope U^{235} of 0.7 percent abundance. The rare isotope U^{234} is present in such small quantities that it never figures in any of our calculations. The resonance capture of slow neutrons is by the abundant isotope U^{238}. Anderson[9] has measured an effective absorption cross section of 5×10^{-21} cm^2 and a resonance energy of 5 volts. If this were due to U^{235}, the cross section would have to be at least $139\times5\times10^{-21}$ cm^2, which is much larger than the maximum possible value $\pi\lambda\!\!\!^{-2}=125\times10^{-21}$ cm^2 for 5 volt neutrons.

The cross section measured is not simply related to σ_o, the cross section at resonance, first because of the spread in energy of the incident neutrons and second because of the Doppler broadening of the resonance width. The Doppler effect results from the vibrational motion of the capturing nuclei, which produces a spread in the *relative* kinetic energy of the neutron with respect to the nucleus even if there is a monochromatic neutron beam. The magnitude of this spread in kinetic energy can be estimated very simply. v, the relative velocity of the neutron, is v_o plus the forward velocity of the nucleus. If E_o is the relative kinetic energy,

$$\triangle E_o = mv_o \triangle v \sim mv_o\sqrt{kT/238m}$$
$$= \sqrt{E_o kT/238}\ .$$

The total broadening is twice this, i.e.

$$\triangle = 2\sqrt{E_o kT/238}\ .$$

If the resonance width Γ is greater than $\triangle$, we have the natural line breadth, and the effective cross section is related to the cross section at resonance σ_o by[10]

$$\sigma_{Eff} = \tfrac{1}{2}\sigma_o\ . \qquad (32)$$

If Γ is smaller than $\triangle$ then[11]

$$\sigma_o = \frac{2\sqrt{2}}{\sqrt{\pi}}\,\frac{\triangle}{\Gamma}\,\sigma_{Eff}\ . \qquad (33)$$

For 5 volt neutrons, $\triangle$ is 0.06 volts, which is comparable but seems somewhat smaller than the values of Γ, 0.1 to 1 volt, estimated from the other heavy nuclei. We shall apply expressions 32 and 33 to see which leads to the most reasonable results. The dispersion

[9] H. L. Anderson, *Phys. Rev.* 57, 566 (1940).

[10] For a complete discussion of the Doppler broadening see H. A. Bethe, *Rev. Mod. Phys.* 9, 140 (1937).

[11] Bohr and Wheeler have used the factor 4 instead of $2\sqrt{2}$, thus differing from Bethe.

formula 22 gives at resonance (i is zero for U^{238})

$$\sigma_0 = 4\pi\lambda\!\!\!^-{}^2 \Gamma_n' \Gamma_\gamma / \Gamma^2$$

$$= 5 \times 10^{-19} cm^2 \Gamma_n' \Gamma_\gamma / \Gamma^2 = \begin{cases} 10^{-20}\, cm^2; & \text{(a) Natural Width} \\ 3.9 \times \dfrac{10^{-22}}{\Gamma(\text{volts})} cm^2; & \text{(b) Doppler Width} \end{cases}$$

Since radiative capture is predominant, $\Gamma \cong \Gamma_\gamma$ and (a) gives

$$\Gamma_n' = \Gamma / 50 \ .$$

If we take the value $\Gamma \sim 0.1$ volts,

$$\Gamma_n' \sim 0.002 \text{ volts} = j\, E^{1/2} \ .$$

So that $j \sim 8 \times 10^{-4}$ volts $^{1/2}$, in agreement with the observations on other elements (see page 106). Choice (b) gives Γ_n' directly as

$$\Gamma_n' \sim 8 \times 10^{-4} \text{ volts},$$

$$j \sim 4 \times 10^{-4} \text{ volts}^{1/2},$$

which seems a little small although it may be reasonable because the large number of particles in U^{238} makes escape of a neutron less probable. Choice (b) is conditioned by $\Gamma < \Delta$ so that at most we can make $\Gamma_n \sim 0.05$ volts.

These values can be used to estimate the capture cross section in uranium at thermal neutron energies, which is important in slow neutron chain reactions. The effective temperature of thermal neutrons as measured by thin absorbers[12] is $(\pi/4)T$, so that $E_{Th} = \pi kT/4 = 0.028$ volt at room temperature. The one level dispersion formula gives at thermal energies

$$\sigma(\text{Thermal}) = \pi\lambda\!\!\!^-{}^2_{Th} \Gamma_n' (\text{Thermal}) \Gamma_\gamma / E_o^2 \ .$$

Using the value obtained from choice (a) we find the natural width to give

$$\sigma(\text{Thermal}) \sim 14 \times 10^{-24} \text{cm}^2 \ .$$

For choice (b):

$$\sigma(\text{Thermal}) \sim 2.8 \times 10^{-24} \text{cm}^2 \ .$$

Actually, there are conflicting experimental results in the literature which do not permit a unique choice of one of the above values. Anderson and Fermi[13] measured 1.2×10^{-24}cm^2 for the absorption cross section, whereas Whitaker,[14] et al., obtained a much larger value. The latter group measured the total cross section for removal of neutrons from the incident beam and then, by changing the position of the absorber, estimated the fraction of this total due

[12] H. A. Bethe, *Rev. Mod. Phys.* **9**, 136 (1937).
[13] Anderson and Fermi, *Phys. Rev.* **55**, 1106 (1939).
[14] Whitaker, Barton, Wright and Murphy, *Phys. Rev.* **55**, 793 (1939).

to scattering and the fraction due to absorption. Their results are:

$$\sigma_{Total} = 23.2 \times 10^{-24} \text{cm}^2 \quad .$$
$$\sigma_a \sim 11 \times 10^{-24} \text{cm}^2 \quad ,$$
$$\sigma_{Sc} \sim 12 \times 10^{-24} \text{ cm}^2 \quad .$$

Anderson and Fermi measured the production of the 23 minute activity producing $_{93}Np^{239}$, so that possibly there is another mechanism of absorption.

With the numbers used here, both (a) and (b) give the same value for the absorption cross section at resonance:

$$\sigma_o \sim 10^{-20} \text{ cm}^2 \quad .$$

It seems justifiable to use the one level formula 22 because, as we shall see below, the level spacing seems to be about 20 volts in the slow neutron region, so that the other levels could contribute only about one-tenth as much as the 5 volt resonance level to the cross section at thermal energies.

The thermal neutron induced fission is due to U^{235}, which has an excitation energy above fission threshold of about 1.6 Mev (see table 7 and figure 24) on capture of a slow neutron. The fission width at this energy is about 100 volts (see figure 30), which is much larger than the level spacing of the compound nucleus U^{236}. Therefore, the many level formula 22 must be applied. Slow neutrons can be captured only with zero orbital angular momentum, so that the only values of J available in the compound nucleus are $i \pm 1/2$, where $i\hbar$ is the angular momentum of the original nucleus U^{235}. (If i is zero, only $J=1/2$ is possible.) In U^{235}, i is certainly not zero, and so equation 24 becomes in this case

$$\sigma_f = \frac{\pi \lambda\!\!\!^{-2}}{2} \Gamma_n' \left(\frac{2\pi}{d} \right) . \qquad (34)$$

This should follow the $1/v$ law. A check has been made by Anderson, et al.,[15] by comparing the activity produced by thermal (cadmium) neutrons with that produced by neutrons absorbed in boron whose mean energy is several volts. They observed a mean cross section for thermal neutrons in uranium of 2×10^{-24} cm². Multiplying by 139, the cross section for U^{235} is

$$\sigma_f(\text{Thermal}) \sim 2.8 \times 10^{-22} \text{cm}^2 .$$

Substituting into equation 34, $\pi\lambda^2 = 23 \times 10^{-18}$ cm² and using the two values: Natural Width (a) Γ_n' (Thermal $\sim 1.3 \times 10^{-4}$ volts
Doppler Width (b) Γ_n' (Thermal) $\sim 0.7 \times 10^{-4}$ volts

[15] Anderson, Booth, Dunning, Fermi, Glasoe and Slack, *Phys. Rev.* **55**, **511** (1939).

we estimate that

$$d = \frac{23 \times 10^{-18}}{2 \times 2.8 \times 10^{-22} \times 2\pi} \times \Gamma_n'$$

$$\sim \begin{cases} 30 \text{ volts; (a) Natural Width} \\ 15 \text{ volts; (b) Doppler Width} \end{cases} .$$

The value $d=20$ volts was used in the calculation of the partial widths Γ_n and Γ_f on pages 107 and 110. This should be about the same as the spacing of the compound nucleus U^{239} when U^{238} captures a slow neutron because of the odd-even fluctuations mentioned in the last section.

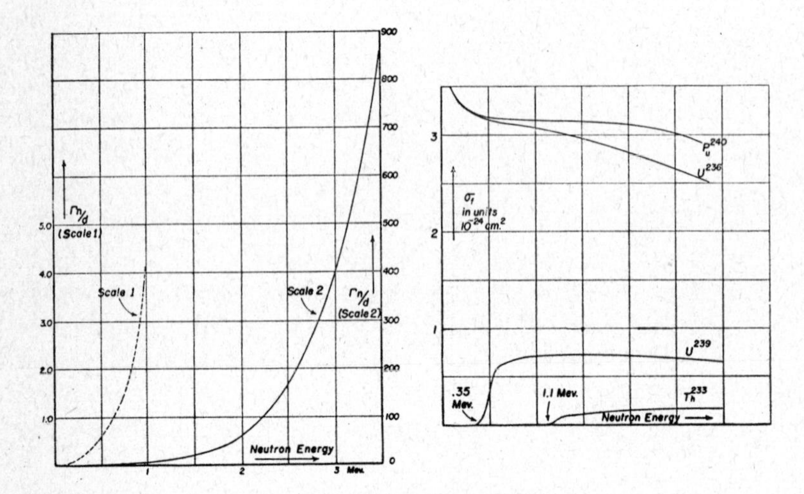

Fig. 30. The quantity Γ_n/d of equation 35 is plotted against the maximum energy of the emitted neutrons.
Fig. 31. The fission cross sections of Th^{232}, U^{235}, U^{238} and Pu^{239} plotted as a function of the energy of the incident neutrons.

The fission cross section for fast neutrons is somewhat simpler to discuss because of the simple form of equation 26. For our purposes it is convenient to write equation 26 in the form

$$\sigma_f = \pi R^2 \frac{\Gamma_f}{\Gamma_f + \Gamma_n} = \pi R^2 \frac{\Gamma_f/d}{\Gamma_f/d + \Gamma_n/d} . \qquad (35)$$

Γ_f/d is plotted in figure 28 as a function of the energy in excess of the critical energy for fission. Γ_f/d is a more convenient quantity than Γ_f itself because of its simple form; also, d depends on the position of the critical energy with respect to the neutron binding energy and Γ_f/d does not. Γ_n/d is plotted in figure 30 against the maximum kinetic energy of the neutron emitted. The details of estimating Γ_n/d were discussed on page 113. For the nuclear radius we take the results of Sherr[16] from the scattering of high energy neutrons on nuclei. The nuclear radius can be written in the form

$$R = r_o A^{1/3} + b \quad ,$$
$$\text{where } r_o = 1.25 \times 10^{-13} \text{ cm}$$
$$\text{and } b = 2.3 \times 10^{-13} \text{ cm.}$$
$$\text{For } U^{238},\ R = 10.15 \times 10^{-13} \text{ cm,}$$
$$\pi R^2 = 3.2 \times 10^{-24} \text{ cm}^2.$$

Figure 31 shows the results of applying equation 35 and the values in figures 28 and 30 to the compound nuclei U^{236}, U^{239}, Th^{233} and Pu^{240}. The essential difference in these four nuclei is the shifting of the zero on the abscissa $E{-}E_c$ in figure 28 with respect to figure 30. The threshold values shown in figure 26 are the newest published data. The plateaus in the cross sections of U^{238} and Th^{232} were first calculated by Bohr and Wheeler.[17] They actually got better agreement with the measurements of Ladenburg etc.,[18] than is the case in figure 31. Ladenburg found little change in the cross section of U^{238} from 2 to 3 Mev neutron energy. The U^{238} cross section was $0.5 \times 10^{-24} \pm 25\%$ cm², which is somewhat smaller than our 0.7×10^{-24} cm². Similarly, they measured Th^{232} in the same region and obtained 0.1×10^{-24} cm² as compared to 0.15×10^{-24}cm² in figure 31. However, Bohr and Wheeler used values for the fission thresholds which are about 0.5 Mev high, so that both the agreement they obtained and the approximate agreement in figure 31 must be considered somewhat fortuitous.

The cross sections for U^{235} and Pu^{239} are more interesting. Both decrease from very large values at thermal energies $\sigma \sim 3 \times 10^{-22}$ cm² until somewhere near $E \sim 0.25$ Mev when $\lambda\!\!\!^{-}$ becomes smaller than the nuclear radius. The further decrease, especially in U^{235}, is caused by the competition from neutron emission, which is negligible only below, say, .25 Mev. For example, in U^{235} the cross section is decreased by 4 percent at neutron energies of 1 Mev. It is evident from

[16] R. Sherr, *Phys. Rev.* 68, 240 (1945).
[17] Bohr and Wheeler, *Phys. Rev.* 56, 429 (1939).
[18] Ladenburg, Kanner, Barschall and Van Voorhis, *Phys. Rev.* **56**, 168 (1939).

the form of Γ_n/d and Γ_f/d, equations 31 and 30, that the neutron competition will increase. Γ_n/d contains the kinetic energy E times the level density, while Γ_f/d contains only N^*, the number of available "nonfission" levels. The competition from neutron emission is not so large in plutonium, so that its cross section will be several percent higher than that of U^{235}; this may be of some advantage in a chain reacting unit.

CHAPTER 9

EARLY WORK ON CHAIN REACTIONS

In chapter 5 we described the experiments of von Halban, Joliot and Kowarski[19] which first conclusively established that more than one neutron was emitted for each neutron absorbed in producing a fission in uranium. In the same paper these physicists suggested that this phenomenon could, under suitable conditions, result in the propagation of a nuclear chain reaction in uranium. They pointed out that a chain reaction would just continue if one of the secondary neutrons produced a fission, and thus another complement of neutrons, and if one of these in turn produced a fission and so on. Since each fission results in the release of about 200 Mev of energy, the energy of reaction in a large mass of uranium would be very large. This suggestion was followed by numerous experiments and calculations to determine the conditions necessary for the establishment of this revolutionary type of nuclear reaction.

The possibility of using the chain reaction for the production of power naturally excited great interest. But the reaction was also intensively studied for its inherent scientific value because it represented a new phenomenon in nuclear physics. In the past, many nuclear reactions had been found in which individual reactions produced a large net amount of energy, but these reactions could not be used for the production of large quantities of energy, because the efficiency of production of an individual reaction was very poor, and once the reaction was produced neither enough energy was produced nor were the right particles emitted to give further reactions. Thus fission opened a new field of nuclear chemistry.

It is interesting to note that a fission chain reaction differs from the more usual chemical chain reactions—the former being dependent on the production of new particles, whereas the latter proceeds by the rearrangement of particles already present in the system.

It is our main purpose in this chapter to review the early work that was done to establish the possibility of a fission chain reaction. We shall see that many of the fundamental scientific considerations that went into the design of the first successful nuclear chain re-

[19] von Halban, Joliot and Kowarski, *Nature* **143**, 470, 680 (1939).

action by the Manhattan District project were developed in this early period.

Qualitative considerations.—The essential problem in the production of a fission chain reaction is to realize such conditions that, after all competing processes have absorbed neutrons, there is one neutron left over from each fission to produce another fission. In any neutron processes involving ordinary uranium, there are four mechanisms competing for the available neutrons.

(a) Fission capture of neutrons by the uranium. (This could be caused either by the capture of thermal neutrons by U^{235} or by the capture of fast neutrons by U^{235} and U^{238}.)

(b) Escape of neutrons from the system.

(c) Nonfission capture of neutrons by U^{238}, leading ultimately to the production of plutonium. This is a resonance process whose peak occurs at about 5 ev.

(d) Capture of neutrons by other materials present, such as impurities or deliberately added materials.

A chain reaction will ensue only if process (a) produces enough neutrons to compensate for these losses, with at least one neutron left over to continue the chain. Since a fission reaction produces only a given number of neutrons, effort must be directed toward minimizing the parasitic (neutron-consuming) processes relative to the fission processes.

These considerations may be expressed quantitatively in terms of a multiplication factor (or reproduction factor), k, for the system. k is defined as the average number of new neutrons ultimately produced by each neutron in the system. For the chain reaction to act, k must be greater than or at least equal to unity. The reaction will proceed steadily if k is kept just equal to unity.

Critical size.—The escape of neutrons from a chain reacting system is a surface effect and varies as R^2 for a sphere, where R is the radius of the sphere; on the other hand, capture processes are volume effects and vary as R^3. Thus the ratio of the rate of escape of neutrons from the system to the rate of capture inside the system varies as R^{-1} and decreases with increasing size of the system. Therefore, if a chain reaction is possible, it can take place only in systems larger than a certain critical size for which the rate of escape is small enough, compared to the rate of capture, for the reaction to proceed. The first calculation of the minimum mass of uranium oxide necessary for the production of a *fast neutron* chain reaction was made by Perrin.[20] Similar calculations

[20] F. Perrin, *Comptes Rendus* **208**, 1394, 1573 (1939).

Table 10 — FISSION DATA

Target substance	$_{92}U^{235}$	$_{93}Pu^{239}$	$_{92}U^{238}$	$_{90}Th^{232}$	$_{91}Pa^{231}$	$_{92}U^{233}$
Compound nucleus	$_{92}U^{236}$	$_{93}Pu^{240}$	$_{92}U^{239}$	$_{90}Th^{233}$	$_{91}Pa^{232}$	$_{92}U^{234}$
Threshold energy for fast neutron fission, in Mev (1)	0	0	0.35 ± 0.1	1.1 ± 0.1	~ 1	0
Energy released per fission, in Mev (2)	200		Estimated to be same as for $_{92}U^{235}$			
Energy of fission neutrons, in Mev (3)	<3.5 (Ave$\sim$ 1)					
Average number of neutrons released per fission, μ (3)	2.3 (2 to 3.5)					
Average number of neutrons released per thermal neutron absorbed, η (3)	1.4					

(1) References are given in table 1.
(2) Bohr and Wheeler, *Phys. Rev.* **56**, 426 (1939).
(3) Zinn and Szilard, *Phys. Rev.* **56**, 619 (1939).

were made by Flügge and by Adler.[21] In these calculations it was assumed that the changes in neutron density accompanying a chain reaction in a mass of uranium could be treated by diffusion theory.

Following Adler, we consider a sphere of uranium oxide (U_3O_8) of radius R in which the concentration N of fast neutrons at any point at some time t is given by the diffusion equation

$$\frac{\partial N}{\partial t} = D\Delta^2 N + KN \quad . \tag{36}$$

This is the usual diffusion equation with the extra term, KN, giving the net extra number of neutrons produced per unit volume per second at any instant at a given point in the sphere. K is given by

$$K = v\,[N_U\sigma_f(\mu - 1) - \sum_i N_i\sigma_{ai}] \ , \tag{37}$$

where v is the average speed of the neutrons, N_U is the concentration of uranium in atoms per cc, σ_f is the cross section for fission by fast neutrons, μ is the average number of neutrons produced per fission, σ_{ai} is the cross section for capture of fast neutrons by any element present and N_i is the concentration of that element. The diffusion coefficient D is given by:

$$D = (1/3)\,\lambda\, v = (1/3)\, v\, (N_U\sigma_{tU} + N_i\sigma_{si})^{-1}, \tag{38}$$

where λ is the mean free path of a neutron, σ_{tU} is the total cross section of uranium for a neutron and σ_{si} is the scattering cross section for neutrons of the other atoms. For diffusion theory to be valid, the mean free path must be much smaller than the radius of the sphere.

Neutrons at the surface of the sphere will escape rapidly, so that the concentration at the surface will be very small; for simplicity we take the concentration at the surface to be zero at all times. (Better boundary conditions will be considered in later chapters.) The solution of equation 36 under these conditions is

$$N(r,t) = \sum_{\nu=1}^{\infty} \frac{A_\nu}{r} \sin\frac{\nu\pi r}{R}\, e^{(K - \nu^2\pi^2 D/R^2)t} \ . \tag{39}$$

The values of A_ν are determined by the initial concentration distribution of the neutrons; for an initial uniform neutron concentration N_o throughout the sphere, $A_\nu = (-1)^{\nu+1}\, 2N_oR/\nu\pi$, while

[21] S. Flügge, *Naturwiss.* 27, 402 (1939); M. F. Adler, *Comptes Rendus* **209**, 301 (1939).

for an initial number Q_o of neutrons concentrated at the origin, $A_\nu = Q_o \nu / 2R^2$.

The concentration of neutrons at any point will increase exponentially in time if one of the coefficients $(K - \nu^2\pi^2 D/R^2)$ is positive. This gives us the two conditions that must be satisfied if a chain reaction is to proceed:

(a) That $K > 0$; this condition may be written as

$$\frac{N_U \sigma_f \mu}{N_U \sigma_f + \Sigma N_i \sigma_{ai}} > 1 \quad , \tag{40}$$

which states that the number of neutrons produced per neutron absorbed in the system must be greater than unity.

(b) That $R > \pi\sqrt{D/K}$. (41)

From equations 37 and 38 this becomes

$$R > \pi (3[N_U(\mu - 1)\sigma_f - \sum_i N_i \sigma_{ai}] [N_U \sigma_{tU} + \sum_i N_i \sigma_{si}])^{-\frac{1}{2}} \quad .$$

The radius $R_c = \pi\sqrt{D/K}$ is called the critical radius. If $R < R_c$, the concentration of neutrons at each point decreases with time; if $R = R_c$, the concentration approaches an asymptotic value; if $R > R_c$, the concentration increases exponentially with time, tending to produce an explosive reaction. Perrin[22] found 140 cm for the critical radius for uranium oxide, for a fast neutron reaction. This corresponds to 40 tons of uranium oxide. The mean free path for a neutron is 10 cm, so that the use of diffusion theory is approximately valid.

The concentration may be integrated over the volume of the sphere to obtain the total number of neutrons within the sphere at any time; this function is plotted in figure 32 for the case of an initial concentration of neutrons at the origin.

It should be noted that the effect of inelastic collisions has been neglected in this calculation. These collisions rapidly reduce the energy of the neutrons and change the values of v and the cross sections. The effect of the impurities produced in the fission reaction has also been neglected; as the reaction proceeds, these impurities increase the value of $\sum_i N_i \sigma_{ai}$ and tend to reduce K to a value less than zero, thereby halting the reaction. However, if the radius is large enough, the mass will be blown apart before these effects enter.

[22] F. Perrin, *Comptes Rendus* **208**, 1394 (1939).

Table 11—CROSS SECTIONS OF FISSIONABLE NUCLEI FOR NEUTRONS (IN UNITS OF 10^{-24} cm^2)

Target Substance	Process	Energy Ranges: Thermal		Resonance		Fast*	
$_{92}U^{235}$	fission	420±100	(1)	30	(6)	2.4	(6)
	scattering	17	(5)	17	(5)	6	(8)
$_{92}U^{238}$	fission	0		0		0.5	(9)
	scattering	17	(3)	17	(7)	6	(8)
	absorption (resonance)	3	(2)	5000**	(12)	0	(10)
ordinary uranium	fission	3 (ave.)	(1)	0.2(ave)	(6)	0.5	(9)
	scattering	17	(3)	17	(5)	6	(8)
	absorption (resonance)	3	(2)	5000**	(12)	0	(10)
$_{94}Pu^{239}$, $_{92}U^{233}$	fission, scattering	assumed same as for $_{92}U^{235}$					
$_{90}Th^{232}$	fission	0		0		0.1	(9)
	scattering	17	(5)	17	(5)	6	(8)
	absorption	8.3	(15)				
$_{91}Pa^{231}$	fission	0		0		3	(11)
	scattering	17	(5)	17	(5)	6	(8)
$_{90}Io^{230}$	fission	0		0		0.3	(15)
	scattering	17	(5)	17	(5)	6	(8)

*Most of the scattering of fast neutrons is inelastic scattering, resulting in large energy losses (as much as 90 percent) (14).
**The resonance peak for U^{238} occurs at approximately 5 ev (12) and is taken to have an effective width of 0.16 (13).

Perrin[23] also suggested that a "neutron-reflecting" layer at the surface could be used to reduce the critical size. Substances like carbon, beryllium and iron have large scattering cross sections and negligible neutron absorption cross sections. Thus, if a chain-reacting mass of uranium were surrounded by a layer of one of these materials, the layer would act to reflect neutrons back into the system which would ordinarily escape from the surface. This has the effect of increasing the number of available neutrons for the chain reaction, and thus the chain can be propagated in a system of smaller size. Perrin calculated that a layer of iron 35 cm thick surrounding a uranium-oxide system would reduce the critical mass of oxide from 40 tons to only 12 tons. We shall discuss the effect of a reflecting layer or tamper in more detail in chapters 10 and 11.

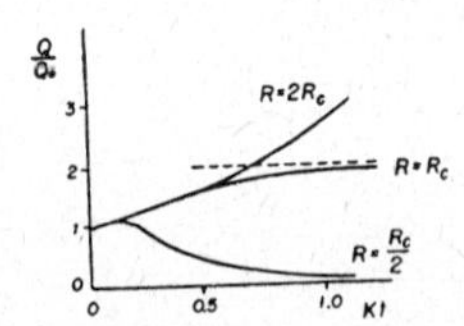

Fig. 32. The total number Q of neutrons in the sphere is shown as a function of kt for spheres of various radii. Q_o is the number of neutrons initially concentrated at the origin. R_c is the critical radius (Adler).

REFERENCES TO TABLE 11

(1) Booth, Dunning and Slack, *Phys. Rev.* **55**, 876 (1939).
(2) This is a theoretical estimate. See page 100.
(3) Whittaker, Barton, Bright and Murphy, *Phys. Rev.* **55**, 793 (1939). Their value of σ_t for ordinary uranium is used.
(5) These values are assumed to be roughly the same as for $_{92}U^{238}$.
(6) See page 89. For fast neutrons σ_f is taken to be πR^2. For neutrons in the resonance range σ_f is obtained by interpolation.
(7) H. A. Bethe, *Rev. Mod. Phys.* **9**, 69 (1937). This estimate is based on Bethe's statement that as a rule σ_s for slow neutrons is independent of the nuclear energy.
(8) H. A. Bethe, loc. cit. This value is obtained by extrapolation from the values of σ_s for other heavy nuclei, given in Bethe's article, on page 151.
(9) Ladenburg, Kanner, Barschall and Van Voorhis, *Phys. Rev.* **56**, 168 (1939).
(10) H. A. Bethe, loc. cit., table on page 160.
(11) Grosse, Booth and Dunning, *Phys. Rev.* **56**, 382 (1939).
(12) H. L. Anderson, *Phys. Rev.* **57**, 566 (1940). The effective height of the peak is listed in the table.
(13) On page 112 the width Γ of the resonance is estimated to be 0.5 ev. The effective width is taken to be $\pi\Gamma$. (H. A. Bethe, loc. cit. p. 141).
(14) H. A. Bethe, loc. cit. p. 158.
(15) Curie and Joliot, *Ann. de Phys.* **19**, 107 (1944).

[23] F. Perrin, *Comptes Rendus* **208**, 1394 (1939).

Use of a moderator.—The neutrons produced by fission have energies ranging up to 3.5 Mev, but these energies are reduced to thermal energies by a succession of elastic and inelastic collisions with uranium. Upon reaching energies in the neighborhood of 5 ev the neutrons become subject to nonfission resonance capture by the uranium. In the resonance region the absorption cross section is very high (about 5000 x 10^{-24} cm^2), so that very few neutrons get through to the thermal energy region where they may cause slow-neutron-induced fissions.

It occurred to many physicists that the effect of resonance capture could be reduced by mixing uranium with another substance which did not strongly absorb neutrons. Then, in slowing down to thermal energies, the neutrons would collide with the moderator instead of the uranium, and the probability of nonfission capture would be reduced. Elements of low atomic weight are particularly suited for this purpose since they cause large reductions in the kinetic energy of a neutron in each elastic collision.

The latter consideration makes water a natural first choice for a moderator, since a neutron loses on the average half its kinetic energy in a head-on elastic collision with a hydrogen nucleus. However, water possesses the disadvantage that its hydrogen absorbs slow neutrons by the reaction

$$_0n^1 + {_1H^1} \rightarrow {_1D^2} + \gamma \quad .$$

Consequently, if too much water is added the hydrogen will absorb appreciable numbers of thermal neutrons, thereby decreasing the possibility of a chain reaction. Many experiments were performed to determine if a slow neutron chain reaction could be propagated in a mixture of uranium and water. Fermi and Szilard (see page 46) used cylindrical rods of uranium-oxide in water and concluded that it was questionable whether a chain reaction could be propagated in such a system. This was followed by a series of experiments by von Halban, Joliot, Kowarski and Perrin,[24] using various concentrations of uranium, uranium-oxide and water and various geometrical arrangements, from which they concluded that a chain reaction was possible in a uranium-water system. However, Turner [25] showed that the data had been incorrectly interpreted, and that they in fact indicated that a chain reaction was definitely impossible in a uranium-water system. However, Turner stated that since about 1.5 neutrons were produced in uranium

[24] von Halban, Joliot, Kowarski and Perrin, *J. de Phys*, ser. 7, **10**, 428 (1939).
[25] L. A. Turner, *Phys. Rev.* **57**, 334 (1940).

for each thermal neutron absorbed, it was possible that a slow neutron chain reaction would take place in a mixture of uranium and some other moderator.

To be useful, a moderator must have a very small absorption cross section for neutrons in addition to being an element of low atomic number. The other substances which received early consideration for use as moderators were heavy water, beryllium and graphite, all of which have far lower absorption cross sections for neutrons than ordinary hydrogen (see table 13). Heavy water was seen to be best, but it was unobtainable in large quantities in 1940.

Isotope separation and plutonium fission.—Thermal neutrons produce fission only in the isotope of uranium of mass number 235 (see page 6). This isotope occurs in natural uranium in a concentration of one part in 140. The cross section for fission by thermal neutrons in U^{235} is about 400×10^{-24} cm^2. It is the abundant isotope U^{238} that has the resonance absorption for neutrons of 5 ev energy.

When these facts became clear it was obvious that the likelihood of producing a chain reaction could be increased by increasing the relative amounts of U^{235} to U^{238} over the concentration ratio

Table 12

CROSS SECTIONS FOR FAST NEUTRONS (1) (IN UNITS OF 10^{-24} cm^2)

Substance	Cross section σ_t
$_1H^1$	1.7
$_1H^2$	1.7
$_3Li^6$	1.8
$_4Be$	1.6
$_5B^{10}$	1.6
$_6C$	1.6
$_7N$	1.8
$_{13}Al$	2.4

(1) H. A. Bethe, *Rev. Mod. Phys.* 9, 69 (1937); table on p. 151.

Table 13 — Cross Sections for Thermal Neutrons (1) (3) (in Units of 10^{-24} cm^2)

Atomic number	Substance	"Average nucleus"		Isotope		
		Absorption σ_a	Total σ_t	Mass number	σ_a	Relative natural abundance
						Percent
1	H	0.30	20 (4)	1	0.31	99.98
1	D	—	4.0(2) (4)	2	0.00065	0.02
2	He	0	1.5	—	—	—
3	Li	—	66.3	6	860	7.5
				7	0.004	92.5
4	Be	—	6.1	9	0.0085	100
5	B	—	703	10	3525	18.4
6	C	0.0045	4.84			
7	N	1.75	11.75			
8	O	0.0016	4.1			
9	F	0.065	4.1			
10	Ne	—	2.8			
11	Na	0.5	4			
12	Mg	0.4	4			
13	Al	0.23	1.7			

occurring in nature. The enrichment of uranium with U^{235} is useful for slow neutron chain reactions with a moderator, because the total amount of fissionable material is increased, thereby decreasing relatively the parasitic effect of the nonfission absorption by the U^{238}. Enrichment in U^{235} (or the use of pure U^{235}) is probably useful for fast neutron chain reactions because the fission cross section of U^{235} for fast neutrons is probably greater than the corresponding cross section of U^{238}.

As described in chapter 6, McMillan and Abelson[26] discovered that the absorption of neutrons in U^{238} resulted in the ultimate formation of a transuranic element of atomic number 94 and mass number 239, which we now call plutonium. The Bohr-Wheeler theory of fission (see chapter 7) predicts that plutonium 239 has fission properties which are similar to the properties of U^{235}. In particular, plutonium should fission under the action of slow neutrons. The production of plutonium in a slow neutron chain reacting system would thus tend to compensate for the depletion of the fission-producing U^{235}.

It was realized that if plutonium could be separated from the uranium of the system, the plutonium could be used instead of U^{235} for enrichment or, in its pure form, for fast neutron chain reacting systems. This separation would be a chemical separation and might be more practicable than the isotopic separation of U^{235} from natural uranium.

Use of a lattice.—Smyth (2.11) reported that Fermi and Szilard suggested the use of a moderator with lumps or rods of uranium imbedded in it, rather than a homogeneous mixture of moderator and uranium. The advantage of such a lattice structure, or "pile," over the homogeneous mixture lies in the shielding that the surface of the uranium lump affords for the interior, in so

REFERENCES TO TABLE 13

(1) Segrè's Chart, 1945, is the source of all these data except σ_t for the deuteron.
(2) H. A. Bethe, *Rev. Mod. Phys.* 9, 69 (1937); table on page 151.
(3) H. A. Bethe, loc. cit., points out that the scattering cross sections are independent of energy for slow neutrons above the thermal range.
(4) H. A. Bethe, loc. cit., section 59, discusses the effect of chemical binding on the scattering cross section at thermal energies. In paraffin σ_s is increased by a factor of 2.8 above the free hydrogen value. A similar argument applied to water (vibrational frequencies 3400, 3600 and 1500 cm^{-1}) yields a factor of 3.4, giving $\sigma_s = 68 \times 10^{-24}$ cm^2. For heavy hydrogen in water a similar argument, assuming the same vibrational frequencies, leads to a factor of 1.9 giving $\sigma_s = 7.6 \times 10^{-24}$ cm^2. This adjustment is for thermal values only, the cross sections in the region of a few electron volts remaining unchanged.

[26] McMillan and Abelson, *Phys. Rev.* 57, 1185 (1940).

far as resonance absorption is concerned. If neutrons in the resonance energy range strike the uranium surface, they penetrate it on the average only to a depth of the order of the mean free path for absorption of such neutrons in uranium. This mean free path λ is given by the formula

$$\lambda = \frac{1}{N_U \sigma_a} , \tag{42}$$

where N_U is the atomic concentration of uranium and σ_a is the absorption coefficient. As we shall see later (page 142), λ is 0.0042 cm for resonance energy neutrons. The shielding effect does not act substantially for thermal and fast neutrons if the dimensions of the lump are less than the mean penetration distance of thermal and fast neutrons in uranium (3.8 cm and 43 cm, respectively).

This arrangement was utilized by Anderson, Fermi and Szilard[27] in early experiments to determine the average number of neutrons produced per thermal neutron absorbed by uranium. This experiment is described in chapter 5 (page 46). They used cylindrical cans 5 cm in diameter and 60 cm in height, filled with uranium-oxide (see figure 18). The ratio of the average atomic concentration of hydrogen to that of uranium was 17 to 1.

Control of a chain reaction.—It is clear that the equilibrium condition of a slow neutron chain reacting system can be varied by changes in the amount of neutron-absorbing material in the system. Insertion or withdrawal of a highly absorbent material like cadmium or boron could thus be used to adjust the reaction to a given stable condition. Such an adjustment would be difficult if the system reacted very quickly to changes in the available neutron density. Fortunately, the propagation of the chain is dependent on the production of delayed neutrons accompanying fission (see chapter 5). Thus, if the neutron density is changed, there will be a time delay before this change is reflected in the number of secondary neutrons produced in the system. As a result, the system will adjust itself slowly to changes in the total neutron absorption and the reaction can be controlled easily.

An interesting possible mechanism for self-stabilization of a chain reaction in the presence of a cadmium absorber was suggested by Adler and von Halban.[28] On page 121 we demonstrated that

[27] Anderson, Fermi and Szilard, *Phys. Rev.* **56**, 284 (1939).
[28] Adler, von Halban, *Nature* **143**, 793 (1939).

the necessary condition which must be satisfied if a chain reaction is to proceed is

$$\frac{N_U \sigma_f \mu}{N_U \sigma_f + \sum_i N_i \sigma_{ai}} > 1. \qquad (43)$$

The absorption cross section of most substances for thermal neutrons is proportional to the reciprocal of the neutron velocity, that is, the probability that a neutron is captured is proportional to the time it spends in the neighborhood of a nucleus. This is called the $1/v$ - law of neutron absorption (see chapter 8). If all the absorbers follow the $1/v$ - law, equation 43 will be independent of v and thus independent of the temperature. However, if we introduce an absorber such as cadmium, which does not obey the $1/v$ - law but has an approximately constant absorption cross section for neutron energies ranging from the thermal energy region to 4 ev, then equation 43 may be written

$$\frac{N_U \sigma'_f \mu}{N_U \sigma'_f + \sum_i N_i \sigma_{ai}' + N_{Cd} \sigma_{aCd}\, v} > 1 \;, \qquad (44)$$

where the primes denote that the $1/v$ dependence has been factored out; and σ_{aCd} is constant in the thermal energy region. Now as the reaction proceeds and the temperature increases, v will increase and tend to reduce the quotient to unity, thereby slowing the reaction. Thus the chain reaction will eventually stabilize itself at some elevated temperature.

Other considerations.—An obvious method of increasing the probability of a chain reaction is to purify the materials used, thus reducing the undesirable neutron absorptions. Smyth discusses the problems which arose in securing adequate quantities of sufficiently pure uranium and graphite for a pile system. One of the troublesome impurities in both cases was boron (Smyth 4.42, 6.10ff, 6.20), whose cross section for absorption of thermal neutrons is 160,000 times the corresponding cross section for carbon (table 14), so that its presence to even a few parts per million adds appreciably to the neutron absorption.

If a self-sustaining chain reaction is successfully established, its maintenance results in changes which affect its further continuance. Thus, the production of large amounts of energy would tend to heat the system unless the energy is removed by some cooling system. (A cooling system, however, adds to the parasitic neutron absorptions.) Also, as the fission products build up they tend to poison

Table 14

CROSS SECTIONS OF SOME FISSION PRODUCTS FOR THERMAL NEUTRONS (1) (UNITS 10^{-24} cm^2)

Atomic Number	Substance	"Average nucleus"		Total listed branching ratios, in percent for elements with long lives (2)	Isotope		
		Absorption σ_a	Total σ_t		Mass number	σ_a	Relative natural abundance (1)
							Percent
35	Br	—	—		79	3.05	50.6
					81	2.25	49.4
36	Kr	—	27		78	0.27	0.35
					86	0.061	17.47
37	Rb	—	12		85	0.724	72.8
					87	0.135	27.2
38	Sr	1.5	11		88	0.005	82.56
39	Y	—	—		89	1.1	100
40	Zr	0.5	15	6	94	0.053	17.0
					96	1.07	1.5
41	Nb	1.0	69	6	93	0.0099	100

42	Mo	3.9	79		98	0.37	24.1
					100	0.23	9.25
51	Sb	4.7	9		121	6.8	56
					123	2.5	44
52	Te	5	10	27	126	0.073	18.7
					128	{ 0.154	31.86
						{ 0.133	
					130	0.008	34.52
53	I	7.8	9.4	36	127	68	100
54	Xe	—	3.5	17			
56	Ba	1.25	9.25	15	138	0.56	71.66
57	La	12	25	15	139	3.1	100
62	Sm	8000	—		149	53000	15.5
63	Eu	2500	—		151	530	49.1
64	Cd	38000	—				

(1) Segrè's Chart, 1945.
(2) Anderson, Fermi and Grosse, *Phys. Rev.* **59**, 52 (1941).

the system by adding to the probability of nonfission capture of neutrons. Depletion of the U^{235} also would tend to halt the reaction, although it is compensated partially by the production of plutonium. These questions will be considered in more detail in chapter 10.

Pertinent data.—Before entering into a more detailed analysis of chain-reacting systems, it is desirable to collect and summarize the data about the processes and materials involved. Though the fundamental data obtained during the war are not yet available, enough in known from published material (or can be estimated from basic physical considerations) to give us confidence in the validity of the main details of our interpretations and conclusions.

The considerations outlined in the preceding chapter indicate the type of data needed for a more detailed analysis. We need to know the details of the fission processes—average number of neutrons released, energy released, fission cross sections, etc., as well as cross sections for all accompanying neutron processes, such as scattering and absorption in uranium, in possible impurities, in possible moderators and in the fission products. We need to know the densities, atomic concentrations and neutron mean free paths, for the moderator and the fissionable materials.

In his review article on nuclear fission, Turner[29] discusses the experimental data available in 1939 concerning the nuclear cross sections for uranium. Though some of the results were uncertain and discrepancies existed among determinations by different investigators, one can select reasonable values for the cross sections with confidence that the precise data when available will not seriously effect the conclusions derived from these values.

The data concerning the fission processes were analyzed in the earlier chapters of this book, and references to the sources are given there. The principal sources of the other absorption and scattering data are Bethe's[30] review article on nuclear physics and a chart of nuclei from E. Segrè, revised May 1945. Where these sources overlap, Segrè's chart has been given precedence.

[29] L. A. Turner, *Rev. Mod. Phys.* **12**, 1 (1940).
[30] H. A. Bethe, *Rev. Mod. Phys.* **9**, 69 (1937).

CHAPTER 10

SLOW NEUTRON CHAIN REACTIONS—PILES

As Smyth points out in his report, the primary purpose of developing the pile systems was the immediate production of plutonium in large quantities. Other obvious uses of piles—the production of energy, of neutrons and of radioactive materials, were of only incidental interest. This purpose naturally directed the course of the development. For instance, the selection of materials (ordinary uranium, with graphite moderator) was governed by considerations of immediate availability, purity and general expediency, with little regard for cost, size and ultimate practicability for peacetime uses. Since we shall use the information given in the Smyth report as a guide in making our analysis and as a check on the validity of our conclusions, we shall restrict the discussion in the first part of the chapter to this type of pile (ordinary uranium with graphite moderator). More general considerations of pile design and application will be set up in the second part.

CARBON-URANIUM PILES

Considerations affecting pile design.—The general considerations of the preceding chapter together with a survey of the Smyth report indicate the possible analysis of the physical factors involved in the development of piles. The following are the considerations which we shall attempt to analyze in greater detail:

(1) The action of a moderator in reducing resonance absorption, including the calculation of optimum proportions of moderator and uranium.

(2) The shielding effect of lumps or rods of uranium and its influence on the proportions of materials.

(3) The relationship of power production, plutonium production and production of radioactive fission products.

(4) The effects of additional materials, such as impurities, cooling system, fission products and control absorbers.

(5) The calculation of neutron lifetimes and neutron densities in a pile.

(6) The determination of the critical size of a pile.

(7) The effect of a reflecting layer.

(8) The inertial effect of delayed neutrons.

We shall make use of the data given at the end of the preceding chapter, in which, as already stated, there are a number of uncertainties. It will be necessary also to make many simplifying assumptions in our analysis. Though the data are not sufficiently reliable to justify calculations to two significant figures we have done so to preserve internal consistency in the calculations, and the general pattern of the analysis should be valid.

Summary of pile data.—We shall summarize the pertinent data given in Smyth's report regarding the piles constructed during the period 1940 to 1945.

FIRST COLUMBIA PILE

This was a graphite cube approximately 8 feet on an edge, containing 7 tons of uranium oxide in iron containers which were distributed at equal intervals through the graphite. It was unsuccessful ($k<1$). A second larger pile gave $k_\infty=0.87$. The uranium oxide contained 2 to 5 percent impurities, including a little boron.

CHICAGO WEST STANDS PILE

This was the first self-sustaining pile. It contained 12,400 pounds of uranium metal, distributed in lumps in a graphite moderator, and an unspecified amount of pressed uranium-oxide lumps. It was an oblate spheroid in shape. Calculated values for k_∞ for the control metal lattice was 1.07; for the two uranium oxide lattices making up the bulk of the rest of the pile, k_∞ was 1.04 and 1.03 In operation the effective k was 1.006. The metallic uranium and the oxide were very pure, the graphite having a neutron absorption 20 percent less than the standard commercial material. Cadmium strips were used for control. The pile was first operated at a power level of 0.5 watt, and later this was increased to 200 watts. A reconstructed version of this pile at Argonne was run at a few kilowatts.

CLINTON PILE

This was a cube of graphite with horizontal channels filled with uranium; it was considerably larger than the West Stands pile. The uranium was in the form of cylindrical rods encased in gas-tight aluminum casings, space being left for air-cooling. It eventually attained a power level greater than 1800 kw. The efficiency of separation of plutonium was increased from the initial 50 percent to about 80 or 90 percent. By February 1944, one-third ton of uranium per day went from the pile to the separation plant. During February several grams of plutonium were delivered.

HANFORD PILES

These are three water-cooled graphite-uranium piles, which use uranium rods sealed in aluminum jackets. Aluminum pipes are used in the cooling system. Smyth estimates, as an example, that a production of 1 kg/day of plutonium corresponds to a power production of 0.5×10^6 to 1.5×10^6 kw. He states that the rise in temperature of the Columbia river is too small to affect fish life.

ARGONNE HEAVY WATER PILE

This pile, which used heavy water as a moderator, was small compared to the uranium-graphite piles. It was run at 300 kw and operated so successfully that some uranium had to be removed. It took several hours to reach equilibrium and could not be shut down as completely or as rapidly as the graphite piles because delayed gamma rays tended to produce additional neutrons from the water. The neutron density at the center was high.

Collision theory for a moderator.—From the elementary theory of elastic collisions of spheres (or collisions of particles with the distribution in angle isotropic in a coordinate system in which the center of mass is at rest)[33] each collision of a neutron of mass m with a nucleus of mass M reduces the neutron energy on the average (arithmetic mean) by a fraction

$$f = \frac{2\,m\,M}{(M+m)^2} \qquad (45)$$

For heavy nuclei this reduces to $2m/M$. The fractional loss has equal probability of being anywhere from zero to twice the average value. The average losses per collision are given in table 12 for such elements as carbon ($f_C = 0.142$) and uranium ($f_U = 0.0083$). The results refer to elastic collisions only; for inelastic collisions the energy losses amount to as much as 90 percent.[34]

If we assume that in each collision a neutron loses exactly this fraction f of its energy, then after q elastic collisions its energy is reduced from an initial value E_o to a final value E, where

$$(1 - f)^q = E/E_o \qquad (46)$$

Reducing the energy of a neutron from, say, 1 Mev to thermal energy (about 0.025 ev) by collisions in carbon would therefore require 115 collisions, while reducing its energy to 5 ev would require 80. Actually the first few collisions would be inelastic, but this would not change greatly the number of required collisions.

In the next section we shall need to know the probability p_1 that a neutron, initially at a high energy E_o, shall at some time have

[33] Condon and Breit, *Phys. Rev.* **49**, 229 (1936).
[34] H. A. Bethe, *Rev. Mod. Phys.* **9**, 69(1937).

been in the resonance region E_r to $E_r—w$,[35] where the width w is small. If we assume that the initial energy E_o is much greater than the resonance energy E_r, that the probability distribution of neutrons in the energy range near E_r is uniform, and that the neutrons suffer exactly the average fractional energy loss f in a collision, then those neutrons which cross the resonance energy value E_r in a collision will have come from the region E_r to $E_r/(1-f)$ and will be uniformly distributed in the region $E_r(1-f)$ *to* E_r. The probability that a neutron, in crossing the resonance value E_r, will be in a region of width w (the resonance width) is

$$p_1 = w/(fE_r) \; .$$

This value is decreased because the probability distribution of neutrons increases with decreasing energy, and because the neutrons crossing the resonance value E_r spread out over a larger energy region than fE_r, viz., $2f\,E_r$. It is increased by the probability that neutrons with energy greater than $E_r/(1-f)$ may drop into the resonance region in one collision. A more exact calculation of p_1 shows that

$$p_1 = Bw/E_r \qquad (47)$$

where B varies from 1 for hydrogen to $1/f$ for heavy elements. For carbon B turns out to be 6.3. For very heavy elements such as uranium itself, where w is equal to or greater than $f\,E_r$, the neutron is certain to land in the resonance region and the probability p_1 must be taken to be unity.

Let us consider the case of a neutron in a hydrogen medium. After one collision its energy has equal probability of lying anywhere from 0 to E_o.[36] The probability of its being in the energy range E to $E+dE$, therefore, is $p_o(E)dE$, where

$$p_o\,(E) = 1/E_o .$$

After a number of collisions the regions below E_o will have greater probability of having at some time contained the neutron. Let $p(E)$ be the ultimate probability distribution, that is, the probability that the range E to $E+dE$ ever contains the neutron is $p(E)dE$. Then, by considering the effect of collisions after the first one,

$$p(E) = \int_E^{E_o} p(E)\frac{dE}{E} + \frac{1}{E_o} \; .$$

[35] For simplicity E_r is taken to be the "top" of the resonance region rather than its center. Since the region is very narrow, this choice does not affect the value of E_r appreciably.

[36] Condon and Breit, *Phys. Rev.* 49, 229 (1936).

Therefore

$$\frac{dp}{dE} = -\frac{p}{E} .$$

The solution of this equation, subject to the initial condition that $p(E_o)$ is $1/E_o$, is

$$p(E)=1/E . \tag{48}$$

The probability p_1 of the neutron's being at some time in the range E_r to E_r-w (where E_r is the resonance energy and w is the width of the resonance region) is, therefore,

$$p_1 = \int_{E_r-w}^{E_r} \frac{1}{E}\, dE = \frac{w}{E_r} \tag{49}$$

if $w \ll E_r$.

Let us consider the problem of a neutron of energy E_o in some other medium of atomic mass M. After one collision its energy has equal probability[37] of being anywhere from $E_o(1\text{-}2f)$ to E_o, where

$$f = \frac{2mM}{(M+m)^2} .$$

The probability of its being in the range E to $E+dE$, therefore, is $p_o(E)dE$, where

$$p_o(E) = \frac{1}{2fE_o}, \qquad E_o(1-2f) < E < E_o$$

$$= 0, \qquad E<E_o(1-2f).$$

Let the probability that the range E to $E+dE$ ever contains the neutron be $p(E)dE$. Then, by considering the effect of collisions, we see that for $E>E_o\ (1\text{-}2f)$

$$p(E) = \int_{E}^{E_o} p(E)\frac{dE}{2fE} + \frac{1}{2fE_o} .$$

Therefore,

$$\frac{d\,p}{d\,E} = \frac{-\,p\,(E)}{2\,fE} .$$

[37] Condon and Breit, *Phys. Rev.* **49**, 229 (1936).

The solution of this equation, subject to the initial condition that $p(E_o)$ is $1/(2fE_o)$, is

$$p(E) = \frac{1}{2f\,E_o}\left(\frac{E_o}{E}\right)^{1/2f}.$$

On the other hand, for $E<E_o(1-2f)$,

$$p(E) = \int_{E}^{\frac{E}{1\text{-}2f}} p(E)\,\frac{d\,E}{2\,fE}.$$

$$\frac{dp}{dE} = \frac{p\left(\frac{E}{1\text{-}2f}\right)}{2fE} - \frac{p(E)}{2fE}.$$

For $E \ll E_o(1\text{-}2f)$, where the disturbing effect of the different behavior of $p(E)$ near $E=E_o$ is small, the solution of this equation is

$$p(E) = B/E \tag{50}$$

B may be evaluated by the condition that the probability that the neutron at some time will cross an energy value E_1 is unity. It a neutron has energy E between E_1 and $E_1/(1\text{-}2f)$ the probability of its crossing the value E_1 in its next collision is $\{E_1—E(1\text{-}2f)\}/2fE$. The condition becomes:

$$\int_{E_1}^{\frac{E_1}{1\text{-}2f}} p(E)\left\{\frac{E_1 - E(1\text{-}2f)}{2fE}\right\}dE = 1 \tag{51}$$

This yields the value

$$B = \frac{2f}{2f + (1\text{–}2f)\ \ln(1\text{–}2f)}. \tag{52}$$

This varies from 1 for $2f=1$ to $1/f$ for $2f \ll 1$. The probability that the neutron will be at some time in the range E_r to $E_r–w$ (where $E_r \ll E_o$) is then

$$p_1 = \int_{E_r-w}^{E_r} \frac{B}{E}\,\text{d}E = \frac{Bw}{E_r}, \tag{53}$$

if $w \ll E_r$.

Equation 50 can also be obtained from equation 51, if the assumption is made that $p(E)$ is independent of E_1/E_o and therefore does not contain E_1 explicitly. This may be seen by substitution in equation 51 of E_1u for E and $q(E_1u)$ for $E_1u\, p(E_1u)$.

It may be noted that if B/E is multiplied by $q\lambda/v$, where q is the number of neutrons produced per second, λ is the mean free path and the neutron velocity $v = \sqrt{2E/m}$, the result gives the density of neutrons as a function of energy under conditions of steady production of monochromatic neutrons. It then agrees with the formula given by G. Placzek.[38]

Proportion of carbon and uranium.—Given a pile of infinite size consisting of a mixture of carbon and ordinary uranium, one problem that concerns us is the determination of the multiplication factor k_∞, that is the average number of new neutrons ultimately produced by each neutron in the system. We shall use the following notation:

μ = number of neutrons released per fission
N_U = atomic concentration of uranium (atoms/cc)
N_C = atomic concentration of carbon (atoms/cc)
ρ = ratio of atomic concentration of carbon to that of uranium
σ_{sx} = scattering cross section for element X
σ_{fx} = fission capture cross section for element X
σ_{ax} = non-fission absorption cross section for element X

The superscripts *th, res* and *fast,* as in $\sigma_{fU}{}^{th}$, refer to the thermal, resonance and fast energy regions, respectively.

We shall need to determine the following quantities:

k_∞ = average number of new neutrons ultimately produced per neutron in the system.
= average number of new fissions ultimately produced by the μ neutrons produced in one fission.

P_f = probability that a fast neutron will be slowed down by collisions to thermal energies and then be captured by the uranium to produce fission.

P_{aU} = probability that a fast neutron will ultimately undergo a non-fission absorption in uranium, to cause the formation of plutonium.

P_{aC} = probability that a fast neutron will ultimately be absorbed in carbon.

p = probability that during the slowing down process the neutron will enter the resonance region near 5 ev and be captured by the uranium.

[38] G. Placzek, *Phys. Rev.* **55**, 1130 (1939).

$\beta_f =$ probability that a thermal neutron will be absorbed in uranium to produce fission.

$\beta_{aU} =$ probability that a thermal neutron will undergo non-fission absorption in uranium to produce plutonium.

$\beta_{aC} =$ probability that a thermal neutron will be absorbed in carbon.

Most of these quantities can immediately be expressed in terms of cross sections and atomic concentrations. Thus, we have:

$$\beta_f = \frac{N_U \sigma_{fU}^{th}}{N_U \sigma_{fU}^{th} + N_U \sigma_{aU}^{th} + N_C \sigma_{aC}^{th}} \tag{54}$$

$$\beta_{aU} = \frac{N_U \sigma_{aU}^{th}}{N_U \sigma_{fU}^{th} + N_U \sigma_{aU}^{th} + N_C \sigma_{aC}^{th}} \tag{55}$$

$$\beta_{aC} = \frac{N_C \sigma_{aC}^{th}}{N_U \sigma_{fU}^{th} + N_U \sigma_{aU}^{th} + N_C \sigma_{aC}^{th}} \tag{56}$$

$$P_f = \beta_f (1-p) \tag{57}$$

$$P_{aU} = \beta_{aU} (1-p)+p \tag{58}$$

$$P_{aC} = \beta_{aC} (1-p) \tag{59}$$

$$k_\infty = \mu P_f = \mu\beta_f (1-p) \tag{60}$$

The probability p is the product of two probabilities, the probability p_1 that a neutron in being slowed down will enter the resonance region, and the probability p_2 that if it is in this resonance region it will be captured by the uranium. The probability p_2 is evidently given by

$$p_2 = \frac{N_U \sigma_{aU}^{res}}{N_C \sigma_{sC}^{res} + N_U \sigma_{aU}^{res}} \tag{61}$$

if we assume that elastic collisions with carbon will always cause a neutron's energy to fall below the resonance region, whereas elastic collisions with uranium always leave it in the resonance region. A 5 volt neutron loses $0.142 \times 5 = 0.720$ ev on the average in a collision with carbon, whereas it loses $0.0083 \times 5 = 0.042$ ev on the average in a collision with uranium. The width w of the resonance region has been taken to be 0.16 ev, so that the relative error for carbon is of the order $(0.16 \times \frac{1}{2})/(0.720 \times 2) = 0.06$.

The probability p_1 was calculated in the preceding section to be Bw/E_r for carbon and unity for uranium. For a mixture of carbon and uranium an average probability must be used. In taking this average, allowance must be made for the fact that the energy regions from which the neutron may cross into the resonance region are proportional in extent to the respective values of the average fractional energy losses f_C (carbon) and f_U (uranium). Therefore the probability of landing in the resonance region is

$$p_1 = \frac{N_C\, \sigma_{sC}^{res}\, f_C\, (6.3\, w/E_r) + N_U\, \sigma_{sU}^{res} f_U (1)}{N_C\, \sigma_{sC}^{res}\, f_C + N_U\, \sigma_{sU}^{res} f_U} \qquad (62)$$

In this discussion we have neglected absorption of high energy neutrons in carbon and uranium, since this absorption is very slight (Smyth 8.11: see also Bethe[39]). We have also neglected the production of fissions by fast neutrons, which would tend to enhance slightly the number of neutrons reaching the lower energy regions per fast neutron produced in the pile. We shall allow for this enhancement later (page 148).

Table 15

AVERAGE FRACTIONAL ENERGY LOSS OF A NEUTRON IN ONE ELASTIC COLLISION (1)

Substance	Average fractional energy loss
H^1	0.500
H^2	0.444
He	0.320
Be	0.180
C	0.142
N	0.124
O	0.111
Bi	0.0095
U	0.0083

(1) Condon and Breit, *Phys. Rev.* **49**, 229 (1936). See also page 135.

If we now substitute the cross section data of chapter 9 into equations 54 to 62, we obtain

$$k_\infty = 2.3\left(\frac{3}{6+0.0045\rho}\right)\left\{1-\frac{5000}{5000+4.84\rho}\left(\frac{0.14+0.134\rho}{0.14+0.688\rho}\right)\right\}$$

$$= \frac{1.15}{1+0.00075\rho}\left\{1-\frac{0.20}{1+0.00097\rho}\left(\frac{\rho+1.0}{\rho+0.20}\right)\right\} \qquad (63)$$

[39] H. A. Bethe, *Rev. Mod. Phys.* **9**, 69 (1937), page 160.

A graph of k_∞ versus ρ is shown in figure 33. From it we see that k_∞ is a maximum for a ratio ρ of atomic concentrations of about 20, at which ratio k_∞ is 0.90. We also notice that the variation of k_∞ with ρ is slight for ρ between 5 and 100.

For a value of ρ of 20, equations 54 to 62 yield the following probabilities for the ultimate disposition of a neutron in the system: P_f (fission) $= 0.39$; $P_{a\text{U}}$ (non-fission absorption in uranium) $= 0.59$ (0.20 of this being absorption in the resonance region and 0.39 being thermal absorption): and $P_{a\text{C}}$ (absorption in carbon) $= 0.012$. For $\rho = 100$ the results are: $P_f = 0.38$; $P_{a\text{U}} = 0.56$ (0.18 in the resonance region); and $P_{a\text{C}} = 0.06$.

Another way of stating the above results is to say that for every 2.3 neutrons created in a fission, 0.90 lead to another fission, 1.36 are absorbed in uranium to cause the formation of plutonium and 0.03 are absorbed in carbon, for $\rho = 20$. A similar statement holds for $\rho = 100$.

Shielding effect of a lattice.—As we indicated in the preceding chapter, the advantage of using a lump or rod of uranium in a moderator rather than a homogeneous mixture is that the surface of the lump shields the interior from resonance absorption. This is true if the dimensions of the lump are larger than the mean distance of penetration of neutrons in the resonance energy range. At the same time the dimensions should not exceed the mean distance of penetration of thermal neutrons or the number of fissions will be reduced.

The mean penetration distance L^{res} for neutrons in the resonance energy range may be taken to be λ, where

$$\lambda = \frac{1}{N\sigma_t} \quad . \tag{64}$$

This is 0.0042 cm for ordinary uranium. For neutrons in the thermal range the problem is complicated since a neutron makes several elastic collisions before being absorbed. For a large number q of such collisions, the mean penetration distance is then

$$\begin{aligned} L^{th} &= \sqrt{q}\lambda_s \\ &= \sqrt{q}/(N\sigma_s) \end{aligned} \tag{65}$$

The number q is given by

$$q = \frac{\sigma_s}{\sigma_a + \sigma_f} + 1 = \frac{\sigma_t}{\sigma_a + \sigma_f} \tag{66}$$

For ordinary uranium q is 3.8. If we assume the formula 65 to be applicable for such small numbers, the mean penetration distance for thermal neutrons, L^{th}, is 2.4 cm. We shall, therefore, take the optimum radius of the lump or rod to be 2.4 cm in order to provide optimum shielding of the uranium from neutrons in the resonance region.

For a spherical lump the shielding effect will reduce the effective uranium atomic concentration N_U for resonance absorption in equations 61 and 62 in the ratio

$$\frac{(4\pi R^2 L^{res})}{4/3\, \pi R^3} = \frac{3\, L^{res}}{R} \tag{67}$$

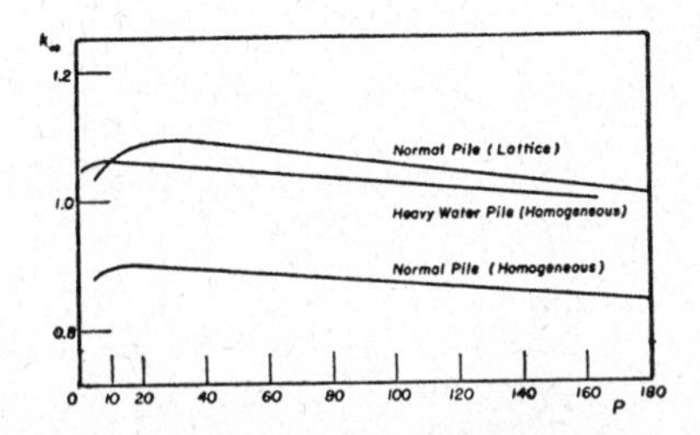

Fig. 33. Multiplication constant k_∞ as a function of atomic concentration ratio ρ, for various pile systems.

For a radius of 2.4 cm this is 0.0052. For a cylindrical rod, the reduction factor is

$$\frac{2\pi\, R\, l\, L^{res}}{R^2\, l} = \frac{2\, L^{res}}{R} \tag{68}$$

which is 0.0035 in our case. For the spherical lumps the effect is to change equation 63 to

$$k_\infty = \frac{1.15}{1+0.00075\rho}\left\{1 - \frac{0.20}{1+0.19\rho}\left(\frac{\rho+0.005}{\rho+0.0001}\right)\right\} \tag{69}$$

ρ now refers to the ratio of the number of atoms of carbon in the pile to the number of uranium atoms rather than to the ratio of atomic concentrations in a homogeneous mixture. A graph of k_∞ versus ρ is shown in figure 33. From it we see the k_∞ is a maximum for a ratio ρ of atomic concentrations of about 40, at which ratio k_∞ is approximately 1.09. We also notice that the variation of k_∞ with ρ is small for ρ between 10 and 100. We may compare our results with the first Columbia pile, described briefly on page 134. From its dimensions (8 feet on an edge) and the fact that it contained 7 tons of uranium oxide, the atomic ratio may be calculated. If the oxide is solid material (density 7.31 gm/cc) the

Table 16

DENSITIES AND NEUTRON MEAN FREE PATHS

Substance	Density (gm/cc)	Concentration N (in units of 10^{23} atoms /cc)	Type of neutron	Mean free path of neutron $\lambda = 1/N\sigma_t$ or $1/\sum_i N_i \sigma_{ti}$ (cm)
C (graphite)	2.25(1)	1.13	thermal	1.9
			fast	5.7
H_2O	1.00(1)	6.68 (H)	thermal	
		3.34 (O)	resonance	
			fast	
D_2O	1.11(2)	6.68 (D)	thermal	
		3.34 (O)	resonance	
			fast	
Be	1.84(1)	1.23	thermal	1.3
			fast	5.1

Bi (liquid)	10.2(1)	0.29	thermal	
			fast	
U (ordinary)	18.7(1)(5)	0.47	thermal	0.93
			resonance	0.0042
			fast	3.3
U^{235}	18.7(1)(5)	0.47	thermal	0.048
			resonance	0.44
			fast	2.5
Th	11.5(1)	0.30	thermal	1.9
			fast	5.5
U_3O_8	7.31(1)	0.16 (U)		
U_3O_8 (powder)	~3.3(4)	0.071 (U)	thermal	14
		0.19 (O)	resonance	0.071

(1) Handbook of Chemistry and Physics, edition **28**, Chem. Rubber Publ. Co. (1944).

(2) Van Nostrand's Scientific Encyclopedia, Van Nostrand (1938).

(4) Anderson, Fermi and Szilard, *Phys. Rev.* **56**, 284 (1939). This value was calculated from the data given therein.

(5) Analysis of the *X*-ray data of Jacob and Warren, *J. Chem. Soc.* **59**, **2588** (1937), indicates a maximum density of 19.1 gm/cm^3 for uranium.

ratio will be 110. If it is in powder form (density 3.3 gm/cc, as given in table 17) the ratio will be 100. For such large ratios the oxygen absorption may be neglected, so that comparison with our results for metal uranium is valid. Our optimum atomic ratio was 40. The check is adequate, therefore, considering the slow variation of k_∞ with ρ and the inaccuracies in our data. Our calculated value of ρ would be increased if the absorption cross section at resonance were increased, if the breadth w of the resonance region were taken to be larger, or if the shielding effect were less pronounced. On the other hand, the increased extraneous absorption due to oxygen and impurities would reduce the calculated value of ρ somewhat. The fast fission enhancement effect would also reduce the optimum ρ.

As we shall see, Smyth's value of 200 (Smyth 8.9) for the number of collisions made by a neutron in carbon in a typical graphite moderated pile, is also in better agreement with an atomic ratio of 100 rather than 40. For this reason we shall use the value $\rho=100$ in our future discussion. For $\rho=100$, k_∞ is 1.06. In view of these uncertainties, the agreement with the calculated value of 1.07 for k_∞ for the central metal lattice of the Chicago West Stands pile (Smyth, Appendix 4) is fortuitous.

As on page 142, it is interesting to calculate the relative probabilities of the various methods of disposal of a neutron in the infinite pile. For $\rho=100$, $P_f=0.46$, $P_{aU}=0.47$ (0.010 in the resonance region and 0.46 in the thermal region), and $P_{aC}=0.069$. If a finite pile is used (together with control absorbers), so that k is unity, then these figures would be reduced by the factor 1/1.06. Therefore, for each 2.3 neutrons formed in a fission, 0.15 are absorbed in carbon, 1.02 are absorbed in uranium to form plutonium (0.022 in the surface layer and 1.0 in the body of the lump), 1.0 is available for further fissions, and 0.14 escape from the pile (or are taken up by the control absorbers).

Though the probability of absorption by uranium of neutrons in the resonance energy region is small, the effect is sufficient to make the surface of the lump far richer in plutonium than the interior. The ratio of body absorption to surface absorption is 1.0/0.022 or 45. However, the plutonium produced by the body absorption is distributed over a volume which for spherical lumps is larger than the volume of the surface layer by a factor $R/3L^{res}$ or 190 (for cylindrical lumps the factor is 290), and it may therefore be advisable to strip the surface layer from the lump before separating out the plutonium. That this is done with the Hanford piles might be concluded from Smyth's statements that "only a few grams of U-238 and of U-235 are used up per day per million grams of uranium present" (Smyth 8.15), and "the problem then is to make a chemical separation at the daily rate of, say, several grams of plutonium from several thousand grams of uranium" (Smyth 8.20). To illustrate, 1,000,000 gm of uranium contain 7,300 gm of U^{235}. If, say, 3 gm of this U^{235} are used up per day for 25 days (making a total consumption of U^{235} of about 1 percent), the plutonium production will be $75 \times 1.02 \times 239/235$ or 78 gm (3.1 gm/day), of which 1.7 gm (0.068 gm per day) are in the surface layer. The 1.7 gm must then be separated from $10^6/190$, or 5300 gm of uranium.

For a more valid application of our calculated results to the Hanford piles, the calculations should be repeated using cylindrical rods of uranium rather than spherical lumps. From equations 67 and 68 we see that the difference is to increase the shielding effect by a factor of 3/2, thus increasing k_∞ slightly but reducing the fraction of uranium that constitutes the surface layer.

With a lattice arrangement the fast neutrons produced in a fission leave the uranium lump and traverse a region of pure carbon before entering a uranium lump again. It is interesting to calculate the average number of collisions a neutron makes with the carbon before entering the uranium. The number of collisions it makes while at thermal energies may be calculated by comparing the probability of absorption by carbon in the thermal region (0.069/1.06 here) with the probability p'' of absorption while traversing a mean free path (for elastic collisions) in carbon. The latter probability is given by

$$p'' = \frac{\sigma_{aC}{}^{th}}{\sigma_{sC}{}^{th}}$$

which is 0.00093. Hence the number of collisions in the thermal range is 0.065/0.00093, or 70. Adding this to the 115 collisions we calculated on page 135 to be required for reduction of a neutron's energy to the thermal range, we get a total of 185. This may be compared with the value of "about 200" quoted by Smyth (8.9) for a "typical" graphite moderated pile. An atomic concentration ratio ρ for carbon to uranium of 40 would yield a lower value of 143 collisions.

It is interesting also to recalculate k_∞ for $\rho = 100$, increasing the absorption cross sections for carbon by 25 percent to allow for impurities. It turns out to be 1.04. This may be compared with the value of 0.87 given for the second Columbia pile (Smyth 4.17), in which the graphite had a neutron absorption 25 percent greater than the graphite used later (Smyth 6.20). The agreement would be much better if the thermal nonfission absorption were larger or if the shielding effect were less pronounced.

We also should have made allowance in our calculations of k_∞ for the fissions by fast neutrons. As we pointed out on page 141, the effect of these fissions is to enhance slightly the number of neutrons reaching the lower energy regions per fast neutron produced in the pile. The enhancement factor ϵ would be

$$\epsilon = 1 + \frac{q\mu N_U \sigma_{fU}^{fast}}{N_U \sigma_{fU}^{fast} + N_U \sigma_{sU}^{fast} + N_C \sigma_{sC}^{fast}} \qquad (70)$$

where q is the average number of collisions that the neutron makes in the region above the fast fission threshold (0.35 Mev). Since inelastic collisions reduce the neutron's energy very rapidly, we shall take q to be approximately 1. The data of chapter 9 then yield

$$\epsilon = 1 + \frac{1.15}{6.5 + 1.6\rho} \qquad (71)$$

For ρ=100, this factor is 1.007. (Actually, this factor should have been included in equation 69 in determining the optimum value of ρ. It would have reduced ρ even further).

Power production and plutonium production.—If each fission produces 200 Mev of energy, we can easily calculate the relationship between power production and plutonium production. Using the results of page 146, in which we estimated that 1.02 atoms of plutonium are formed per fission, we find that a production of 1 kg/day of plutonium corresponds to

$$\frac{10^3 \text{ gm Pu}}{\text{day}} \times \frac{1 \text{ mol}}{239 \text{ gm}} \times \frac{6.03 \times 10^{23} \text{ atoms}}{1 \text{ mol}} \times \frac{1 \text{ fission}}{1.02 \text{ atoms}}$$

$$\times \frac{200 \text{ Mev}}{\text{fission}} \times \frac{1.6 \times 10^{-13} \text{ joules/sec}}{1 \text{ Mev}} \times \frac{1 \text{ kw}}{1000 \text{ joules/sec}} \times \frac{1 \text{ day}}{86400 \text{ sec}}$$

$$= 0.92 \times 10^6 \text{ kw.}$$

Smyth (6.32) states that 1 kg/day of plutonium corresponds to between 0.5×10^6 kw and 1.5×10^6 kw.

A rate of production of plutonium of this magnitude would require a very large pile. On page 147 we assumed that 10^6 gm of uranium produce 3.1 gm of plutonium per day. A production of $\frac{1}{3}$ kg/day/pile, as an example, would then require 1.1×10^8 gm of uranium. (We are neglecting the differences involved in the use of cylindrical rods of uranium rather than spherical lumps.) An atomic ratio ρ of 100, corresponding to a mass ratio of 5.0, leads to a total volume of

$$\frac{1.1 \times 10^8}{18.7} + \frac{5.5 \times 10^8}{2.25} = 2.5 \times 10^8 \text{ cc} \tag{72}$$

which is the volume of a cube of edge 630 cm or 21 feet.

To dissipate such large powers (as at Hanford, where plutonium production was the primary objective) requires an extensive cooling system. If we assume that water at 20° C is raised to 80° C, we find that 0.92×10^6 kw (0.31×10^6 kw/pile) corresponds to

$$0.92 \times \frac{10^9 \text{ joules}}{\text{sec}} \times \frac{60 \text{ sec}}{1 \text{ min}} \times \frac{1 \text{ cal}}{4.2 \text{ joules}} \times \frac{1 \text{ cc}}{60 \text{ cal}} \times \frac{1 \text{ cu. in.}}{16.4 \text{ cc}} \times \frac{1 \text{ gal.}}{231 \text{ cu. in.}} = 60000 \frac{\text{gal.}}{\text{min.}}.$$

This may be compared with the capacity of the Columbia River, whose mean flow at the Grand Coulee Dam is 211,000 cu. ft/sec, or 95×10^6 gal/min.[41] It is of course considerably less at Hanford. A power production of 0.92×10^6 kw would raise the temperature of the river at the dam at mean flow by only 0.04°C.

A rough calculation may be made of the size of the cooling system. If we assume a pile 630 cm on an edge, containing a mass of 1.1×10^8 gm of uranium in the form of rods 2.4 cm in radius, the number of tubes in the pile will be 520. With a total water flow of 20,000 gal/min (per pile) each pipe must carry 38 gal/min, or 2300 cc/sec; this is easily attainable. For example, a clearance of 2 mm between the uranium rods and the pipe would require a flow of 25 ft/sec, which could be obtained by maintaining a pressure of about 12 lb/in^2.[42] Such an amount of water would have an absorption relative to the carbon of

$$2 \times \frac{520 \times 630 \times 2\pi\,(2.4)\,(0.2) \text{ cm}^3}{18 \text{ gm mol}^{-1}} \times 0.30 \text{ cm}^2 \times \frac{12 \text{ gm mol}^{-1}}{2.4 \times 10^9 \text{ cm}^3 \times 0.0045 \text{ cm}^2} = 0.12$$

[41] United States Government Report, 1941.

[42] H. Lamb, Hydrodynamics, 4th ed. pp. 579, 571; Cambridge 1916.

The factor of 2 is introduced since there are 2 atoms of hydrogen per molecule of water. This would reduce the value of k_∞ from our previously determined value of 1.06 to about 1.05.

If we had assumed that the figure of 1kg/day of plutonium referred not to total production but to production in the surface layers, the calculated volume of the pile, the power dissipated in the pile and the temperature rise of the river would have been multiplied by a factor of 45 for spherical lumps of uranium, or 70 for cylindrical lumps. (See page 143.)

Fission products. It is clear from Smyth's report (4.27, 6.36, 7.31, 7.35, 8.15) that neutron absorption by fission products is an important factor in the operation of piles. Thus, in section 7.31, Smyth states that "the gradual disappearance of the U^{235} and the appearance of fission products with large neutron absorption cross sections tend to stop the reaction"; and, in section 8.15. "However, other fission products are being produced also. These consist typically of unstable and relatively unfamiliar nuclei so that it was originally impossible to predict how great an undesirable effect they would have on the multiplication factor. Such deleterious effects are called poisoning."

There are four evident criteria for a fission product to have a serious poisoning effect:

(a) It should have a very large cross section for absorption of thermal neutrons.

(b) It should have a large branching ratio.

(c) It should have a mean life for radioactive decay long compared to, or at least of the order of, the mean life for decay by absorption of thermal neutrons. This ensures that the effect of the fission product be predominantly absorption of thermal neutrons.

(d) It should build up to its equilibrium concentration in a time smaller than, or at most comparable with, the time of operation of the pile.

We can compare the neutron absorption of the fission product with that of the carbon moderator under equilibrium conditions, if we adopt the values calculated on page 146. We found there that each fission leads ultimately to 1.02 atoms of plutonium, and that of the 2.3 neutrons emitted, 0.15 are absorbed in carbon. But each fission produces b atoms of the fission product, where b is called the branching ratio. These b atoms may be removed from the system either through natural decay (if the product is unstable), or by absorption of neutrons. If the latter process is the dominant one, then the ratio of neutron absorption by the fission product to that by carbon is $b/0.15$. This gives an upper limit for the absorption by the fission product. We are here neglecting the effect of the fission product absorption on the carbon absorption calculated previously.

Smyth (8.17) states that "About twenty different elements are present in significant concentration. The most abundant of these comprises slightly less than 10 per cent of the aggregate". This implies a maximum branching ratio of about 20 percent. Anderson and Fermi[43] list branching ratios as high as 12 percent, and specify a number of fission products with relatively long half-lives. If any of these have both large absorption cross sections and large branching ratios, they will contribute a substantial amount of poisoning.

A better estimate of the absorption may be obtained as follows: since each fission produces about 200 Mev energy and releases about 2.3 neutrons, operation of a pile at a power level of 0.31×10^6 kw (page 148) corresponds to $(0.31\times10^6\times10^{10}) \div (200\times1.6\times10^{-6})$ or 0.97×10^{18} fissions per second, with a release of 2.2×10^{19} neutrons per second. If N is the number of atoms of the fission product present when equilibrium is established, then, neglecting neutron losses above the thermal range,

$$\alpha=\lambda N+\left(\frac{N\sigma}{\Sigma_i N_i\sigma_i}\right)\times2.2\times10^{19} \qquad (73)$$

where

α=production rate of the fission product
 $=0.97\times10^{19}\ b$ atoms per second
λ=time constant of the radioactive decay, in sec^{-1}
σ=absorption cross section of the fission product for thermal neutrons
N_i=number of atoms of any element present
σ_i=absorption cross section of element i for thermal neutrons.

We may rewrite equation 73 in the form

$$\alpha = \lambda N+\lambda' N, \qquad (73a)$$

where λ' is the time constant (reciprocal mean lifetime in sec^{-1}) for the decay of the fission product by neutron absorption, and is given by

$$\lambda'=\frac{\sigma(2.2\times10^{19})}{\Sigma_i N_i\sigma_i}\cong\frac{\sigma(2.2\times10^{19})}{N_U(\sigma_{aU}+\sigma_{fU})+N_C\sigma_{aC}}$$

[43] See footnote 2, table 14.

The contribution of the fission product $N\sigma$ to the denominator is assumed to be negligible. The cross sections are given in chapter 9. From equation 72 we see that

$$N_U = 1.1\times10^8\times6.03\times10^{23}/238 = 2.8\times10^{29}$$
$$N_C = 5.5\times10^8\times6.03\times10^{23}/12 = 2.8\times10^{31}$$

This yields a relation between λ' and the cross section σ for absorption of thermal neutrons by the fission product, viz.,

$$\lambda' = 1.2\times10^{13}\sigma. \qquad (74)$$

Also the rate of absorption of neutrons by the fission product, $\lambda' N$, may be obtained from equation 73a

$$\lambda' N = \frac{a}{1+(\lambda/\lambda')} = \frac{0.97\times10^{19}b}{1+(\lambda/\lambda')}$$

The rate of absorption of neutrons by carbon is $0.97\times10^{19}\times0.15$ atoms per second, since 0.15 are absorbed in carbon for every fission. Hence the ratio of absorption of neutrons by the fission product to that by carbon is

$$\frac{b/0.15}{1+(\lambda/\lambda')}$$

It is thus clear that for a large poisoning effect the branching ratio b should be large (criterion (b)). Also the reciprocal mean lifetime λ for radioactive decay should be smaller than, or at most comparable with, the time constant λ' for decay by absorption of thermal neutrons (criterion (c)).

Equation 74 yields some interesting results. In the first place, it leads to criterion (a), that the cross section σ of a highly poisonous fission product for absorption of thermal neutrons must be very large. If the fission product is to build up to the neighborhood of its equilibrium concentration in less than, say, 25 days (the time assumed for operation of the pile), $(\lambda+\lambda')$ must be greater than $1/25\times86{,}400$ or 4.6×10^{-7} sec^{-1}. Since criterion (c) requires that λ' be at least of the order of magnitude of λ, λ' would have to be at least about 2×10^{-7} sec^{-1}. Equation 74 would then require that σ be at least $17{,}000\times10^{-24}$ cm^2.

A further consequence is that only long-lived fission products are likely to give serious poisoning effects. For example if λ' is 1.2×10^{-5} sec^{-1} (about 1 day^{-1}) equation 74 gives a value of σ of $1{,}000{,}000\times10^{-24}$ cm^2. Values of λ' larger than this (corresponding to a larger upper limit for λ and hence a shorter mean life for radioactive decay than 1 day) would require even larger values of the cross-section σ. The quantum mechanical upper limit for σ for thermal neutrons is πa^2 (where a is the reduced wave length given by de Broglie's formula) and this limit has the value 26×10^{-18} cm^2.

A recalculation of k_∞ for a carbon uranium pile, increasing the thermal neutron absorption by carbon by, say, 50 percent to allow for poisoning reduces the value of 1.06 (page 146) to 1.02. The absorption by the cooling system reduces it further to 1.01 and the fast fission enhancement increases it to 1.02.

Neutron lifetime and neutron density.—The lifetime τ of a neutron is the sum of the time τ_1 that it spends in slowing down to thermal energies and the time τ_2 that it spends making collisions in the thermal energy region before being absorbed. τ_2 is given by

$$\tau_2 = q' \frac{\lambda_s^{th}}{v} = q' \lambda_s^{th} \, (2M/E)^{\frac{1}{2}} \; , \qquad (75)$$

where q' is the number of collisions in the thermal range, m is the neutron mass, and E is the thermal energy (0.025 ev at $T = 300°A$). v is 2.0×10^5 cm/sec. and τ_2 is $1.25\times10^{-5}\,q'$ sec. On the other hand, since the energy after i collisions is given by

$$E_i = E_o \, (1 - f)^i \; ,$$

the time τ_1 may be calculated by evaluating the sum

$$\tau_1 = \sum_{i=1}^{q} \frac{\lambda_i}{v_i}$$

$$= \lambda_o \sqrt{\frac{2m}{E_o}} \sum_{i=o}^{q-1} (1-f)^{-i/2}$$

$$= \lambda_o \sqrt{\frac{2m}{E_o}} \left\{ \frac{(1-f)^{-q/2} - 1}{(1-f)^{-1/2} - 1} \right\} ,$$

where λ_o is the mean free path for elastic collisions at high energies. We have assumed that this mean free path remains constant as the energy decreases. Since it actually decreases and since inelastic collisions serve to reduce the required time, our calculation gives an upper limit to τ_1. If the total energy drop is large $(1-f)^{-q/2}$ is very large compared to unity. Also $E_o\,(1-f)^q$ is the final thermal energy E. Hence, finally, we obtain

$$\tau_1 = \sqrt{\frac{2m}{E}} \left\{ \frac{1}{(1-f)^{-1/2} - 1} \right\} . \qquad (76)$$

In carbon, for neutrons with an initial energy of one Mev. τ_1 is 4.7×10^{-4} sec, which is equivalent to the time required for 38 collisions in the thermal range.

On page 147, we found that an atomic concentration ratio of 100 implied 70 collisions in the thermal range, giving a total time of $108 \times 1.25 \times 10^{-5}$ or 1.4×10^{-3} sec for the lifetime of a neutron in the pile. This average lifetime enables us to make a rough estimate of the neutron density in a pile operating at a power level of, say, 0.31×10^6 kw (corresponding to a production of $\frac{1}{3}$ kg per day of plutonium as discussed on page 149). Each fission produces about 200 Mev energy and releases about 2.3 neutrons, so that operation at a power level of 0.31×10^6 kw corresponds to the production of $(0.31 \times 10^6 \times 10^{10} \times 2.3) \div (200 \times 1.6 \times 10^{-6})$ or 2.1×10^{19} neutrons per second. Since each neutron lives approximately 1.4×10^{-3} seconds, the number of neutrons present at any time is 2.9×10^{16}. For a pile of edge 630 cm (page 149) the mean density is then

$$\frac{2.9 \times 10^{16}}{2.5 \times 10^{8}} = 1.2 \times 10^{8} \text{ neutrons/cc}$$

If the distribution outward from the center of the pile varies roughly as $(\sin \pi r/R)/(\pi r/R)$, the density at the center is greater than the average density by a factor

$$1 \div \left\{ \frac{\int_0^R (\sin \pi r/R)(R/\pi r) 4\pi r^2 \, dr}{\int_0^R 4\pi r^2 \, dr} \right\} = \frac{\pi^2}{3}.$$

Hence it is 3.9×10^8 neutrons/cc at the center of the pile. Since 760 mm of mercury corresponds to 2.6×10^{19} molecules/cc at 80°C, this corresponds to a pressure of 1.1×10^{-8} mm of mercury.

Critical size—on page 120 the diffusion theory was applied to a fast-neutron-induced chain reaction in a sphere of uranium oxide. The critical radius R_o was given by

$$R_o = \pi \sqrt{D/K} \quad , \qquad (77)$$

where the diffusion constant D is

$$D = \tfrac{1}{3} \lambda v = \tfrac{1}{3} v \{N \sigma_t + \sum_i N_i \sigma_{si}\}^{-1} \qquad (78)$$

and the quantity K, which appears as an extra term (KN) in the diffusion equation 36, is

$$K = v\{N(\mu\text{-}1)\,\sigma_f - \sum_i N_i \sigma_{ai}\} \quad . \tag{79}$$

These results also apply to slow neutron chain reactions if the cross sections are suitably averaged over the energy ranges occupied by neutrons during their lifetime in the pile.

The use of the boundary condition that N vanishes at the surface is adequate in this case because the mean free path λ is very small compared to the critical radius. (This point is discussed on page 175.)

Since KN is the net extra number of neutrons produced per unit volume per second, K is the time rate of production of extra neutrons per neutron. It appears as e^{Kt} in the time dependent part of the expression for neutron concentration. Since each neutron has a lifetime τ, the number of extra neutrons produced per neutron in the pile is $K\tau$. But this number is also $k_\infty - 1$.

$$K = (k_\infty - 1)/\tau \quad , \tag{80}$$

or

$$k_\infty = 1 + K\tau \quad .$$

Therefore the critical radius is

$$R_c = \pi \sqrt{\frac{\lambda v \tau}{3(k_\infty - 1)}} \tag{81}$$

Since most of the collisions are in carbon and are due to thermal neutrons, λ and v may be taken to be the mean free path and the velocity of thermal neutrons in carbon. If we take $k_\infty = 1.06$ (page 146) and $\tau = 1.4 \times 10^{-3}$ seconds (page 154), the critical radius R_c turns out to be 170 cm or 5.6 ft. This result would apply approximately to the Chicago West Stands pile which consisted of lattices with values of k_∞ of 1.01, 1.04 and 1.03. (Smyth, appendix 4.)

The dominant term in equation 39 for the neutron concentration is

$$\frac{A_1}{r} (\sin \pi r/R) e^{(K - \pi^2 D/R^2)t} \quad . \tag{82}$$

Hence, the effect of a finite radius R is to reduce the effective value of K to $K' = K - \pi^2 D/R^2$, and of $k_\infty = 1 + K\tau$ to $k = 1 + K'\tau$. If R is the critical radius, $K' = 0$ and $k = 1$. If R is three times the critical value, $K' = 8K/9$ and $k = 1 + 8Kt/9$. A pile with $k_\infty = 1.05$, therefore, has an effective k of 1.04 if its radius is three times the critical radius.

The reduction in K can be interpreted in terms of the mean time required for escape of a neutron from the center of the pile. If a neutron makes q collisions, it travels a radial distance of $\sqrt{q}\lambda$. Hence the number of collisions made in traveling a distance R is R^2/λ^2. The time required for this passage is

$$\frac{q\lambda}{v} = \frac{R^2}{\lambda v} = \frac{R^2}{3D}, \tag{83}$$

which is of the order of magnitude of the reciprocal of $\pi^2 D/R^2$.

Smyth (appendix 4) discussed the problem of extrapolation to the critical sizė during the construction of a pile. The quantity R^2_{eff}/A is plotted against the number of layers, where R_{eff} is the effective radius of the incomplete pile obtained from geometrical considerations, and A is the activity of a neutron detector placed at the center of the pile. The discussion may be clarified by investigating the diffusion of neutrons in a spherical pile in which neutrons are produced at a constant rate S (by cosmic rays or other processes). The diffusion equation is then

$$\frac{\partial N}{\partial t} = D\nabla^2 N + KN + S\ . \tag{84}$$

If we assume that $N = 0$ at the surface, then the steady state solution for the spherical case $K = 0$ ($k_\infty = 1$) is

$$N = \frac{S}{6D}(R^2 - r^2)\ . \tag{85}$$

The activity A of any neutron detector placed at the center of the pile is proportional then to R^2, so that R^2/A remains constant as the size of the pile is increased. On the other hand, for $K>0$ ($k_\infty > 1$) the steady state solution of the diffusion equation is

$$N = \frac{S}{K}\left\{\frac{R}{\sin(\sqrt{K/D}R)}\ \frac{\sin(\sqrt{K/D}\,r)}{r} - 1\right\}. \tag{86}$$

The activity A at the center is the proportional to

$$\frac{S}{K}\left\{\frac{\sqrt{K/D}\,R}{\sin(\sqrt{K/D}\,R)} - 1\right\}, \tag{87}$$

which becomes infinite as R approaches the critical value R_c (at which sin $\sqrt{K/D}\,R$ vanishes). In this case, a plot of R^2/A against the number of layers approaches zero as R approaches the critical size. The critical number of layers may thus be determined by plotting R_{eff}^2/A against the number of layers and extrapolating the curve to the point where it crosses the abscissa (Smyth, appendix 4).

Reflecting layer—As we saw on page 155, the effect of a non-infinite radius R is to reduce the effective value of K to $K' = K - \pi^2 D/R^2$ and the neutron multiplication factor, $k = 1 + K\tau$, to $k_{eff} = 1 + K\tau$. A reflecting layer will compensate partially for this loss. If we consider a layer of inactive, nonabsorbing material or infinite extent surrounding the spherical pile of radius R, the equations to be solved are

$$\frac{\partial N}{\partial t} = D\nabla^2 N + KN\,, \qquad 0<r<R; \tag{88}$$

$$\frac{\partial N}{\partial t} = D'\nabla^2 N, \qquad R<r,\ ; \tag{89}$$

subject to the boundary conditions

$$N = 0 \text{ at } r = \infty\ ;$$

$$N \text{ and } D\frac{\partial N}{\partial r} \text{ continuous at } r = R. \tag{90}$$

The interesting solutions are those in which the neutron concentration is steady or increases with time. The dominant term in the solution in these cases is

$$N = A/r \sin \alpha r\, e^{(K-\alpha^2 D)t}\ , \qquad r<R\ ; \tag{91}$$

$$N = B/r\, e^{-\beta r}\, e^{\beta^2 D' t}\ , \qquad r>R\ ; \tag{92}$$

where

$$\beta^2 D' = K - \alpha^2 D = \pi^2 D/R_c^2 - \alpha^2 D \tag{93}$$

and

$$D\{\alpha R \cot \alpha R - 1\} = -D'\{\beta R + 1\}\ , \tag{94}$$

where R_c is the critical radius without the reflector. (If the reflecting layer were of finite extent with outer radius R', the exponential in equation 91 would have become $\sinh \beta(R'-r)$ and the term βR in equation 94 would have become $\beta R/\tanh\beta(R'-R)$. However, the hyperbolic tangent is very close to unity for $R'-R$ of the order of magnitude of R_c or greater and D' of the order of magnitude of D so that equation 93 is not altered much.)

Consider now a pile which is at the critical size without any reflector, i.e. $R = R_c$; it will be much beyond critical if a reflector is added. For simplicity, we take $D' = D$; then equations 93 and 94 reduce to

$$\frac{\sin \alpha R_c}{\alpha R_c} = \frac{1}{\pi} . \qquad (95)$$

This gives $\alpha R_c = 2.31$ and $\beta R_c = 2.12$, and the reflector thus changes the steady state into one increasing exponentially with time as

$$e^{(2.12)^2 Dt/R_c^2} .$$

This result is comparable with an infinite pile which should have the time factor (neglecting fast neutron effects)

$$e^{\pi^2 Dt/R_c^2} .$$

If the reflector had a lower diffusion coefficient than the pile (say, a larger scattering cross section for thermal neutrons), αR_c would be smaller but βR_c and $\beta^2 D'$ would be larger and the reflector more effective. Actually, such a general statement cannot be made since the effect of different degrees of absorption in the reflector has not been considered.

Another special case is interesting; the effect of the reflector when R is already greater than R_c. For example, if $R = 3\ R_c$ and $D' = D$, the solution of equations 93 and 94 gives $\alpha R = 2.83$ and $\beta R = 8.99$, corresponding to a value of K' of $0.91K$. In the absence of a reflector, αR is π and $K' = 8/9\ K = 0.89\ K$. The effect of the reflector is small because the density of neutrons at the surface of a large pile is small compared to the density in the interior.

It is interesting to calculate the reflection coefficient p at the surface of a pile which has a reflector. This is defined to be the ratio at the surface of the inward flow to the outward flow and is given by [44]

$$p = \left\{ \frac{\frac{v}{4}N + \frac{D}{2}\frac{\partial N}{\partial r}}{\frac{v}{4}N - \frac{D}{2}\frac{\partial N}{\partial r}} \right\}_{r=R} \qquad (96)$$

[44] L. B. Loeb, "The Kinetic Theory of Gases", p. 259, McGraw-Hill 1934.

In our problem this becomes

$$p = \left\{ \frac{\left(\frac{Rv}{4} - \frac{D}{2}\right)\frac{\tan \alpha R}{\alpha R} + \frac{D}{2}}{\left(\frac{Rv}{4} + \frac{D}{2}\right)\frac{\tan \alpha R}{\alpha R} - \frac{D}{2}} \right\} \tag{97}$$

For a pile $Rv \gg D$ since D is $\frac{1}{3}\lambda v$, so that the reflection coefficient is very nearly unity. This can be understood by considering the history of a particular neutron which has just left the surface of the pile. Although it may ultimately progress far away from the pile, the chance of doing so without once reentering the pile is very small.

If there is no reflector the boundary condition is that the reflection coefficient is zero. That is,

$$\frac{\tan \alpha R}{\alpha R} = \frac{-\frac{D}{2}}{\frac{Rv}{4} - \frac{D}{2}} . \tag{98}$$

Since for a pile $Rv \gg D$ this is very nearly equivalent to the requirement that $\alpha R = \pi$, that is, to the condition that the neutron density N vanish at the boundary.

Effect of delayed neutrons.—In solving the diffusion equation on pages 120 and 157 we neglected the fact that about 1 per cent of the neutrons are delayed. The diffusion equation is

$$\partial N / \partial t = D \Delta^2 N + KN, \tag{99}$$

where K is the time rate of production of extra neutrons per neutron in the pile, and is related to k_∞ by the relation

$$k_\infty = 1 + K\tau \tag{100}$$

where τ is the lifetime of a neutron in the pile. We found τ to be approximately 1.4×10^{-3} seconds for a graphite-uranium pile with an atomic concentration ratio ρ of 100.

The dominant term in the solution of equation 99 is

$$N(r,t) = \frac{A}{r} \sin (\pi r / R)\, e^{(K - \pi^2 D / R^2)t} , \tag{101}$$

so that the effect of a noninfinite radius is to reduce the effective value of K to $K' = K - \pi^2 D/R^2$ and the effective k to

$$k_{eff} = 1 + (K - \pi^2 D/R^2)\,\tau\ . \tag{102}$$

When K' is zero the system is in a steady state.

Now for simplicity let us assume that a small fraction ϵ of the neutrons are delayed by a time t_o. Then equation 99 becomes

$$\frac{\partial N(r,t)}{\partial t} = D\nabla^2 N(r,t) + K_1 N(r,t) + K_2 N(r,t-t_o) \tag{103}$$

where

$$\frac{K^2\tau}{1+K_1\tau+K_2\tau} = \epsilon. \tag{104}$$

The solution for the steady state $(N(r,t) = N(r,t-t_o))$ is unaltered if K is taken to be $K_1 + K_2$. Equation 104 may then be solved to give

$$K_2 = \epsilon(1+K\tau)/\tau = \epsilon k_\infty/\tau\ . \tag{105}$$

However, for a nonsteady state equation 99 has the solution whose dominant term is

$$N = (A/r)\sin(\pi r/R)\,e^{at}\ , \tag{106}$$

where α satisfies the relation

$$\alpha = K_1 + K_2 e^{-t_o} - \pi^2 D/R^2\,; \tag{107}$$

or, using equation 105,

$$\alpha = (K - \pi^2 D/R^2) - \epsilon k_\infty/\tau(1 - e^{-at_o})\ . \tag{108}$$

This constitutes an implicit equation for α. The effect of the delay is to reduce the effective time factor K' by an amount $\epsilon k_\infty/\tau \times (1 - e^{at_o})$. If αt_o is small this is approximately $\epsilon k_\infty(\alpha t_o/\tau)$. Actually the delayed neutrons are distributed in their delay times with four main decay periods (Smyth appendix 3). Let the fraction delayed by a time t_o to $t_o + dt_o$ be $F(t_o)dt_o$, where

$$F(t_o) = \sum_{i=1}^{4} B_i e^{-b_i t_o}\ , \tag{109}$$

so that the total fraction delayed is

$$\epsilon = \int_0^\infty F(t_o)\,dt_o = \sum_{i=1}^{4} \frac{B_i}{b_i}\ ; \tag{110}$$

and the fraction delayed at least a time T is

$$\epsilon_T = \int_T^\infty F(t_o)dt_o = \sum_{i=1}^{4} \frac{B_i}{b_i} e^{-b_i T} \quad . \quad (111)$$

From the values of b_i and (B_i/b_i) given in Smyth and the fact that $\epsilon_{0.01} = 0.01$ (Smyth appendix 3; also, chapter 5 of this book) we find that the values of B_i are 1.1×10^{-3}, 3.7×10^{-4}, 9.0×10^{-5} and 5.0×10^{-6} for $b_i = 0.28$, 0.099, 0.029 and 0.012 $\sec^{-1}$ respectively; and the fraction ϵ of delayed neutrons is 0.011.
Equation (100) then becomes

$$\frac{\partial N(r,t)}{\partial t} = D^2N(r,t) + K_1N(r,t) + h\sum_{i=1}^{4}\int_0^\infty B_i e^{-b_i t_o} N(r,t-t_o)dt_o \quad (112)$$

where h is a proportionality factor to be evaluated.
The equation for the steady state is unaltered if we now interpret K to be

$$K = K_1 + b\sum_i \int_0^\infty B_i e^{-b_i t_o} dt_o = K_1 + h\epsilon \quad . \quad (113)$$

For this steady state case, just as in equation 104,

$$\epsilon = \frac{\tau(h\epsilon)}{1+K\tau} = 0.011, \quad (114)$$

and

$$1 + K\tau = k_\infty \quad . \quad (115)$$

Hence

$$h = k_\infty/\tau \quad . \quad (116)$$

In the nonstationary case, the dominant term in the solution is again

$$N(r,t) = \frac{A}{r}\sin\frac{\pi r}{R} e^{at} \quad , \quad (117)$$

where substitution into equation 112 gives

$$\alpha = K_1 + h\sum_{i=1}^{4}\int_0^\infty B_i e^{-b_i t_o} e^{-\alpha t_o} dt_o - \frac{\pi^2 D}{R^2}$$

$$= K_1 + h \sum_{i=1}^{4} \frac{B_i}{b_i + \alpha} - \frac{\pi^2 D}{R^2}$$

$$= (K - \frac{\pi^2 D}{R^2}) - \alpha h \sum_{i=1}^{4} \frac{B_i}{b_i(b_i + \alpha)} \qquad (118)$$

This equation must then be solved for α. As in equation 80 the effective multiplication factor then becomes

$$k_{eff} = 1 + \alpha\tau \, . \qquad (119)$$

The relaxation time t_r, defined as the time required for the neutron concentration to double, is given by the relation

$$t_r = \frac{0.693}{\alpha} \, . \qquad (120)$$

We thus see that the effect of the delay is to reduce the effective time factor K' by an amount $\alpha h \sum_{i=1}^{4} B_i / b_i(b_i + \alpha)$, and the effective multiplication factor k by this factor times τ. The relaxation time is increased correspondingly.

To illustrate the delay effect let us use some of the data already calculated. If k_∞ is 1.07 and τ is 1.4×10^{-3} sec, then h is 760. Let the system be adjusted to the steady state so that $K' = K - \pi^2 D/R^2$ is zero. Suppose now that some sudden change occurs in the system so that the neutron concentration starts to build up with a relaxation time of one hour. From equation 117 α is 0.00019. From equation 115 the effective K' in the absence of the delay effect would have been $0.00019\{1 + 760 \times 0.011\}$ or 0.0018 sec^{-1}, and the relaxation time would have been decreased from one hour to 6.5 minutes. Similarly, a relaxation time of 1 minute would have been decreased to 7.3 seconds in the absence of the delay effect.

PILE SYSTEMS

The self-replenishing pile.—As we remarked in the introduction, the carbon-natural uranium pile was designed for the specific purpose of producing plutonium in large quantities as soon as possible for military purposes. In considering the possible types and potentialities of pile systems many factors enter. The primary considerations are the availability of the components, the size desired and the purpose intended.

In connection with the availability of pile components, the considerations of the normal uranium-carbon pile suggest an interesting possibility. For the pile considered on page 146, for each 2.3 neutrons formed in a fission, 1.02 are absorbed in U^{238} to form plutonium. Since the plutonium itself is fissionable it can serve to compensate for the depletion of the U^{235} if it is not removed from the pile. Such a pile, in which each fission results in the ultimate production of at least one plutonium nucleus, will thus preserve or increase the supply of fissionable material. In this case U^{238} becomes the principal source material. Such a pile might be called a self-replenishing (or perhaps regenerative) pile. It has the effect of increasing the available supply of atomic energy by a factor of 140, since U^{238} is the abundant isotope. In fact, a further increase in available source material is immediately suggested by using thorium plus an initial amount of U^{235} or plutonium. The capture of a neutron by thorium eventually forms U^{233} by the reactions shown below.[45]

$$_{90}Th^{232} + \underset{\text{slow}}{n} \longrightarrow {}_{90}Th^{233} + \gamma \ (\sigma_a{}^{th}{=}8.3\times10^{-24}\ \text{cm}^2;\ \text{resonance at} \sim 2\ \text{ev}).$$

$$_{90}Th^{233} \xrightarrow[\text{25 min}]{} {}_{91}Pa^{233} + {}_{-}\beta$$

$$_{91}Pa^{233} \xrightarrow[\text{25 days}]{} {}_{92}U^{233} + {}_{-}\beta(0.23\ \text{Mev}) + \gamma(0.3\ \text{Mev}).$$

U^{233} probably fissions like U^{235} and consequently could serve to keep the pile running. Such a thorium pile then increases the available supply of atomic energy since thorium is present to twelve parts per million in the earth's crust whereas uranium is only four parts per million (Smyth 2.24).

This self-replenishing pile could also be made to run indefinitely if the fission products were chemically removed periodically from the fissionable material and more U^{238} or thorium were added.

The steady state.—A useful feature of many pile systems is that during their operation the proportions of fissionable material and source material (i.e., source of new fissionable material) tend to approach a stable relationship. To illustrate, consider a pile containing, say, α atoms of U^{235} or plutonium to each atom of U^{238}. Let us suppose that it is imbedded in a moderator which has negligible absorption for neutrons and which reduces the neutron energies to thermal values so rapidly that the nonfission absorption in the resonance region may be neglected. The rate R_1

[45] Meitner, Strassmann and Hahn, *Zeits. für Phys.* **109**, 538 (1938); L. Meitner, *Phys. Rev.* **60**, 58 (1941).

of depletion of fissionable material due to fissions to the rate R_2 of production of fissionable material by neutron absorption in the thermal range is

$$\frac{R_1}{R_2} = \frac{\alpha \sigma_{fU235}{}^{th}}{\sigma_{aU238}{}^{th}} = 140\alpha \tag{121}$$

with the data we are using. If the atomic ratio α is 1/140 the supply of fissionable material will be maintained if source material (U^{238}) is periodically added to the system to replace that used up. If the ratio α is greater than 1/140 (it is 1/140 for normal uranium), the depletion rate for the fissionable material exceeds the production rate, and α will decrease. On the other hand, if α is less than 1/140 the production rate will exceed the depletion rate, and α will increase. Thus a stable ratio will tend to be set up if the process continues long enough.

For such a system k_∞ may easily be determined in terms of the ratio α. It is

$$k_\infty = \frac{2.3 \times 420\,\alpha}{420\,\alpha + 3} = 1 + \frac{180\alpha - 1}{140\,\alpha + 1}\,. \tag{122}$$

When α is 1/140 this gives $k_\infty = 1.15$. (This is the maximum value k_∞ may have for a self-replenishing system.) When α is 1/70 k_∞ becomes 1.52, and it becomes still larger for larger proportions of fissionable material.

The presence of some absorption by the moderator (and other materials) will not affect the ratio of production to depletion rate, but will reduce k_∞ to the value

$$k_\infty = \frac{2.3 \times 420\,\alpha}{420\,\alpha + 3 + \sigma_a{}'\rho}\,, \tag{123}$$

where ρ is the ratio of atomic concentration of moderator to that of U^{238}, and $\sigma_a{}'$ is its absorption cross section for thermal neutrons in units of 10^{-24} cm^2.

The operation of the pile now results in the production of neutron-absorbing fission products which tend to reduce the value of k_∞ to less than unity and thus stop the chain reaction. Also, if the supply of source material is not replenished the atomic ratio α will continue to maintain itself at 1/140 (or to adjust itself toward this value if it is not originally so). The amount of both fissionable and source material, however, will decrease, thereby increasing the relative effects of the neutron absorbers (moderator and fission products) and reducing the effective k to below unity more rapidly. On the other hand, if k_∞ were originally made large enough to require the use of extra control absorbers in the system which could be withdrawn gradually, the system would continue to operate until all the absorbers were removed. At that time, cleaning out the system (chemical removal of the fission products) and perhaps replenishment of the source material would permit reintroduction of control absorbers and further operation of the system.

The existence of a small amount of nonfission absorption of neutrons in the resonance energy range will not effect the general nature of the above discussions but will alter the stable ratio α and the equations for k_∞. Enough absorption will increase the ratio of the production to the depletion of fissionable material until normal uranium is self-replenishing.

If thorium were used as the source material the fissionable material which it ultimately produces is U^{233} and the ratio of the depletion rate of fissionable material to the production rate would be

$$\frac{R_1}{R_2} = \frac{\alpha\, \sigma_{U233}{}^{th}}{\sigma_{aTh}{}^{th}} = \frac{420\ \alpha}{8.3} = 51\alpha \qquad (124)$$

Hence, for a self-replenishing pile with negligible absorption by the thorium of other than thermal neutrons, the ratio α of fissionable material to source material must be less than 1/51. If it is 1/51, and if we assume the same value 2.3 neutrons per fission for the ultimate product U^{233}, then k_∞ is 1.15 (or less if absorption of neutrons by the moderator is taken into account).

Our discussion has been oriented toward the use of the pile as a source of energy or of neutrons. By adjusting the pile so that the production rate of fissionable material is greater than the depletion rate and then gradually introducing more neutron absorbers to keep the effective k equal to unity, it could be used to enrich the system in fissionable material, which could later be removed.

Heavy water piles.—Except for the problem of obtaining it in sufficient quantities, heavy water is a better moderator than graphite. The average fractional energy loss of a neutron in elastic collisions with deuterium is 0.444 as contrasted with 0.142 for carbon (table 15) so that much less heavy water is needed and the pile may be made smaller. Also the absorption of thermal neutrons is much less than for carbon (0.0045 for carbon, 0.0016 for oxygen and 0.00065 for deuterium[46], as given in table 13). Hence the multiplication factor k_∞ may be made larger than for the carbon-uranium pile.

The neutron collision cross section of oxygen for energies above the thermal range is not given by Bethe (see table 13) but is probably very close to the corresponding cross section for deuterium. Hence, when a neutron is slowed down in heavy water from 1 Mev to thermal energy, it will make q collisions with oxygen and $2q$ collisions with deuterium, where

$$(1-0.111)^q \ (1-0.444)^{2q} = 0.025/10^6 \qquad (125)$$

The solution for q is 19. To slow the neutron down to 5 ev would require q to be 13 (26 collisions with deuterium and 13 with oxygen).

From equation 52 the numerical factor B needed for computing the probability p_1 that a neutron will be at some time in the resonance region, turns out to be 1.4 for deuterium and 8.4 for oxygen. If we let N_W represent the molecular concentration of heavy water, equation 62 for this probability then becomes

$$p_1 = \frac{2N_W\sigma_{sD}^{res}f_D\left(\frac{1.4w}{E_r}\right) + N_W\ \sigma_{sO}^{res}f_O\left(\frac{8.4w}{E_r}\right) + N_U\,\sigma_{sU}^{res}\,f_U\ (1)}{2N_W\ \sigma_{sD}^{res}\,f_D + N_w\ \sigma_{sO}^{res}f_O\ N_U\ \sigma_{sU}^{res}\ f_U}, \qquad (126)$$

where the quantities f, w, E_r and σ_s are defined on page 136. The other equations, 54 to 71, may be applied to the heavy water case if in all of them $N_C\sigma_C$ is replaced by $N_W\ (2\sigma_D + \sigma_O)$ for the appropriate cross sections, and ρ now signifies the ratio N_W/N_D. We shall assume in this case that the probability p' of absorption of a fast neutron before reaching the resonance region is negligible. Equation 69 then becomes, for normal uranium homogeneously distributed in heavy water,

$$k_\infty = \frac{1.15}{1 + 0.00048\rho}\left\{1 - \frac{1}{1 + 0.00024\rho}\left(\frac{1 + 1.8\rho}{1 + 27\ \rho}\right)\right\} \qquad (127)$$

We are again disregarding neutron losses in the higher energy ranges. A graph of k_∞ versus ρ is shown in figure 33. We see that k_∞ has its maximum value 1.06 for a value of ρ of approximately 10, corresponding to a mass ratio of 0.83 and a volume ratio of 14. This value of k_∞ varies very slightly for ρ between 1 and 50. Furthermore, for $\rho = 10$ we find that, of each 2.3 neutrons produced in a fission, 0.01 on the average are absorbed in heavy water, 1.22 are absorbed in uranium to form plutonium (1.06 in

[46] A low capture cross section for deuterium was early predicted by L. I. Schiff on theoretical grounds. L. I. Schiff, *Phys. Rev.* **52**, 242 (1937).

the thermal range and 0.16 in the resonance energy range) and 1.06 are available for further fissions. Use of a finite volume together with added absorbers to reduce k to 1.0 would reduce the above numbers to 0.01, 1.15, and 1.0 respectively, leaving 0.14 neutrons to escape from the pile or to be taken up by the control absorbers or other added materials. Such a pile would be self-replenishing.

The addition to the system of some other moderator having very small neutron absorption but which is less efficient than heavy water in getting a neutron safely past the resonance region (e.g., beryllium oxide, carbon, or carbon dioxide or materials of greater atomic mass) would increase the amount of absorption in the resonance region relative to the other processes and make the pile self-replenishing for normal uranium.

Enriched piles.—If the self-replenishing feature of the pile is not required it is clear that the use of uranium enriched with an additional amount of U^{235} or plutonium would be an improvement over the "normal" uranium pile. The moderator could be either graphite or heavy water. Such a pile, having a very large value of k_∞, could be made quite small and still furnish large amounts of energy and large neutron densities. As an extreme case let us consider a pile consisting of pure U^{235}, or plutonium, homogeneously distributed in a heavy water moderator. For such a system

$$k_\infty = \frac{2.3\,N_U\,\sigma_{fU235}{}^{th}}{N_U\,\sigma_{fU}{}^{th} + N_w(2\sigma_{aD}{}^{th} + \sigma_{aO}{}^{th})} = \frac{2.3}{1+0.0000069\rho} \cong 2.3 \qquad (128)$$

Assuming a ratio of heavy water molecules to uranium atoms of 1,000, this should be sufficient to allow for neglect of the fast fission process in view of the tremendously greater density of thermal neutrons in the pile.

As on page 147 and page 154, we can calculate the number of collisions that are made by a neutron in the pile before it is absorbed and thereby obtain its average lifetime as well as the average neutron density in the pile. From equation 128, the probability of absorption of a neutron by heavy water is 0.0069. If the neutron makes on the average q' elastic collisions with oxygen in the thermal range before being absorbed, then

$$\frac{q'_D}{q'_O} = \frac{2\sigma_{sD}{}^{th}}{\sigma_{sO}{}^{th}}, \qquad (129)$$

and

$$q'_D \frac{\sigma_{aD}{}^{th}}{\sigma_{sO}{}^{th}} + q'_O \frac{\sigma_{aO}{}^{th}}{\sigma_{sO}{}^{th}} = 0.0069 \ . \qquad (130)$$

Solving, we get q_O' to be 10 and q_D' to be 36. The time required for these collisions is

$$\tau_2 = q'_O \lambda_{sO}{}^{th}/v \ , \qquad (131)$$

which is 3.7×10^{-4} sec. To this must be added the time τ_I required to slow the neutron down to thermal energies. We found before that this requires 19 collisions with oxygen and 38 with deuterium. Neglecting the effect of the former we can apply equation 76, which yields 1.3×10^{-4} sec. The total lifetime τ in the pile is therefore 5.0×10^{-4} sec.

By using equation 81, we may now make a rough estimate of the critical size for a pile operating under the above conditions. If λ, v and D are calculated for thermal energies, R_c turns out to be 20 cm, corresponding to a sphere of volume 1,700 cc containing 2,900 gm of U^{235} and 1,300 gm of heavy water. If the pile were operated at a power level of, say, 100 kw, then at 200 Mev per fission the time required to use up the U^{235} is

$$\frac{2900 \times 6.02 \times 10^{23} \times 200 \times 1.6 \times 10^{-6}}{235 \times 3 \times 10^{5} \times 10^{7} \times 86400 \times 365} = 77 \text{ years} \ .$$

Application of the treatment mentioned on page 154, enables us to make a rough estimate of the neutron density in such a pile operating at 100 kw. This corresponds to a production of 7.2×10^{15} neutrons per second, each with an average lifetime of 5.0×10^{-4} sec. Hence the average neutron density is $7.2 \times 10^{15} \times 5.0 \times 10^{-4}/1700$, or 2.1×10^{9} neutrons/cc; at the center of the pile it is 7.0×10^{9} neutrons/cc.

In actual practice such a pile would be built somewhat larger than the critical size, and would contain some control absorbers like cadmium or boron. It would also be surrounded by a reflecting layer of a substance like beryllium or graphite to reduce the neutron losses, as well as by a thick shield of concrete or some other substance.

Conclusion.—Table 18 summarizes the main types of possible piles and their distinguishing characteristics. One additional possible type not mentioned above is the bismuth pile, in which bismuth is used as a cooling fluid in order to permit operation of the pile at much higher temperatures than water-cooled piles, thus giving more efficient power production.

Table 17—PILE TYPES

Type	Distinguishing characteristics	Main reasons for design	Fissionable material	Source material	Moderator	Coolant
Normal	Normal U	availability of materials	U^{235}	U^{238}	graphite (usually)	Water
Enriched	Enriched U^{235} or Pu	smaller size	U^{235} or Pu	U^{238}	graphite	Water
Thorium	Th as principal source	increased availability	U^{235} or Pu	Th		Water
Heavy water	Heavy water as moderator	smaller size	U^{235} or Pu	U^{238}	heavy water	Water
Bismuth	Bi as coolant	higher operating temperature, for more efficient power production	U^{235} or Pu	U^{238}	graphite, etc.	Bi

In conclusion, attention should be called to some of the simplifying assumptions made in this chapter.

(1) We assumed that the cross sections and mean free paths for various neutron processes remain constant in the energy ranges considered (thermal, intermediate and fast). Actually they have in general a rather complicated dependence on energy, as is illustrated in the discussion of the absorption cross section for cadmium in chapter 9 (see figure 12).

(2) We neglected fission by fast neutrons except in so far as it necessitated a substantial proportion of moderator to fissionable material in order to reduce this effect.

(3) We applied rather crude statistical arguments in the analysis of the shielding effect of a lattice arrangement, such as the disregard of the actual spacing of lumps or rods in the lattice and the use of approximate expressions for the distances of penetration of neutrons into lumps .

(4) In calculating the critical size of a pile, we applied diffusion theory to the lattice as though the uranium were homogeneously distributed through the system.

We have tried to present a semiquantitative treatment of some of the fundamental physical considerations involved in controlled slow-neutron-induced chain reaction processes and their application to the design of piles. It should be stressed again that the numerical values are based on inadequate data and that the analysis is intended only to illustrate the possible type of treatment.

CHAPTER 11

FAST NEUTRON CHAIN REACTION

Possibility of a fast neutron chain reaction. Mathematical theory.—The use of a fast neutron chain reaction to produce large neutron pulses and to release large amounts of energy in a short time has many interesting applications. The feasibility of such reactions can best be studied by considering the diffusion equation first discussed on page 120.

While the mean free path of the neutrons is comparable with the dimensions of fissionable material used, limiting the validity of the calculation, nevertheless analysis will show what quantities are important and will suggest their orders of magnitude.

The neutron density at any point will vary with time as a result of diffusion of the neutrons, and also because of absorption of neutrons by capture and the production of new neutrons by fission. The appropriate differential equation for this variation is

$$\frac{\partial N}{\partial t} = D \nabla^2 N + KN \tag{132}$$

These quantities have been previously defined on page 120, but their dependence on other more physical quantities should be recalled. The diffusion coefficient is

$$D = \frac{\lambda v}{3} \tag{133}$$

where v is the velocity of the neutrons. The mean free path, λ, of the neutrons is given by

$$\lambda = \frac{1}{\sigma_s N_U + (\text{other scattering processes})}, \tag{134}$$

where σ_s is the cross section for scattering and N_U the concentration of the main constituent of the reacting material.

$$K = v\sigma_f N_U(\mu - 1) - (\text{other processes which use up neutrons}), \tag{135}$$

where σ_f is the cross section for fission and μ the number of new neutrons released per fission.

We will assume a sphere of radius R and investigate the solution subject to the initial condition:

$$N\,(r,0) = N_o,$$

and the boundary condition

$$D \frac{\partial N}{\partial r} + aN = 0,$$

at $r = R$. This second equation provides for continuity in the normal component of the neutron current at the surface of the sphere. a is related to the rate at which neutrons leave the surface per unit area per unit concentration.

The solution of this problem is given in Byerly[47] and may be written as

$$N(r,t) = \frac{1}{r} \sum_{i=0}^{\infty} A_i \exp \left[\left(K - \frac{Dx_i^2}{R^2}\right) t \right] \sin \frac{x_i r}{R} \qquad (136)$$

where x_i is the i'th root of the equation

$$\tan x = \frac{x}{1 - \dfrac{aR}{D}} \qquad (137)$$

If the concentration of neutrons is to increase with time, the coefficient of the time in one of the exponentials must be positive. If x_o is the smallest root of equation 137, a chain reaction must proceed when

$$R \geqq x_o \sqrt{\frac{D}{K}} \qquad (138)$$

Before we can determine x_o we must find the value of a.

Solving equation 96 for $\left\{ -D \dfrac{\partial N}{\partial r} / N \right\}_{r=R}$ we find that

$$a = \tfrac{1}{2}\, v \frac{(1-p)}{(1+p)}$$

where p is the fraction of neutrons reflected back.

[47] Byerly, "*Fourier's Series and Spherical Harmonics*," pp. 117-122, Ginn and Co. (1895).

From the previous definitions

$$\tan x_o = \frac{x_o}{(1-ax_o/\sqrt{DK})} \qquad (139)$$

$$\frac{a}{\sqrt{DK}} = \frac{(1-p)/(1+p)}{\sqrt{4(\mu-1)\,\sigma_f/3\sigma_s}} \qquad (140)$$

$$\sqrt{K/D} = N_U\,\sqrt{3\,(\mu-1)\,\sigma_f\,\sigma_s} \qquad (141)$$

In this discussion we have assumed that fission is the only important process and have neglected the other absorbing processes mentioned in equations 134 and 135. We have also assumed that the velocity of all neutrons is the same and that the cross sections are independent of neutron velocity.

Possibility of a fast neutron chain reaction. Calculations for U^{235}.—The equation for the critical radius expressed in terms of measurable physical constants is

$$R_c = \frac{x_o}{N_U(3(\mu-1)\sigma_f\sigma_s)^{\frac{1}{2}}} \qquad (142)$$

To get as small a critical radius as possible x_o should be as small as possible and the other quantities as large as possible. The value of x_o can be decreased by use of a tamper which will reflect as many neutrons as possible. To increase N_U the material should be a pure isotope, such as U^{235} or plutonium. μ is taken as 2.3 and assumed to be the same for all fissionable material. The value of σ_s for fast neutrons probably does not vary as much for various nuclei as does σ_f. However, when σ_f is large, σ_s will also be large since as far as the diffusion equation is concerned, σ_s is the sum of all processes which scatter neutrons. In fission, although one neutron disappears and μ new neutrons are actually produced, we regard the process as consisting of the scattering of one neutron plus the production of μ-1 new ones.

The ideal material for producing a chain reaction is one in which every neutron entering the nucleus will produce a fission. The existence of competitive processes, such as absorption and gamma ray emission, decreases both σ_f and σ_s. An additional contribution to the "scattering" cross section is inelastic scattering, in which the neutron is absorbed and another neutron with a lower energy reemitted. As far as calculation of the critical size is concerned it is not necessary to know the relative contribution of the various scattering processes. Measurements of σ_s for nonfissioning heavy nuclei have been made by Dunning et al.[48] The fact that reemission of a neutron after capture is far more likely than a radiative transition,[49] simplified their measurements. σ_s showed a regular increase proportional to $A^{2/3}$ and would be about 6×10^{-24} cm^2 for uranium and plutonium. In U^{235}, fission replaces neutron emission as the main process but as mentioned above, this does not change the effective σ_s.

The average energy of the neutrons being considered is of the order of 1 Mev, and for them σ_f is not simply the πR^2 mentioned on page 71. Their effective wave lengths are of the order $\lambda \sim 10^{-12}$ cm which is comparable with the nuclear radius R as determined by α particle scattering. Rabi[50] has shown that the measured cross section may be several times larger than πR^2, which accounts for the large value of σ_s.

The exact value of σ_f depends on the neutron energy. We can estimate an approximate value of 3×10^{-24} cm^2, or about half the total scattering cross section for U^{235}. A survey of the published cross sections of heavy nonfissioning nuclei indicates that potential scattering accounts for about half the total scattering. The other half of the fast neutron scattering cross section is contributed by inelastic cross section effects, the most likely one being fission in those nuclei which will fission with thermal neutrons. This was discussed on page 111.

Inaccuracies in the values of σ_f and σ_s seriously change our estimate of the critical mass because of the rapid change in mass with radius. Smyth's description (12.32) of the extensive fast neutron cross section measurements at Los Alamos emphasizes the importance of these cross sections. The calculation presented here should be considered as tentative since in a report to the National Academy in November 1941 (Smyth 4.99) the critical mass could only be fixed at between 2 and 220 pounds, mainly on account of the uncertainties in the cross sections.

In the following calculations μ is taken as 2.3 neutrons per fission and

$$N_U = \rho \frac{N_a}{M} = \frac{19 \text{ gm cm}^{-3} \times 6 \times 10^{23} \text{ atom (gm mol)}^{-1}}{235 \text{ gm (gm mol)}^{-1}}$$

$$= 4.8 \times 10^{22} \text{ gm cm}^{-3}$$

[48] Dunning, Pegram, Fink and Mitchell, *Phys. Rev.* **48**, 265 (1935).
[49] H. A. Bethe, *Rev. Mod. Phys.* **9**, 160 (1937).
[50] I. I. Rabi, *Phys. Rev.* **43**, 858 (1933).

If all the neutrons reaching the surface escape, the reflection coefficient is zero and x_0 is 2.10. It should be noted that Adler's[51] calculation assumed that the concentration at the surface of the sphere was zero and for this case the value of x_0 is π. This assumption is good in systems which are so large that $\lambda \ll R$. In our system $\lambda \sim R$, and the fact that we use the more exact calculation results in a reduction in the critical mass by a factor of 3.4.

The critical radius with no tamper is $R_c = 5.15$ cm which corresponds to a mass of about 23 pounds of U^{235}. The mean free path is given by

$$\lambda = \frac{1}{\sigma_t N_U}$$

and for U^{235}, it is 3.4 cm. This value is comparable with the critical size; therefore our use of the diffusion theory might lead to considerable error. If we compare the solution for the one dimensional random walk problem[52] with the diffusion equation solution, it appears that the probability of escape is different from that calculated for diffusion. This simple consideration indicates that replacing the diffusion theory calculation by a more accurate statistical treatment of the neutron paths would give a different estimate of the critical radius.

Effect of a tamper.—The critical size may be somewhat reduced by the use of a tamper. A value for the reflection coefficient can be found by solving the differential equation for two concentric spherical media. The inner sphere of U^{235} has the values of K and D discussed above. The second sphere is the tamper with about the same value of D, but with $K = O$. The region surrounding it consists of empty space. Since a detailed calculation (see page 158) is hardly worth while at this point, a value for the reflection coefficient will be estimated very crudely.

In our approximation the value of the reflection coefficient p for a tamper of given thickness cannot be calculated until the critical size is known. Consequently a preliminary value of p will be used to get a preliminary value of R_c. This value of R_c will be used to redetermine p and to calculate a more accurate value of R_c. For a tamper of high density the mean free path in the tamper will be about the same as that in the fissionable material, 3.4 cm, since the fast neutron scattering cross sections of materials of large atomic number are about the same. We assume tentatively that

[51] M. F. Adler, *Comptes Rendus* **209**, 301 (1939).
[52] S. Chandrasekhar, *Rev. Mod. Phys.* 15, 1 (1943).

this is of the same order as the radius of the fissionable material. If neutrons leave the core perpendicularly as shown in figure 34, then on the average they will suffer a collision at a distance λ from the surface of the sphere, and will be scattered in all directions. The mean probability of being scattered into the solid angle ω subtended by the core is $\omega/4\pi$, where

$\frac{\omega}{4\pi} = \frac{1}{2}(1 - \cos\theta)$, and $(\sin\theta)_{\text{ave}} = \frac{R}{R+\lambda}$. Therefore, p is roughly given by

$$p = \frac{1 - \sqrt{(1-R^2)/(R+\lambda)^2}}{2} \tag{143}$$

This estimate neglects all multiple scattering processes. However these effects tend to compensate, since the neutrons which eventually return to the core after several scatterings would increase the coefficient, while neutrons which are scattered into the solid angle ω may be scattered out of it before reentering the core. This would decrease p. Of course, all the neutrons do not leave the surface of the core normally, as we have assumed, but this complication does not change the order of magnitude of the reflection coefficient and it is neglected here.

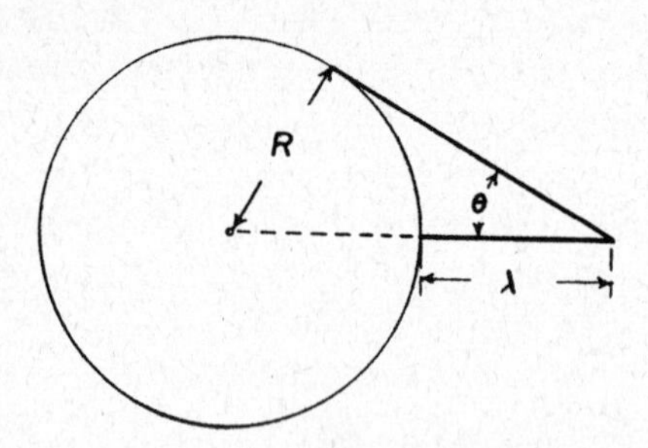

Fig. 34. Diagram of core of fissionable material showing the coordinates pertinent to the reflection of neutrons by tamper.

For the preliminary critical radius, $R_c \approx 5.15$ cm and the reflection coefficient p is 0.10. The new critical radius is $R_c = 4.65$ cm, corresponding to a mass of 16 pounds. The consistency of the calculations could be improved by repeating the calculation for a new value of the reflection coefficient based on the new critical radius.

A more indirect method of estimating the critical size can be made by using remarks in the Smyth report (6.39). The theoretical studies of Manley, Oppenheimer, Serber and Teller indicated that the energy release in a fast neutron chain reaction could be made greater than that estimated in the third report of the National Academy. It was indicated there (Smyth 4.99) that between 1 and 5 per cent of the fission energy should be released at a fission explosion.

A war department release stated that a typical fission explosion contained the explosive equivalent of about 20,000 tons of T.N.T. The fission of 8 pounds of U^{235} will produce this amount of energy. If the chain reaction has an efficiency of between 10 and 30 percent the total mass of U^{235} lies between 24 and 80 pounds. It appears then that substantially more than the critical amount of material was used in these explosions.

If the U^{235} is not pure, the critical radius will be somewhat larger because of the decrease in K. The fission cross section for U^{238} is about one fifth that of U^{235}, and neutron absorption will use up some of the neutrons. In fact, pure uranium metal will not produce a chain reaction, even for an infinite sphere (Smyth 12.10). The diffusion coefficient D will vary little with concentration. The percentage increase in the critical radius is probably about half the percentage concentration of U^{238} in U^{235}. In the same manner, the separation of Pu^{239} from U^{238} need not be taken to completion. No estimate of the critical size for Pu^{239} will be made except to indicate that the value of σ_s and σ_f are both probably a little larger than for U^{235}, making the critical radius somewhat smaller. The energy release per fission may also be a little larger since the electrostatic forces are increased.

We now make an estimate of the rate at which the neutron density will build up in a spherical mass of fissionable material which is larger than the critical size. From equation 136 the time necessary for the neutron density to increase by a factor $e = 2.718$ is

$$T \cong \frac{1}{K - \frac{Dx_o^2}{R^2}} = \frac{1}{\frac{Dx_o^2}{R^2}\left[\left(\frac{R}{R_c}\right)^2 - 1\right]} \qquad (144)$$

If we let $R = R_C + \Delta R$, with $\Delta R \ll R_C$, then

$$T = \frac{R_C}{2K \Delta R} = \frac{1.7 \times 10^{-9} R_C \text{ sec}}{\Delta R}, \qquad (145)$$

since for 1 Mev neutrons in U^{235}, K is of the order of 3×10^8 sec $^{-1}$.

Production of controlled neutron pulses. If a large controlled pulse of neutrons is desired for experimental purposes it can be obtained by combining several pieces each of which is smaller than the critical size, and disassembling then in a time comparable with the time for the neutron density to double.

As the neutron density builds up a pressure will develop which tends to blow the material apart. This pressure at the surface of the sphere has two principal contributions, the gamma radiation and the neutrons. The range of the beta rays and fission products is so short that their kinetic energy is soon shared with other nuclei. This energy is propagated as a shock wave at a velocity which is small in comparison with that of the neutrons and gamma rays. We can make a crude estimate of the neutron pressure P in terms of a gas composed of the 1 Mev neutrons and having a density N_O.

$$P = 1/3\, N_o E$$

where E is the neutron energy. This places an upper limit on the pressure since the neutrons are not reflected at the boundary as in a gas but travel several cm into the tamper before undergoing a collision.

If the tamper will withstand a pressure wave of short duration of 10^4 atmosphere, we can tolerate a neutron density of 10^{16} per cm^3. If the original neutron density were of the order of 1 per cm^3, the density would have to increase by a factor of e^{37} in order to reach a density of 10^{16} per cm^3. The total time required is

$$T^* = \log_e 10^{16} \times 1.7 \times 10^{-9} \frac{R_C}{\Delta R} = 37 \times 1.7 \times 10^{-9} \frac{R_C}{\Delta R} \text{ sec}$$

and if $R_C / \Delta R$ is of the order of 1000, the time becomes about 5×10^{-5} sec. If controlled neutron pulses are to be produced, a mechanical motion of ΔR must be achieved within this time or the reacting material could not be disassembled before an explosion took place. For a radius of 5 cm the velocity of the moving parts must be the order of 100 cm/sec if the critical radius is exceeded by 0.1 percent. The velocity required of the moving parts varies inversely as the square of the fraction by which the critical radius is exceeded.

As mechanisms with parts moving near the speed of sound are feasible, neutron pulses of several microseconds duration and with an intensity of 10^{24} neutrons per second, appear possible. The existence of such pulses would allow the application of many of the timing techniques of the radar art to a large number of problems. However, the possibility of a mechanical failure while the material is over critical is not attractive.

A rough estimate of gamma radiation pressure can be made by assuming that a one Mev gamma ray is given off for every neutron. The momentum of the gamma ray is roughly 1/50 that of a neutron of the same energy, so their contribution to the pressure is of secondary importance. The gamma radiation however, reaches the tamper material first so it may be of importance in a detailed calculation.

In the discussions in this chapter it is assumed that the neutrons are emitted instantaneously during a fission. It is also possible to construct assemblies which are overcrititcal for delayed neutrons but undercritical for fast neutrons; hence the time constant is considerably longer. Smyth (12.46) describes the experiments of this type performed at Los Alamos.

Production of single pulses with maximum number of neutrons.—If the material is over critical and is not intentionally disassembled, the reaction will continue until the material is consumed, the fission products "poison" the reaction by absorption or the material is blown apart. In contrast to piles, there is probably little "poisoning" by the fission products since their most likely reaction with fast neutrons is scattering rather than absorption. If the critical radius has been exceeded by an amount ΔR and there is no motion of the material or capture of the neutrons by the fission products, then the reaction will continue until the concentration of U^{235} has decreased by about twice $\Delta R/Rc$. In most cases the kinetic energy of the fission products will disperse the material long before it has been consumed.

The number of neutrons required to produce fission products of the same energy as 20,000 tons of TNT is of the order of 10^{25} neutrons, or a density of the order of 10^{22} neutrons per cm^3. The pressure developed by a neutron gas in which the average energy of the neutron is of the order of 1 Mev may be about 10^{10} atmospheres. If pressures of a billion atmospheres are to be developed, extremely short times for building up the neutron density are required in order that inertia can be used to hold the material together effectively. The time required for the neutron pressure to build up from 10^4 to 10^{10} atmospheres is

$$1.7\times10^{-9}(R_c/\Delta R)\times(2.3 \log 10^6) \text{ seconds or about}$$
$$2\times10^{-8}R_c/\Delta R=10^{-7}/\Delta R \text{ sec.}$$

To get an idea of the order of magnitudes involved let us assume that the tamper is 4 cm thick and composed of a dense material. The time required to move it through a distance ΔR under the influence of a static pressure P is approximately

$$t = \sqrt{\frac{8\,\Delta R\,\rho}{P}}$$

where ρ is the density of the tamper. For a pressure of a billion atmospheres this time caculated for U^{235} is approximately $10^{-7}\sqrt{\Delta R}$. Thus, if the time for the neutron density to build up is to be less than that required to overcome inertial forces,

$$\frac{10^{-7}}{\Delta R} \leqq 10^{-7}\sqrt{\Delta R}$$

or ΔR is approximately 1 cm. In other words, the critical radius must be exceeded by 20 percent, and thus the total mass increased from 16 to 28 pounds.

CHAPTER 12

SEPARATION OF ISOTOPES

Introduction[53].—In the preparation and use of fissionable material it is desirable to obtain more or less pure samples of a particular isotope. The reason for this is that in the reactions of interest (e.g., slow and fast neutron chain reactions) the isotopes not entering into the reactions either poison or dilute them. Since in nature elements usually occur as mixtures of various isotopes, and since the isotopes of interest often have the smallest concentration in the mixture, it is necessary to use rather complicated separation methods in order to obtain the desired isotope in a sufficiently pure form.

The main difficulty of the isotope separation methods is that isotopes have identical nuclear charges and differ only in their nuclear mass. Consequently, they have but slightly different electronic structures. The most likely property of isotopes, therefore, which can be used in their separation is only the difference in nuclear mass.

Since the kinetic energy of molecules in a gas or liquid depends only on the temperature, the average velocity of two isotopes will be inversely proportional to the square root of their mass. This fact is used in the separation of isotopes by gaseous diffusion. When isotopes are ionized and accelerated in an electric field, their kinetic energies again depend only on the difference of potential traversed (and the ionic charges). The resulting difference of velocity of the isotopes can then be used to separate them by electromagnetic methods.

Isotopes can also be subjected to gravitational and thermal fields. In the latter case, the motion of the molecules does not depend only on the fields and the masses of the isotopes, but also on inter-molecular forces which act differently on different isotopes. Separation effects of this kind are produced by centrifuge and by thermal diffusion methods.

As was previously remarked, the electronic structures of isotopic molecules are so similar that their chemical properties, which depend directly on the outer electron structure, are also much alike.

[53] For a complete summary of isotope separation methods in use before 1939, see W. Walcher, *Erg. d. exakt. Naturwiss.* **18**, 155 (1939); H. C. Urey, *J. App. Phys.* **12**, 270 (1941).

However, the rates of chemical reactions differ slightly for different isotopes since the molecular, kinetic and vibrational energies which determine the reaction rates, depend on the nuclear mass. This permits an isotope separation by "chemical exchange" reactions.

This chapter will discuss in detail the physical nature of some of the isotope separation methods indicated above. Since the present interest of isotope separation lies mainly in the large scale separation of U^{235} and deuterium (in the form of heavy water), an effort is made to obtain numerical values for some of the interesting parameters occurring in the separation of these elements by various processes. It should always be borne in mind that the numerical values of the parameters obtained indicate only orders of magnitude.

Definitions.—In this discussion we shall assume that the desired isotope has to be separated or enriched from an initial mixture of only two isotopes, which we shall call heavy (S) and light (L). The *molecular mass* of the heavy isotope is denoted by M_S and that of the light one by M_L.

We assume that in the original mixture the mole fraction of the heavy isotope is σ_o and that of the light isotope is λ_o. The ratio σ_o/λ_o, or its inverse ,is then called the *mole ratio* of the original material. It is evident that

$$\sigma_o + \lambda_o = 1 \tag{146}$$

As we said in the introduction, the purpose of the isotope separation is to obtain one isotope in a more or less pure form with a final mole ratio σ_F/λ_F. Supposing for example, that starting with ordinary water: $\sigma_o\,(D)/\lambda_o\,(H) \cong 1/5000$, it is desired to obtain 99 per cent pure heavy water. The final mole ratio of the water has to be

$$\sigma_F(D)/\lambda_F(H) = 99/1$$

and the original isotope has to be enriched

$$\frac{\sigma_F/\lambda_F}{\sigma_o/\lambda_o} \cong 5 \times 10^5 \text{ times.}$$

If this enrichment were produced by one separation process, the quantity

$$E = \frac{\sigma_F/\lambda_F}{\sigma_o/\lambda_o} \tag{147}$$

would be called the *overall enrichment factor* (for the heavy isotope).

In the case of uranium, it is desired to enrich the lighter isotope. It is convenient therefore to define an overall enrichment factor for the light isotope, called E', where

$$E' = \frac{\lambda_F/\sigma_F}{\lambda_0/\sigma_0} = \frac{1}{E} \tag{148}$$

(In ordinary uranium $\lambda_0(U^{235})/\sigma_0(U^{238}) \cong 1/140$.)

In an isotope separation it is desirable to obtain as large an enrichment as possible. A particular separation apparatus though, is usually characterized by a separation factor rather than by an enrichment factor.

If a given amount of material is processed in a separation apparatus, after a certain length of time there will appear at the two ends of the apparatus (denoted by I and II), material, whose original mole ratio has been changed from σ_0/λ_0 to σ_I/λ_I and σ_{II}/λ_{II} respectively. It is clear that if the material at end II is enriched in the heavy isotope (i.e., $\sigma_{II}/\lambda_{II} > \sigma_0/\lambda_0$) then the material at end I must be enriched in the light isotope (i.e, $\sigma_I/\lambda_I < \sigma_0/\lambda_0$) since we have assumed that the total amount of material in the apparatus was constant.

The ratio Q

$$Q = \frac{\sigma_{II}/\lambda_{II}}{\sigma_I/\lambda_I} \tag{149}$$

or its inverse, whichever is greater than unity, is called the *overall separation factor* of the apparatus for the heavy isotope. This quantity Q is generally independent of the original mole ratio, σ_0/λ_0, and is determined solely by the nature of the separation apparatus. The overall separation factor for the light isotope, Q', is

$$Q' = \frac{\lambda_{II}/\sigma_{II}}{\lambda_I/\sigma_I} = \frac{1}{Q} \,. \tag{150}$$

If the separation apparatus consists of several stages, it is convenient to define *single stage separation factors* (greater than unity)

$$q = \frac{\sigma_{k+1}/\lambda_{k+1}}{\sigma_k/\lambda_k} = \frac{1}{q'} \tag{151}$$

where k and $k+1$ refer to two adjacent stages. In most multistage apparatus q is the same for all stages.

From the remarks preceding equation 149 it should be clear that in an apparatus processing a fixed quantity of material, the enrichment factor is less than the separation factor. If the desired enriched isotope (assumed heavy) appears at end II of the apparatus so that $\sigma_F=\sigma_{II}$ and $\lambda_F=\lambda_{II}$ are the mole fractions of the heavy and light isotopes at that end, then

$$Q=\frac{\sigma_F/\lambda_F}{\sigma_I/\lambda_I}>E=\frac{\sigma_F/\lambda_F}{\sigma_o/\lambda_o} \text{ (constant amount of material). (152)}$$

In order to increase the enrichment of the heavy isotope, (i.e., to increase σ_F/λ_F) one has to increase σ_I/λ_I, since Q is constant for a given apparatus. This is possible if, instead of processing only a fixed quantity of material in the apparatus, one supplies at end I a theoretically infinite amount of original material (mole ratio σ_o/λ_o), so that even upon completion of the separation, the mole ratio at end I is still the original mole ratio (i.e., $\sigma_I/\lambda_I=\sigma_o/\lambda_o$). It follows from equation 152 then that

$$\sigma_F/\lambda_F=Q\cdot\sigma_o/\lambda_o$$

so that

$$E=Q \text{ (infinite supply).} \qquad (153)$$

In every separation process the *yield* of the desired isotope is evidently of great importance. In general a process with a high separation factor has a low yield of the enriched isotope. In an apparatus processing a fixed amount of material, the yield is the total amount of material collected at the end of the apparatus at which the desired isotope is enriched. In an apparatus with an infinite supply it is better to give the yield of the isotope in terms of rate of production, (i.e., in terms of the rate at which the isotope can be removed at the "enriched" end of the apparatus). This removal of the enriched material causes a decrease in separation factor, so that in an actual separation process a compromise between production and enrichment of the desired isotope has to be made.

In most processes which separate isotopes by differences in their average properties (statistical separation methods), the enrichment of the desired isotope increases slowly until a steady condition or equilibrium value is reached. (The previously used expression of separation factor (Q) and enrichment factor (E) refer to equilibrium values of these quantities). Although the steady condition is approached only in an asymptotic fashion, the approach to equilibrium is characterized by what is known as *start-up* or *equilibrium time.* This time can be defined in whatever manner is most convenient to the process under consideration.

The total amount of material being processed at a given time in a separation apparatus is called its *hold-up per stage.* Hold-up is most conveniently expressed in moles of material.

Individual separation processes; mass spectrometer.—Separation processes are most conveniently divided into those which depend on the behavior of single molecules (individual separation methods) and those which depend on the average statistical behavior of many molecules (statistical separation methods). There are many more methods of the latter kind than of the former. As examples of the individual separation method we shall describe the mass spectrometer and the isotron (Smyth 11.24). Only the mass spectrometer has been used for the large scale production of enriched U^{235} (Smyth 11.1) but the isotron is of interest since it has not been described elsewhere in detail.

The mass spectrometer is an "individual" separator since the behavior of each ion is individually determined, within certain limits, by the experimental conditions. Since the mass spectrometer is fundamentally a device which separates isotopes, it can evidently be used for the production of isotopes, but is limited in this by the small ion currents ordinarily obtained. In fact, the usual ion current of 1 microampere is equivalent to a transport of 10^{-5} micromoles of singly charged ions per second. In ordinary uranium ($\lambda(U^{235})/\sigma(U^{238}) = 1/140$) this corresponds to about $1/16$ microgram of U^{235} per hour.

In order to use an increased ion current, the following problems must be solved:

1. Production of large quantities of gaseous ions.
2. Use of greatest number of ions in the ion beam.
3. Elimination of space charges in the magnet chamber.

A large ion current evidently should not entail too great a decrease in enrichment due to defocussing by mutual repulsion of the ions, so that this problem has to be investigated also. Furthermore, a good yield depends on the ability to collect all the ions which arrive at the collector. This might be quite difficult when ions enter the collector in large quantity and with high velocities.

Prior to 1939 several attempts had been made to overcome some of the above mentioned difficulties. Ion currents up to about 100 microamps were obtained, although not with uranium.

The successful large scale separation of uranium was first reported by Smyth (11.1). Probably because of its simplicity, a Dempster type mass spectrometer was used for this separation.

The most important features of this spectrometer as well as paths for heavy and light isotopes are indicated in figure 35.

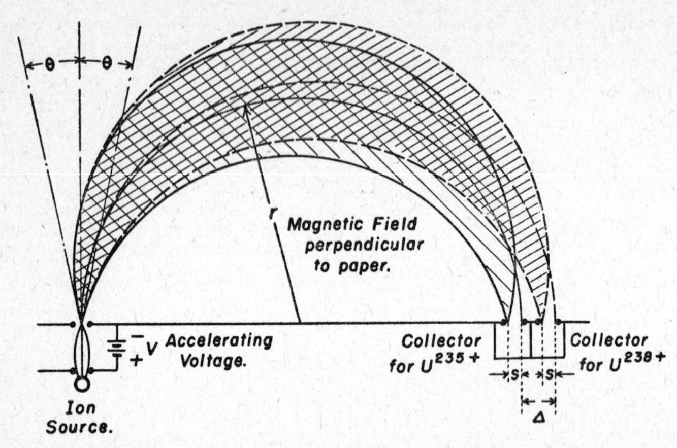

Fig. 35. Separation of isotopes by a Dempster type mass spectrometer. The ions (U^{235+} and U^{238+}) leave the ion source and after acceleration by a voltage V, enter a magnetic field H. The ions are approximately focussed after a deflection of 180° by the magnetic field, the heavier ions further away from the entrance slit than the lighter ions, and are collected by separate collectors.

A very convenient ion source to use for uranium is an adaptation of a cyclotron arc type ion source; (see figure 36 and Smyth 11.20). Such ion sources give currents up to several milliamperes[54]. The use of line sources of this type would probably enable the production of ion beam currents of over 10 milliamperes. Smyth (11.1 and 11.5) reports that A. O. Nier used uranium bromide vapor in the isotope separation of uranium and that E. O. Lawrence found mostly U^+ ions in the ion beam (not necessarily using the same uranium salt vapor).

[54] M. S. Livingston, *J. App. Phys.* **15**, 15 (1944); Alkazov, Mescheryakov and Chromochenko, *J. Phys. USSR* 8, 56 (1944).

One disadvantage of the arc type ion sources is the rather appreciable variation in energy (often about one hundred volts)[55] of the ions which enter the beam. This can produce a considerable defocussing at the collector. Furthermore, these sources require large pumps to retain a sufficently good vacuum in the magnet chamber.

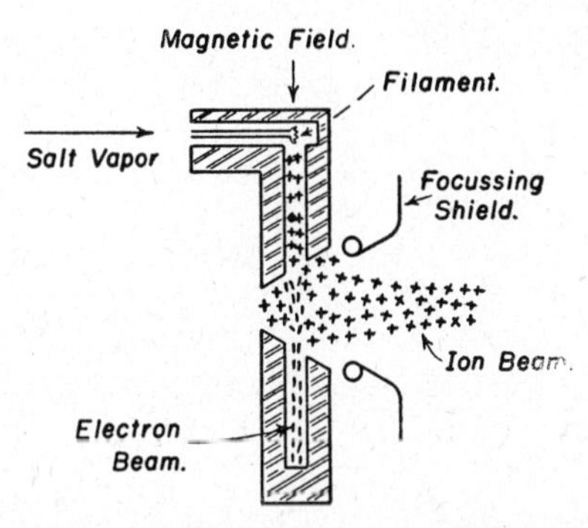

Fig. 36. Schematic cross-section of a possible arc type ion source. The salt vapor to be ionized flows down a capillary and is ionized by the beam of electrons which traverses the same capillary. At the opening of the capillary the mutual repulsion of the ion pushes them out of the arc so that by suitable focussing shields they can be drawn away from the source.

One of the previously mentioned difficulties connected with large ion currents (i.e., the space charge in the magnet chamber) can be overcome by ionization of the residual gas in the magnet chamber (Smyth 11.5). It appears then that by proper construction of the ion source and accelerating system, a large number of the ions produced can be brought into a beam of small angular variation. On figure 35 this angular variation is denoted by θ. In some mass spectrometers: $\theta \cong 5^0$. If the angular variation becomes less than this value, the intensity of the beam is noticeably decreased, while a much greater value of θ causes an appreciable spread at the collector. In fact it is well known that an ion beam (of one isotope) traversing a magnetic field H with a radius r, has a spread at the collector due to the angular variation θ of

$$S_\theta \cong r\theta^2 \tag{154}$$

where

$$r = \frac{c}{H}\sqrt{\frac{2mV}{e}} \tag{155}$$

[55] For example, Lamar, Buechner and Van de Graaff, *J. App. Phys.* **12**, 132 (1941).

In this equation c is the velocity of light, H is the magnetic field (in gauss), m is the mass of the isotope ion, e is its charge and V is its accelerating potential. Let S_1 be the width of the entrance slit of the ion beam (at the accelerating system) and S_v the spread due to the fluctuations ΔV in the energy of the ions, where

$$S_v = r\frac{\Delta V}{V} \tag{156}$$

Then the total spread of the ion beam at the collector is approximately

$$S = \sqrt{S_\theta^2 + S_V^2 + S_1^2} \quad . \tag{157}$$

(This assumes that each of the spreads has a gaussian shape.)

In order to have an appreciable separation of two isotopes of weights M_S and M_L, it is necessary to have the dispersion Δ due to the mass difference greater than the total spread S. Now

$$\Delta = \frac{M_S - M_L}{M} r \ , \tag{158}$$

where M is the average mass of the two isotopes. For U^{238} and U^{235}, $\dfrac{M_S - M_L}{M} \cong 1/80$, so that for a reasonable dispersion ($\Delta \cong 2$ cm) a radius r of at least 160 cm is necessary. With the previously given angular variation $\theta = 5°$, the spread due to this variation (equation 154) becomes approximately $S_\theta = 1.4$ cm. With an energy fluctuation $\Delta V = 100$ volts and an accelerating voltage $V = 10$ kilovolts, the total spread S (equation 157) is increased to about 2.2 cm. With $V = 40$ kilovolts, the total spread S is reduced to less than 1.5 cm which is more desirable. (These considerations neglect the entrance slit width S_1 which, if much larger than 0.5 cm, will affect the total spread.) It might be noted that according to equation 155, an accelerating voltage $V = 40$ kilovolts and a radius $r = 160$ cm require a magnetic field strength H of approximately 3000 gauss.

Figure 37 indicates the number of ions collected per second as a function of the distance from the entrance slit. The mass dispersion Δ is about 2 cm and the width S of each peak is about 1.5 cm. In order to have an appreciable yield the collector entrance for U^{235} should have a width of the order of S. If this gives an insufficient separation, it may be necessary to run a given sample of enriched U^{235} through an electromagnetic separator several times in order to achieve a sufficient purity.

Smyth (11.37) mentions that it is advantageous to start with enriched uranium. This is evidently in line with the above remarks. Furthermore, if the initial material is, for example, doubly enriched in U^{235} ($\lambda_0(U^{235})/\sigma_0(U^{235})=1/70$), twice as much U^{235} is produced at twice the purity since the total amount of U^{238} collected is roughly independent of the initial enrichment of U^{235} (as long as this is small) and depends on the total ion current. In terms of figure 37, this means that the height of the U^{238} peak depends mainly on the ion current while the height of the U^{235}

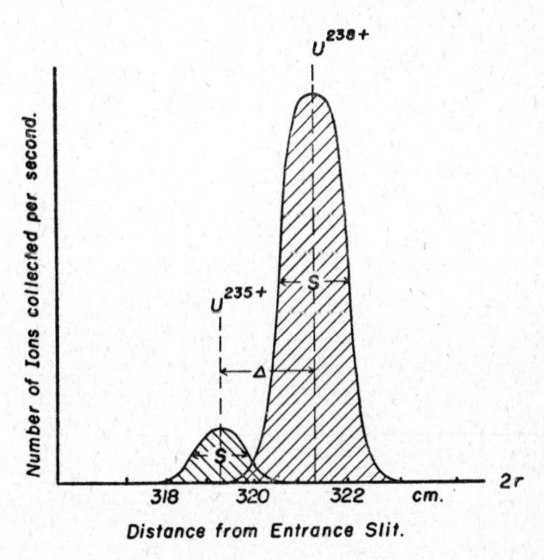

Fig. 37. Ion current distribution near the collectors. Because of the angular width of the ion beam at the entrance slit, the ions are not focussed perfectly at the collectors, but each isotope beam has a spread *S*. The overlap of the ion beams makes a perfect isotope separation impossible without a tremendous decrease in yield. For simplicity the distribution of the ion beams is assumed to be approximately Gaussian.

peak is approximately equal to the product of the initial mole ratio $\lambda_0(U^{235})/\sigma_0(U^{238})$ and the height of the U^{238} peak (since the peak widths are practically the same).

From the above remarks we can calculate the number of separators necessary to produce say 1 mole of 90 percent pure U^{235} per day, starting with material which is for example five times enriched ($\lambda_0(U^{235})/\sigma_0(U^{238})=1/22$). Assuming an ion current of 10 milliamperes and a single run in each separator, the total number of separators required is calculated as follows: 1 mole of 90% U^{235} per day divided by 10^4 microamps per separator, divided by 10^{-11} moles of U per sec per microamp, divided by 8.8×10^4 sec per day, multiplied by 22×0.9 moles of U per mole of 90% U^{235}, which equals 2×10^3 separators. (With double the ion current evidently only one half of the separators are needed.) In order to build such a large number of separators, it is advantageous to use many

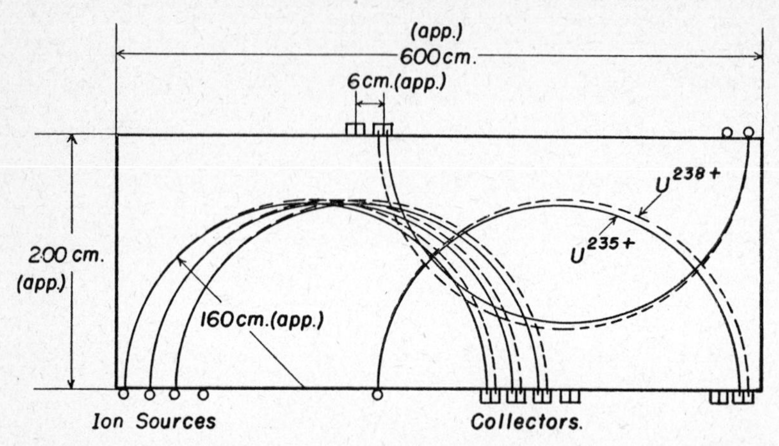

Fig. 38. Possible layout of a magnetic separator. Ion sources and collectors are placed along the long sides of a rectangular pole gap of a magnet in order to utilize the magnetic field to the best advantage and to have easy access to all parts. By proper dimensioning of the apparatus it should be possible to use about 100 ion sources per magnet.

separators in one magnetic field (Smyth 11.15). An arrangement, similar to the one shown in figure 38 with many separators along each side of the pole gap of a rectangular magnet, is very convenient since it permits easy access to ion sources and collectors. If 100 separators can be used per magnet perhaps 20 magnets of the above type might be sufficient to produce quite pure U^{235} in appreciable quantities per day.

Isotron.—Smyth (11.24) reported a new isotope separator, first suggested by R. R. Wilson of Princeton, which permits the use of an extended ion source. This separator is based on the same principle as a velocity modulated electron oscillator (klystron). Ions from an extended source are first accelerated by travelling through a constant high intensity electric field of total potential V, then further accelerated by a low intensity electric field varying at radio frequency $(1/\tau)$ in a saw tooth manner with maximum total potential V'. The result of the intense field is to give the ions a high velocity v which is inversely proportional to the square root of their mass m:

$$v = \sqrt{\frac{2Ve}{m}} . \qquad (159)$$

On the other hand, the saw tooth modulated field produces small periodic variations in the velocity of the ions, varying the velocities of the ions from v (equation 159) to a maximum velocity v':

$$v' = v(1+V'/V)^{\frac{1}{2}} \quad . \tag{160}$$

Figure 39 shows, on a so-called Applegate diagram (distance travelled by ions plotted as a function of time), the bunching which results from the velocity modulation. It is easily shown that, for the above conditions, bunching occurs a time t after the ions have passed the modulator, where

$$t = \tau(2V/V'). \tag{161}$$

This equation assumes $V'/V \ll 1$. It is interesting to note that with this approximation, t is independent of the mass of the ions and furthermore that the saw tooth modulation gives perfect bunching of each isotope. (An ion leaving at a time $\Theta\tau(0 \leqq \Theta \leqq 1)$ receives a voltage modulation $\Theta V'$ so that the bunching time $t_\Theta = \Theta\tau \dfrac{2V}{\Theta V'}$ is independent of Θ).

If the bunching of the light ions takes place at a distance h from the modulator, where

$$h = v \cdot t \,, \tag{162}$$

then the heavy ions bunch a distance $\triangle h$ nearer to the modulator where

$$\triangle h/h = \tfrac{1}{2}(M_S - M_L)/M \,, \tag{163}$$

M_S being the mass of the heavy ions, M_L the mass of the light ions and M their average mass. For U^{235} and U^{238} $\dfrac{\triangle h}{h} \cong \dfrac{1}{160}$. In order to collect only light ions, deflecting plates are placed at $h - \triangle h$. The voltage on the deflecting plates is synchronized with the modulator and is adjusted with a time delay so that at the moment when the light ions pass the point h, the heavy ions are deflected out of the beam. (A time delay can be avoided if the constants of the apparatus are so made that $t = n\tau$ where n is an integer.)

In order to give some reasonable values to the constants of the apparatus we note that the spread in h, $\triangle h'$, with voltage fluctuation $\triangle V$ of the ions coming off the ion source is given by

$$\frac{\triangle h'}{h} = \frac{3}{2}\frac{\triangle V}{V} \,. \tag{164}$$

This spread has to be less than the dispersion Δh due to mass difference (equation 163), for separation to occur. If $\Delta V \cong 5$ volts (this evidently depends on the ion source) the above condition re-

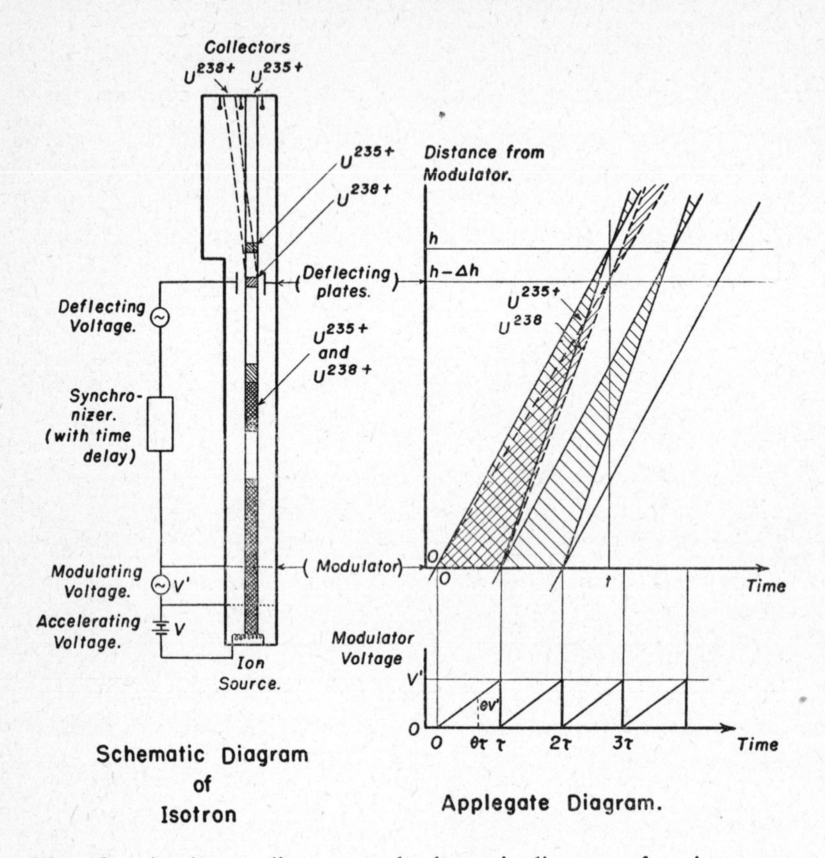

Fig. 39. Applegate diagram and schematic diagram of an isotron. The ions (U^{235+} and U^{238+}) from an extended ion source are accelerated with a voltage V and velocity modulated with a "saw tooth" voltage of maximum amplitude V'. The light ions (U^{235+}) bunch at a distance h from the modulator and the heavy ions (U^{238+}) at a distance $h - \Delta h$. At the latter point deflecting plates deflect the heavy ion bunches out of the ion beam so that the isotopes can be collected in separate collectors.

quires $V > 1200$ volts in the case of uranium. Taking $V = 1000$ volts, $V' = 10$ volts, $\tau = 10^{-7}$ sec, the modulator to collector distance is approximately $h = 60$ cm and the mass dispersion $\Delta h =$ .3 cm. To use convenient deflecting plates it is necessary to increase the distance Δh to 1 to 2 cm, so that h has to be increased by decreasing the modulator frequency $(1/\tau)$ or increasing the voltage V.

Just as in the case of the mass spectrograph the yield of U^{235} depends on the initial enrichment and on the total ion current. Too large ion currents might cause debunching, due to space charge, and make a good separation difficult.

Statistical separation processes; single stage Rayleigh processes; electrolysis.—As we said in the introduction, statistical separation processes depend on the average behavior of molecules of different mass. Since the behavior of individual molecules can vary between wide limits, the separation factors obtainable in a single operation will generally be much smaller than in the case of the individual separation processes.

Two main classes of these separation processes are now in use. These will be called Rayleigh processes and equilibrium processes. In the Rayleigh processes, the mixtures of isotopes (one of which is always enriched in the desired isotope) into which the original material is divided by the isotope separation process, are never allowed to be in contact for any length of time so that no process opposing the dividing process has an appreciable chance to develop. On the other hand in the equilibrium processes the dividing process and the process opposing this are allowed to develop sufficiently so that some kind of equilibrium state is reached.

A good illustration of the difference between a Rayleigh and an equilibrium process is the process of evaporation from a liquid. A vapor is said to be in equilibrium with a liquid when the rate of molecules entering the liquid from the vapor is equal to the rate of molecules going from the liquid to the vapor. This equilibrium state is reached when the liquid is placed in a closed vessel. If the liquid consists of two isotopes, the mole ratio of the isotopes in the liquid ($(\sigma/\lambda)_{liq}$) is different from the mole ratio of the isotopes in the vapor ($(\sigma/\lambda)_{vap}$). This is an example of an equilibrium separation process which can be shown to have a separation factor

$$\frac{(\sigma/\lambda)_{liq}}{(\sigma/\lambda)_{vap}} = \frac{P_{oL}}{P_{oS}} \tag{165}$$

where P_o is the vapor pressure of the pure isotope (S-heavy, L-light). On the other hand, if the vapor molecules are removed so rapidly from the liquid-vapor interface that no appreciable number of molecules returns to the liquid, then we have an example of a Rayleigh process. This process can be shown to have an instantaneous separation factor (see page 195):

$$q = \frac{P_{oL}}{P_{oS}}\sqrt{\frac{M_S}{M_L}}\ , \tag{166}$$

where M is the molecular weight of an isotope.

In general it is more convenient to put a certain physical process, such as evaporation, into either a Rayleigh type process or an equilibrium type process although often it is possible to use them in both classes. This chapter describes only the conventional use of some of the physical separation processes.

Both the Rayleigh and the equilibrium processes can be used in a single stage and in multiple stages (cascade). A Rayleigh type process can be put in cascade by using recycling while an equilibrium process can be cascaded by using countercurrent flow. This will be described later in more detail.

We shall now describe the Rayleigh process in a single stage, stressing those methods which are particularly suited to separate U^{235} and deuterium. We shall illustrate this process by a description of electrolysis which has been used in Sweden, for example, to produce large quantities of heavy water. (Another important Rayleigh process is gas diffusion through a porous barrier which, in cascade, was so successful in separating U^{235} by the use of uranium hexafluoride or a similar gas (Smyth 10.1).)

It was first discovered by Washburn and Urey[56] that the electrolysis of aqueous solution heavy water was enriched in the remaining solution.

In order to examine this effect let us consider a given volume V of solution containing n ions of the element under investigation, σn ions of the heavy isotope and λn ions of the light isotope. The rate at which ions of the isotopes are electrolysed at the electrodes will be proportional to the number of ions of each kind present and inversely proportional to the total volume V of the solution, so that at any time t

$$\frac{d(\sigma n)}{dt} = -a\frac{\sigma n}{V} \text{ and } \frac{d(\lambda n)}{dt} = -b\frac{\lambda n}{V}\ . \tag{167}$$

In these equations a and b are constants depending on the mass of the ions in a fashion to be explained later. These equations assume that at all times the molecules in the volume V are perfectly mixed so that no complicating concentration gradients occur.

[56] E. W. Washburn and H. C. Urey, *Proc. Nat. Acad. Amer.* **18**, 496 (1932).

It is convenient to define as the *instantaneous separation factor* q for the heavy isotope, the ratio

$$q = \frac{\sigma n / \lambda n}{d(\sigma n)/d(\lambda n)} = \frac{b}{a} \quad . \tag{168}$$

This is the ratio of the mole ratio in the remaining solution to the mole ratio of that part of the solution which is instantaneously removed from it at any time t. As can be seen, q is independent of time.

In order to obtain the overall separation and enrichment factors, we integrate equation 167 to obtain

$$\ln \frac{\sigma n}{\sigma_o n_o} = -a \int_o^t \frac{dt}{V} \text{ and } \ln \frac{\lambda n}{\lambda_o n_o} = -b \int_o^t \frac{dt}{V} \quad .$$

It follows, using equation 168, that

$$\lambda n / \lambda_o n_o = (\sigma n / \sigma_o n_o)^q$$

where the index zero refers to the initial ions. In the case of water, the remaining solution contains the desired enriched isotope (deuterium) so that in accordance with equation 149 we can call

$$\sigma = \sigma_{II}, \ \lambda = \lambda_{II} \text{ and } n = n_{II} \quad ,$$

so that from above

$$\lambda_{II}/\lambda_o = (\sigma_{II}/\sigma_o)^q (n_{II}/n_o)^{q-1} \quad . \tag{169}$$

Since initially there were $\lambda_o n_o$ light ions and $\sigma_o n_o$ heavy ions, the electrolysis must have removed

$$\lambda_o n_o - \lambda_{II} n_{II} \equiv \lambda_I \ (n_o - n_{II}) \tag{170}$$

light ions and

$$\sigma_o n_o - \sigma_{II} n_{II} \equiv \sigma_I \ (n_o - n_{II}) \tag{171}$$

heavy ions, where the index I refers to the total amount of gas released during the electrolysis. Making use of equations 169, 170 and 171 and keeping in mind that $\sigma + \lambda = 1$, it is possible to work out the overall separation factor $Q = \dfrac{\sigma_{II}/\lambda_{II}}{\sigma_I/\lambda_I}$, and the overall enrichment factor $E = \dfrac{\sigma_{II}/\lambda_{II}}{\sigma_o/\lambda_o}$, for this process. Both of these quantities are functions of the ratio n_{II}/n_o and increase with the decreasing ratio n_{II}/n_o. This can be seen from figure 40 which shows the variation of σ_{II} and λ_{II} for the electrolysis of water with the instantaneous separation factor $q = 5$. (Here $\sigma =$ mole fraction of deuterium and $\lambda =$ mole fraction of hydrogen.) For ordinary water σ_o (D)/λ_o (H) $= 1/5000$. In order to get heavy water which is 80 per cent enriched in deuterium, its initial volume (proportional to n_o) has to be decreased approximately 10^5 times which means that this process has a very low yield for such high enrichment.

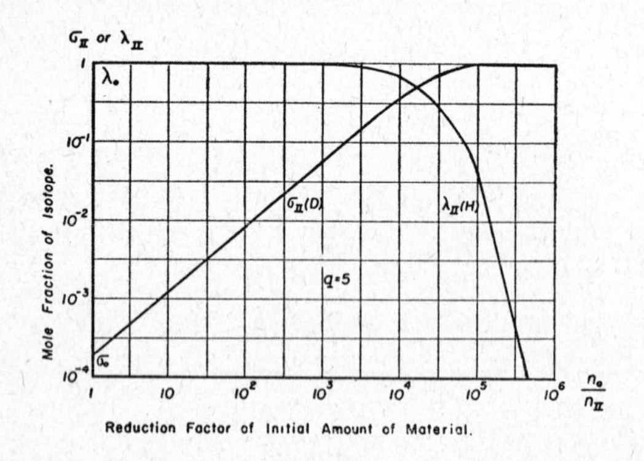

Fig. 40. Isotope separation by electrolysis. The mole fraction of the heavy isotope increases with decreasing volume of the substance in the fashion shown. The case illustrated is for water with $q=5$. (From W. Walcher, *Erg. d. exakt. Naturwiss* **18**, 155 (1939).)

These considerations are quite general for any Rayleigh type process, since this process is characterized by equation 167. Equation 169 is known as the Rayleigh formula. Processes other than electrolysis usually have instantaneous separation factors q which are not much larger than unity. In that case, equation 169 immediately yields the overall enrichment factor

$$E \cong (n_o/n_{II})^{q-1} \qquad (q \cong 1) \quad . \qquad (172)$$

If in addition $(q\text{-}1)\ ln \dfrac{n_o}{n_{II}} < 1$, the overall enrichment factor becomes

$$E \cong 1+(q-1)\ ln \frac{n_o}{n_{II}} \quad . \qquad (173)$$

In order to explain the difference between the coefficients a and b in equation 165, we must consider the process of electrolysis in some detail. The theory of this process is by no means in its final form,[57] but it is known that the following phenomena could be determining in the electrolysis of water:

1. Mobility of the ions in the electrolyte.
2. Neutralization of the ions at the cathode.
3. Combination of atoms to form molecules and their liberation from the cathode.
4. The exchange reaction

$$H_2O + HD = HDO + H_2.$$

The first process does not enter into the separation of isotopes because the mobility of ions is determined primarily by their electronic configuration. It might be noted, though, that under high electric fields the mobility of ions can serve to separate isotopes to some extent (Smyth 9.31).

Of the processes 2 and 3, the slowest one will be most important since it will determine the rate at which ions are transformed into liberated molecules. It is generally believed that the neutralization of ions at the cathode is this rate determining process. On the cathode there appear to be more or less fixed layers of OH, H and perhaps H_2O (see figure 41) which form a barrier for any H^+H_2O and D^+H_2O ions coming to be discharged. In order to have neutralization, electrons from the metal have to effectively overcome the so-called over voltage V which is the potential between the null point energy level of the hydrated ion and the Fermi level in the metal. The rate of discharge of the heavy ions is therefore

$$a = C \exp(-\alpha V_S / kT) \quad ,$$

where C is the same for different isotopes and depends mostly on the current density and α takes care of the fact that right after neutralization the atom (H or D) might still be at a different potential for an electron in the neighborhood than the H_2O molecule it carried along. V_S is the over voltage for the heavy ion and is equal to a constant voltage plus the null point energy, E_{OS}. Since the null point energies of different isotopes are not similar, one finds for the instantaneous separation factor

$$q = b/a \cong \exp\{\alpha(E_{os} - E_{OL})/kT\}, \tag{174}$$

where E_{OL} is the null point energy of the light isotope.

It is quite clear that the above picture is insufficient since q actually depends on the electrode metal, the electrode surface, the

[57] Eyring, Glasstone and Laidler, *J. Ch. Phys.* **7**, 1053 (1939); Kimball, Glasstone and Glassner, *J. Ch. Phys.* **9**, 91 (1941); J. A. V. Butler, *J. Ch. Phys.* **9**, 279 (1941).

current density (q generally increases with current density) and the concentration ratio of the isotopes.

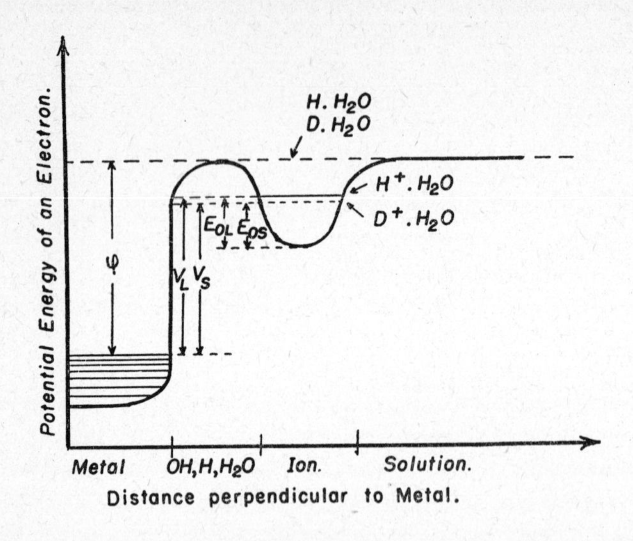

Fig. 41. Electron potential near a metal surface during electrolysis. A hydrated ion which approaches a cathode surface in order to be discharged is held at some distance from the surface by layers of H, OH and H_2O molecules. The discharge can take place only by electrons leaving the metal and effectively overcoming the over voltage V (W. Walcher, *Erg. d. exakt. Naturw.* **18**, 155 (1939).)

For water at 20°C., q varies from about 3 to 15. The lowest value of 3 seems to occur with so-called poisoned cathodes on which

Table 18

SOME INSTANTANEOUS SEPARATION FACTORS FOR ELECTROLYSIS

Isotopes	Cathode	Separation factor q	Reference
Li^6, Li^7	Hg	1.06	(1)
O^{16}, O^{18}	Fe	1.008	(2)
Cl^{35}, Cl^{37}	Pt	1.006	(3)

(1) Johnston and Hutchinson, *J. Chem. Phys.* 8, 869 (1940).
(2) Johnston, *J. Am. Chem .Soc.* 57, 484 (1935).
(3) Johnston and Hutchinson, *J. Chem. Phys.* **10**, 469 (1942).

the gas is retained long enough for the exchange reaction

$$H_2O + HD \rightleftharpoons HDO + H_2$$

to come to equilibrium. In fact the equilibrium constant

$$K = \frac{[HD0]/[H_20]}{[HD]/[H_2]}$$

which is approximately equal to 3, is exactly the instantaneous separation factor q as defined by equation 168. Therefore if the gas (H_2 and HD) is retained long enough on the cathode for the exchange equilibrium to set in, q has to be equal to K, as happens in the case of poisoned cathodes.

In order to show that electrolysis probably cannot be used to enrich uranium isotopes, the instantaneous separation factors q of some known isotopes are given in table 18. It can be seen that the separation factor decreases with the increasing mass of the isotopes.

Gas diffusion.—The separation of isotopes by gaseous diffusion through a porous wall was first suggested by Lindeman and Aston.[58] In order to investigate the separation process in some detail let us consider a cylinder filled with a mixture of two isotopes. One end of the cylinder is closed by a porous barrier of area A and thickness L and the other end consists of a movable piston. As the gas diffuses through the barrier, the pressure inside of the piston is kept at a constant value P_1 by moving the piston while the pressure outside of the cylinder is kept at a much lower constant value P_2 (by suitable pumping for example). The flow of gas through the porous barrier can be calculated if we idealize the pores to be cylindrical tubes of length L and radius r and assume that the pores are arranged in a square array over the area A of the barrier and have touching sides. The number of pores per square centimeter is then $(1/2r)^2$.

If the radius r of the pores is of the order of one tenth or less of the mean free path of the gas molecules at the pressure P_1 (at atmospheric pressure the mean free path of molecules is approximately 10^{-5}cm) and if the length L of the pores is several orders of magnitude larger than that mean free path, the flow through the pores is almost purely diffusive (Smyth 10.14).[59] This means that the molecules flowing down the pores collide only with the pore walls and do not interact among themselves. It has been shown[60] that under these conditions the net rate of molecules flow-

[58] F. A. Lindeman and F. W. Aston, *Phil. Mag.* 37, 523 (1919).
[59] de Bethune and Present, *Bull. Amer. Phys. Soc.* 21, 19 (Jan. 1946).
[60] L. B. Loeb, "The Kinetic Theory of Gases," p. 305 McGraw-Hill Co. (1934).

ing out of each pore is

$$\frac{4\sqrt{2\pi}}{3} \frac{n_A}{\sqrt{MRT}} \frac{r^3}{L} (P_1 - P_2) ,$$

where n_A is Avogadro's number, M the molecular weight, R the gas constant and T the absolute temperature. Referring back to the assumed arrangement of the pores, the total number of molecules per second flowing through a porous barrier of area A and thickness L is

$$\frac{dn}{dt} = - \frac{\sqrt{2\pi}}{3} \frac{n_A}{\sqrt{MRT}} \frac{r}{L} (P_1 - P_2)\, A \tag{175}$$

Assuming $r = 10^{-6}$ cm, $L = 10^{-1}$ cm, $P_1 = 1$ atm, $P_2 \cong 0$, equation 175 shows that for uranium hexafluoride (M=350) the rate of flow through a barrier is approximately

$$\frac{1}{A} \frac{dn}{dt} \cong - 1.6 \times 10^{23} \text{ molecules/cm}^2\text{/day} . \tag{176}$$

If the process described is to be a Rayleigh type process, there must be no appreciable back diffusion (i.e., $P_2 \ll P_1$). If P_1 then denotes the total pressure in the piston, the partial pressure of the heavy isotope will be σP_1 and equation 175 must be rewritten for the heavy isotope as

$$\frac{d(\sigma n)}{dt} = - \frac{\sqrt{2\pi}}{3} \frac{n_A}{\sqrt{M_S RT}} \frac{r}{L} \frac{P_1}{n} A \cdot \sigma n , \tag{177}$$

where n is the total number of molecules in the piston at any time. A similar equation obtains for the light isotope. Comparison of equation 177 with equation 167 and 168 indicates that gas diffusion in one direction is a Rayleigh process with an instantaneous separation factor q

$$q = \sqrt{\frac{M_S}{M_L}} \tag{178}$$

For uranium hexafluoride gas ($U^{235}F_6$ and $U^{238}F_6$) $M_S = 352$ and $M_L = 349$ so that

$$q = 1.0043 . \tag{179}$$

Equation 173 shows that by letting half the original number of uranium hexafluoride molecules escape through a barrier, an enrichment factor $E = 1.003$ is obtained.

Single stage equilibrium processes; thermal diffusion.—As we mentioned previously, equilibrium processes are differentiated from the Rayleigh processes by the fact that the process dividing the original mixture of isotopes into two mixtures (one of which is enriched in the desired isotope) is allowed to come to equilibrium with the process opposing the division. In addition, in equilibrium processes the overall separation factor does not depend on the total number of molecules present. It can be worked out directly from physical considerations without going through a formula such as the Rayleigh formula (equation 169).

The equilibrium processes which will be described here are the thermal diffusion method and the chemical exchange method. The former has been used to enrich liquid uranium hexafluoride in $U^{235}F_6$ (Smyth 11.37) and the latter enters into the large scale production of heavy water (Smyth 9.37). These methods are also characteristic of two main classes of equilibrium processes: those in which the two isotope mixtures mentioned are physically not separated (i.e. are both gases or liquids) and those in which they are physically separated (i.e. one is a gas and the other a liquid). In the first class are thermal diffusion and centrifugation, while the second class contains chemical exchange, distillation and similar processes. The fundamental equations governing the processes in each class are quite similar, but the scope of this chapter does not permit a detailed presentation.

The mechanism of thermal diffusion.—If a temperature gradient is initially established in a gas or liquid at a uniform temperature and concentration, a movement of molecules occurs along the direction of the temperature gradient (neglecting any convection currents). The theory of this effect has been worked out for gases by Enskog and Chapman[61] and for liquids by Wirtz and Hiby.[62]

Consider the simple one dimensional case of a gas or liquid between two infinite, plane heat conducting walls. The fluid at a uniform temperature is subjected to a temperature gradient producing a flow of the light isotope in the fluid, which is perpendicular to the conducting walls. This can be expressed by

$$\lambda v_L = -D \frac{\partial \lambda}{\partial x} + \frac{D_T}{T} \frac{\partial T}{\partial x} , \tag{180}$$

where $\lambda = \lambda (x, t)$ is the mole fraction of the light isotope at the plane which is a distance x from the cold wall and at the time t ($t = O$ when the temperature gradient is established), v_L is the flow velocity of the light molecules, D is the coefficient of self dif-

[61] For all pertinent references see W. H. Furry, R. C. Jones and L. Onsager, *Phys. Rev.* 55, 1083 (1939).
[62] K. Wirtz, *Ann. d. Physik* 36, 295 (1939); K. Wirtz and J. W. Hiby, *Phys. Zeit.* 44, 369 (1943).

fusion of the isotopes, D_T is the thermal diffusion coefficient and T the absolute temperature. (The thermal diffusion coefficient is sometimes defined in a different manner.)

When equilibrium is reached, the flow of molecules stops ($v_L = O$) and one obtains from equation 180

$$D \frac{\partial \lambda}{\partial x} = \frac{D_T}{T} \frac{\partial T}{\partial x} \tag{181}$$

In order to integrate this equation, the ratio D_T/D must be known.

For a gaseous mixture of isotopes, Furry, Jones and Onsager[63] give

$$\frac{D_T}{D} = \alpha \sigma \lambda \tag{182}$$

where

$$\alpha = .35 \frac{M_S - M_L}{M_S + M_L}, \tag{183}$$

$\sigma(x, t)$ is the mole fraction of the heavy isotope at the point x and the time t and M_S and M_L are the molecular weights of the heavy and light isotopes respectively. The numerical coefficient in equation 183 has been chosen semi-empirically by considering the isotope molecules as hard spheres. Its value for very soft molecules, such as UF_6, may be much smaller than .35.

Assuming that one of the isotopes, say the lighter, is present only in small quantity so that $\sigma \cong 1$ at all times, integration of equation 181 gives for gases

$$\lambda_h / \lambda_c = (T_h / T_c)^a \quad . \tag{184}$$

The index h refers to the hot wall and the index c to the cold wall of the separator. We define as overall separation factor of the apparatus (for the lighter isotope)

$$Q' = \frac{\lambda_h / \sigma_h}{\lambda_c / \sigma_c} \cong \frac{\lambda_h}{\lambda_c}$$

since $\sigma_h \cong \sigma_c \cong 1$. From equation 184

$$Q' = 1 + \alpha \, ln \frac{T_h}{T_c} \tag{185}$$

because generally $\alpha \, ln \dfrac{T_h}{T_c} \ll 1$. If $\triangle T \equiv T_h - T_c < T_c$,

[63] Furry, Jones and Onsager, *Phys. Rev.* 55, 1083 (1939).

$$Q' = 1 + \alpha \frac{\Delta T}{T_{av}} \tag{186}$$

where $T_{av} = \frac{1}{2}(T_h + T_c)$. For gaseous uranium hexafluoride $M_S = 352$ and $M_L = 249$ so that $\alpha = .0015$. At one atmosphere pressure the lowest temperature of the cold wall must be about 60° C[64] ($T_c = 333°$ K) and the hot wall can have a temperature $T_h = 2T_c$ (i.e. about 666° K (393° C)) without dissociating the uranium hexaflouride. In that case equation 186 yields

$$Q' = 1.00094 \ . \tag{187}$$

For liquids one can obtain from Wirtz and Hiby's calculations[65]

$$D_T/D = (\epsilon/kT)\sigma\lambda \ . \tag{188}$$

where ϵ is an energy necessary for the light molecules to pass from one position in the "lattice" of the heavy molecules to a neighboring position and k is Boltzmann's constant. Integration of equation 181 then yields for liquids (assuming again $\sigma \cong 1$)

$$\lambda_h/\lambda_c = \exp\{(\epsilon/k)(1/T_c - 1/T_h) \ , \tag{189}$$

so that, as before, the separation factor with respect to the light isotope is approximately

$$Q' = 1 + \frac{\epsilon}{k\,T_{\text{mean}}} \frac{\Delta T}{T_{\text{mean}}} \tag{190}$$

where $\qquad T_{\text{mean}} = \sqrt{T_h\,T_c}$ and $\Delta T = T_h - T_c$

From experiments done by Korsching[66] with heavy water (HDO) and heavy benzine (C_6D_6), it can be estimated that for water at room temperature

$$\frac{\epsilon}{kT} \cong 0.015 \text{ for HDO in } H_2O$$

and

$$\frac{\epsilon}{kT} \cong 0.2 \text{ for } C_6D_6 \text{ in } C_6H_6 \ .$$

This corresponds to a separation factor for the heavy isotope of

$$Q \cong 1.0015 \text{ (HDO)}$$
$$Q \cong 1.02 \ (C_6D_6)$$

[64] W. Krasny-Ergen, *Nature* **145**, 742 (1940).
[65] K. Wirtz and J. W. Hiby, *Phys. Zeit.* **44**, 369 (1943).
[66] H. Korsching, *Nature*, **31**, 348 (1943).

with $\Delta T \cong 30°C$ and $T_{\text{mean}} = 300°\ K$. How the energy ϵ varies with mass and size of the isotopes is not known but thermal diffusion has been used to enrich liquid uranium hexafluoride in $U^{238}F_6$ (Smyth 11.40).

The characteristic time τ_d necessary for the simple thermal diffusion process described in equation 180 to come to equilibrium can be calculated rather easily[67] and turns out to be of the order of

$$\tau_d = \frac{a^2}{D} \tag{191}$$

a is the distance between the hot and cold walls and D is the coefficient of self diffusion of the isotopes.

In the case of gases one generally chooses the distance $a \cong 0.5$ cm and, since $D \cong 0.1\ \text{cm}^2/\text{sec.}$ the equilibrium time $\tau_d \cong 2$ sec. For liquids it is best to make $a \cong 10^{-2}$ cm and with $D \cong 10^{-5}$ cm^2/sec, the equilibrium time $\tau_d \cong 10$ sec. These results are of course very rough and are only meant to indicate orders of magnitude of equilibrium time.

Chemical exchange.—In the earlier discussion of electrolysis (page 199) we stated that the exchange reaction

$$HD + H_2\,O \rightleftharpoons H_2 + HDO$$

has an equilibriuum constant K which is approximately equal to 3 at room temperature and which could serve to separate the hydrogen isotopes. In general, if an exchange reaction is of the form

$$LA + SB \rightleftharpoons LB + SA\ , \tag{192}$$

where L and S represent the light and heavy isotope respectively and A and B any radicals, the equilibrium constant is defined by

$$K = \frac{[LB]\ \ [SA]}{[LA]\ \ [SB]}\ , \tag{193}$$

where the square brackets denote concentration in moles per liter. If σ_A denotes the mole fraction of the heavy isotope S combined with the A radical and λ_A the same for the light isotope ($\sigma_A + \lambda_A = 1$) then evidently

$$\frac{[SA]}{[LA]} = \frac{\sigma_A}{\lambda_A}$$

[67] P. Debye, *Ann. d. Physik* **36**, 284 (1934).

A similar relation holds for the B molecules, so that the equilibrium constant K can be written

$$K = \frac{\sigma_A / \lambda_A}{\sigma_B / \lambda_B} \tag{194}$$

If now the A molecules and B molecules can be separated by physical means, the exchange reaction is effectively a separation process with an overall separation factor Q

$$Q = K \quad . \tag{195}$$

It should be noted that if more than one atom of each exchanging isotope is involved in the exchange reaction the relation 195 no longer holds[68] and must be modified.

Physically, the difference of the equilibrium constant (K) of exchange reactions from unity expresses a difference in reaction rates of isotopes. This is due to a variation with mass of the number of collisions between molecules and to a variation in energy required to form new molecules if a collision leads to such a formation.[69] Urey and Greiff[70] calculated equilibrium constants for various exchange reactions by using the fact that the equilibrium constant K for a reaction is given by the ratio of the product of the partition functions, f, of the resultants to that of the reactants:

$$K = \frac{f_{LB} \cdot f_{SA}}{f_{LA} \cdot f_{SB}} \tag{196}$$

Each partition function is given by

$$f = M^{3/2} \Sigma \, (2\pi)^{3/2} \, (kT)^{5/2} / (h^3 n_A{}^{3/2}) \; , \tag{197}$$

where M is the mass of the molecule, k is Boltzmann's constant, T is the absolute temperature, h is Planck's constant and A is Avogadro's number. Σ is the partition sum,

$$\Sigma = \sum_{v,j} P_j \exp\{-E(I,v)/kT\} \; , \tag{198}$$

where $E(I,v)$ is the energy of the state of the molecules represented by the rotational quantum number I and the vibrational quantum number v and P_j is a weighting factor for the degeneracy of the state. Urey and Grieff have calculated the equilibrium constants K for a number of exchange reactions by using spectroscopic

[68] Urey and Greiff, *J. Am. Chem. Soc.* 57, 321 (1935).

[69] C. N. Hinshelwood, "The Kinetics of Chemical Change," Clarendon Press (1940).

[70] H. C. Urey and L. Greiff, *J. Am. Chem. Soc.* **57**, 321 (1935).

data to calculate vibration frequencies and moments of inertia of the molecules. Table 19 lists some of the calculated separation factors. It should be noted that a system containing hydrogen and water (with deuterium) generally does not have a simple expression for a separation factor.[71] Table 19 shows a decreasing trend of the separation factor with an increase in mass of the exchanging isotope. A separation of uranium isotopes by this method for example is highly unlikely.

The time of equilibrium of exchange reactions could be obtained by a consideration of reaction kinetics but is outside the scope of this presentation. It is worth while to remark, though, that the time to equilibrium will be reduced by having large areas of contact between reacting phases and by the use of catalysts.

Multiple stage recycling process; gas diffusion.—With the exception of electrolysis for hydrogen and centrifugation at very low temperatures, the separation coefficients of the single stage processes are all very close to unity. In order to increase the overall separation factors of the various methods it is necessary to put the single stage processes in cascade.

In a cascade there is usually a continuous flow from one stage to the next. Each stage divides the material it receives into two parts, one of which is enriched in the desired isotope and serves to feed the next higher stage while the other, although impoverished in the desired isotope, is not removed (stripped), but serves to feed the next lower stage. In this way a given quantity of material is recycled many times, resulting in an increased extraction of the desired isotope over what it would be in a single stage separation.

The theory of cascade operation is very complicated, especially when one cascade serves to enrich an initially low mole ratio material (e.g., $\lambda_0/\sigma_0 \ll 1$ if the light isotope is desired) to a final high mole ratio material (e.g. $\lambda_F/\sigma_F \gg 1$). This is because the flow in the initial stages is determined primarily by the undesired material (e.g., heavy isotope) while the flow in the final stages is determined largely by the desired material (e.g. light isotope). To obtain the most efficient flow of desired material the net flow must be inversely proportional to the concentration of the desired material. If the light isotope has a mole fraction λ_k in the kth stage, the net flow of material through that stage must be proportional to λ_F/λ_k.

Since Rayleigh type processes can be put in cascade by recycling, it turns out that the above flow conditions can be arranged

[71] Crist and Dalin, *J. Chem. Phys.* 2, 735 (1934).

Table 19

SOME CALCULATED SEPARATION FACTORS FOR CHEMICAL EXCHANGE REACTIONS

Reaction	Separation factor		Reference
	Q(20°C)	Q(100°C)	
$HD+H_2 \rightleftharpoons H_2+HDO$	3.8	2	(1)
	Q(25°C)	Q(300°C)	
$H_2+2DI \rightleftharpoons 2HI+D_2$	1.082	1.117	(2)
$H_2+2DCl \rightleftharpoons 2HCl+D_2$	0.707	0.895	(2)
	Q(25°C)	Q(325°C)	
$Li^6H+Li^7 \rightleftharpoons Li^7H+Li^6$	1.025	1.008	(3)
$C^{13}O_2+C^{12}O_3^- \rightleftharpoons C^{12}O_2+C^{13}O_3^-$	1.012	0.997	(3)
$N_2^{14}+2N^{15}O \rightleftharpoons N_2^{15}+2N^{14}O$	1.015	1.007	(3)
$CO_2^{16}+2H_2O^{18}(g) \rightleftharpoons CO_2^{18}+2H_2O^{16}(g)$	1.054	1.014	(3)
$CO_2^{16}+2H_2O^{18}(l) \rightleftharpoons CO_2^{18}+2H_2O^{16}(l)$	1.039		(3)
$Cl_2^{35}+2HCl^{37} \rightleftharpoons Cl_2^{37}+2HCl^{35}$	1.003	1.00015	(3)
$Br_2^{79}+2HBr^{81} \rightleftharpoons Br_2+2HBr^{79}$	1.0004	0.99997	(3)

(1) A. and L. Farkas, *Proc. Roy. Soc. A* **144**, 467 (1934).
(2) Urey and Rittenberg, *J. Chem. Phys.* **1**, 137 (1933).
(3) Urey and Greiff, *J. Am. Chem. Soc.* **57**, 321 (1935).

by proper construction of the apparatus. In the case of equilibrium processes, the cascading is accomplished by counter current flow in which recycling occurs naturally. It is usually not possible to proportion the flow conditions artificially as desired and, consequently, a single counter current cascade cannot be used efficiently to produce very large separations, although several single counter current cascades can be connected by artificial recycling to increase their overall efficiency.

We shall first describe recycling processes, using gas diffusion as an illustration since it has been actually used in the large scale separation of U^{235} from U^{238} (Smyth 10.1).

The physical nature of gas diffusion has been described on page 199. In order to put several single stages in cascade, an

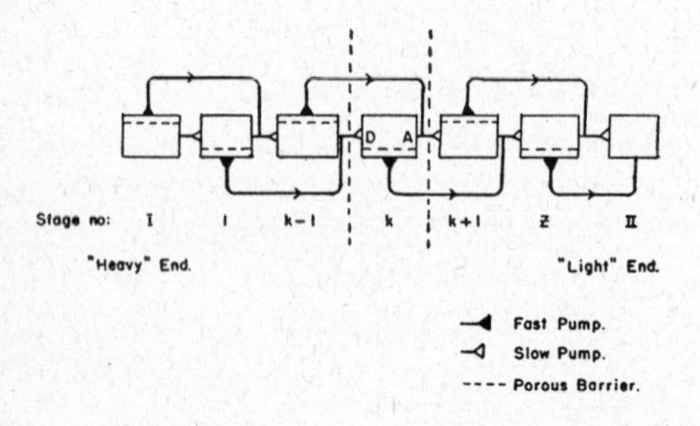

Fig. 42. Schematic diagram of a recycling cascade. As the gas flows through any stage the light isotope diffuses preferentially through the porous barrier and is fed to the next higher stage while the heavy isotope flows preferentially to the next lower stage. In actual practise there is a continuous supply of original material available at the "heavy" end and the enriched light isotope is removed at the "light" end, or *vice versa.*

arrangement such as that shown in figure 42 can be used. A typical stage k is shown enlarged in figure 43a and a possible actual construction of such a stage is sketched in figure 43b.

It is evident that each stage consists of two compartments which are separated by a porous wall (barrier). A pressure difference is established between the two compartments so that part of the gas diffuses through the barrier as it travels along it. If we denote by $N_{x,k}$ the number of molecules per second flowing past the point x in stage k, then $\sigma_{x,k}\, N_{x,k}$ will be the number of heavy molecules per second flowing past the same point and $\lambda_{x,k}\, N_{x,k}$ the number of light molecules per second. Considering figure 43b, it can be shown from considerations similar to those made on page 201 that

$$\sigma_{D,k} N_{D,k} = \sigma_k N_k \exp\{-\frac{\sqrt{2\pi}}{3}\frac{n_A}{\sqrt{M_S RT}}\frac{r}{L} P_k \frac{A_k}{N_k}\} \quad (199)$$

and similarly for light molecules. In equation 199 we have replaced $\sigma_{A,k} = \sigma_k$ and $N_{A,k} = N_k$ as a matter of convenience; n_A is Avogadro's number, M_S is the molecular weight of the heavy isotope, r, is the radius of the cylindrical pores in the barrier, L is the thickness of the barrier, A_k its area and P_k is the pressure in the gas at point (A,k). Equation 199 neglects the effect of any back diffu-

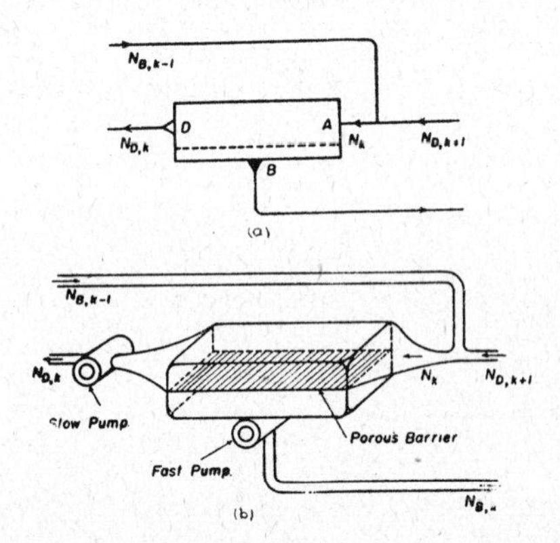

Fig. 43 (a,b). Schematic diagram and possible construction of a stage. The fast pump at B establishes a pressure difference across the porous barrier so that the gas diffuses through the barrier. The slow pump at D serves to overcome the pressure drop across the stage.

sion due to the molecules which are pumped off the barrier and which leave at point (B,k). In general the pressure P_k is made the same for all stages and is conveniently put at 1 atmosphere. Furthermore the fraction f of heavy molecules which diffuse through the barrier is kept constant, so that

$$ln(1-f) = -\frac{\sqrt{2\pi}}{3}\frac{n_A}{\sqrt{M_s RT}}\frac{r}{L} P \frac{A_k}{N_k} \quad (200)$$

For the barrier used to illustrate equation 176 for uranium hexafluoride, one finds

$$A_k \cong 0.6 \times 10^{-23}\, ln \frac{1}{(1-f)} N_k \quad (201)$$

where N_k is expressed in molecules per day. Equation 201 serves to estimate the barrier area for each stage k.

The operation of the separation apparatus is such that the cascade is initially filled with materials of mole ratio σ_0/λ_0 (=1/140 in the case of ordinary uranium hexafluoride) and each stage k sends material enriched in the light isotope to the stage $k+1$ and impoverished material to the stage $k-1$. No further separation of material will occur when each stage k receives from the stage $k+1$ the same number of each isotope as it transmits to that stage. In terms of the notation of figure 43a, this means that at equilibrium

$$\sigma_{D,k+1} N_{D,k+1} = \sigma_{B,k} N_{B,k} \text{ and } \lambda_{D,k+1} N_{D,k+1} = \lambda_{B,k} N_{B,k} \quad (202)$$

This equation assumes that there is no "production" of the desired (light) isotope (i.e., there is no net flow of light molecules toward the light end). Calculations of separation factor with continuous production are too complicated to be considered here.

By using equations 199, 200 and 202 we find that at equilibrium

$$\sigma_{k+1} N_{k+1} (1-f) = \sigma_k N_k f$$
$$\lambda_{k+1} N_{k+1} (1-f)^{\mu} = \lambda_k N_k [1-(1-f)^{\mu}] \quad (203)$$

where the σ and N refer to the entrance (A) of each stage, f is the fraction of heavy molecules diffusing through the barrier in each stage and μ is the square root of the ratio of the mass of the heavy isotope to that of the light one

$$\mu = \sqrt{M_S/M_L} \quad (204)$$

(compare with equation 178).

Equation 203 yields for the separation factor q' per stage for the light isotope

$$q' = \frac{\lambda_{k+1}/\sigma_{k+1}}{\lambda_k/\sigma_k} = \frac{1 - (1-f)^{\mu}}{f(1-f)^{\mu-1}} \quad . \quad (205)$$

By using the proper boundary conditions we find for the apparatus shown in figure 42

$$\sigma_{II} N_{II} = \left(\frac{f}{1-f}\right)^{Z} \sigma_I N_I \quad (206)$$

$$\lambda_{II} N_{II} = \left[\frac{1-(1-f)^{\mu}}{(1-f)^{\mu}}\right]^{Z} \lambda_I N_I \quad ,$$

where the index II refers to the light volume, the index I to the heavy volume and Z is the total number of complete stages. The overall separation factor for the light isotope is

$$Q' = \frac{\lambda_{II}/\sigma_{II}}{\lambda_I/\sigma_I} = q'^{Z} \quad . \quad (207)$$

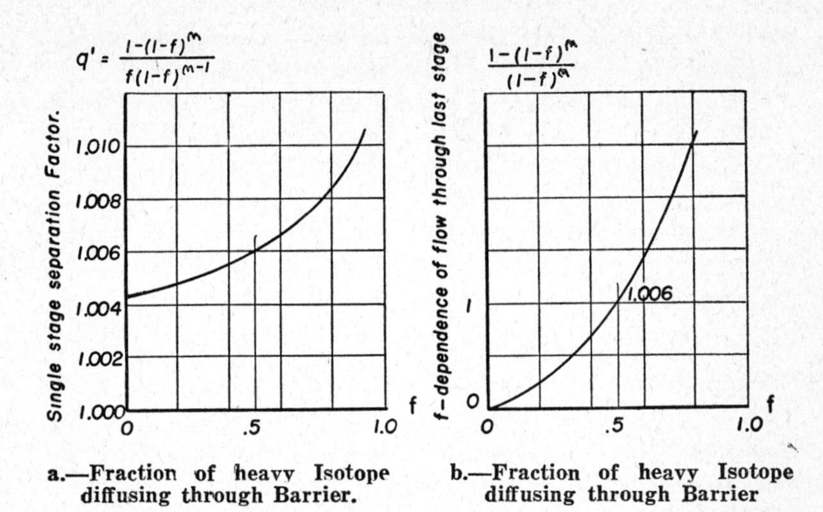

a.—Fraction of heavy Isotope diffusing through Barrier.

b.—Fraction of heavy Isotope diffusing through Barrier

Fig. 44 (a,b). Dependence of separation factor and flow through last stage on the fraction of heavy isotope diffusing through the barrier. Although the single stage separation factor q' becomes infinite when $f=1$ this is actually not the case because of back diffusion in the stage which acts in opposition to the separation. The most efficient flow seems to occur when $f \cong .5$. (see text).

One consideration which so far has been ignored is the fact that since the total number of molecules used in the apparatus (excluding the end volumes) is constant, there exists a relation between each λ_k and λ_o. The number of molecules in each stage is proportional to the volume of that stage which in turn depends on the barrier area A_k and therefore on the flow N_k through the stage (see equation 201). The above mentioned relation therefore gives an additional relation between the λ's and the N's which would permit the solution of equations 203 and 206 for the N's. Since the calculations indicated are complicated, we shall proceed from assumed values for the N's and calculate back some of the interesting constants of the apparatus.

The yield in light isotope of the apparatus can be expressed very roughly by the quantity $\delta\ \lambda_{II}\ N_{II}$, where δ is the fraction of material flowing into the light end of the apparatus which is drawn off continuously. δ generally can not be much larger than 1/1000 without decreasing greatly the overall separation factor Q', which is a function of δ. Analogous to the derivation of equations 224 and 229, it can be shown that $\delta = \theta\ (q'-1)$ where θ is called the production factor. Curves of θ versus Q' for thermal diffusion separation are shown in figure 46. The curves for gas diffusion separation should be similar.

In order to get maximum enrichment of the lighter isotope, we assume that there is an infinite supply of initial material (mole ratio λ_o/σ_o) so that $\lambda_I = \lambda_o$ and $\sigma_{II} = \sigma_o$ at all times. (See remarks leading to equation 153 (page 184)). In that case the overall enrichment factor for the lighter isotope is

$$E' = q'^Z = \left[\frac{1-(1-f)^\mu}{f\,(1-f)^{\mu-1}} \right]^Z \tag{208}$$

and the yield of enriched material is

$$\delta \lambda_{II} N_{II} = \delta \left[\frac{1-(1-f)^\mu}{(1-f)^\mu} \right]^Z \lambda_o N_I \tag{209}$$

The dependence of both these quantities on f is shown in figure 44 for the case of uranium hexafluoride (μ=1.0043).

Suppose it is desirable to obtain 1 mole of 90 percent pure $U^{235}F_6$ per day starting from unenriched uranium hexafluoride Then

$$E' = \frac{\lambda_{II}/\sigma_{II}}{\lambda_o/\sigma_o} = \frac{0.9/0.1}{1/140} = 1260 \; . \tag{210}$$

Smyth (10.7) notes that $f \cong 1/2$ is an efficient arrangement, although it is not known whether the stages were connected as shown in figure 42. As marked on figure 44, $f=1/2$ gives approximately $q'=1.006$ for the single stage separation factor and therefore for the total number of stages

$$Z=1200 \; . \tag{211}$$

In order to calculate the flow N_I through the first stage we set the square bracket in equation 209 equal to unity [72] and since we want

$$\delta\lambda_{II}N_{II} = 0.9 \text{ moles/day}$$

we obtain for the flow N_I, if $\delta=1/1000$ and $\lambda_o=1/140$,

$$N_I=126000 \text{ mole/day} \; . \tag{212}$$

Now in the first 400 stages of the apparatus (k=1 to 400) the initial material ($\lambda_o/\sigma_o=1/140$) is only enriched by a factor of approximately 10 ($\lambda_{400}/\sigma_{400} = 1/14$) so that the flow N_k through these stages must be approximately the same as that given for N_I in equation 212. (In the rest of the stages the flow N_k decreases as λ_{II}/λ_k until N_{II}=1000 mole / day). Equation 201 then requires a barrier area for each of the first 400 stages

$$A_k = 3 \times 10^6 \text{ cm}^2 \quad k = 1 \ldots .400 \tag{213}$$

[72] For most efficient flow, see page 208.

(while for the last stage, $A_{II} \cong 0.23 \times 10^4$ cm^2). The total barrier area in the apparatus would then be approximately 3 acres, which compares favorably with Smyth (10.13).

The actual variation of concentration at each stage with time is very complicated. Calculations have been made by Sherr[73] for the case of Hertz pumps which can also be represented by the schematic diagram shown in figure 43a, assuming the same holdup and flow for each stage and assuming that only a small fraction of the lighter isotopes is present all the time.

In general the concentration or mole fraction at any point in the apparatus (such as the light end in our case) increases quite rapidly at first and then approaches its equilibrium value asymptotically. Huffmann and Urey[74] give a rather simple derivation for the characteristic time of equilibrium assuming that the initial rapid increase in enrichment continues until the maximum concentration ratio is reached. Urey's calculations are for fractional distillation and assume what would be constant holdup and flow in the present case. The results can be transposed though and yield for the equilibrium time, assuming $\lambda_o \ll 1$.

$$\tau = \frac{n_o}{\ln q'} \frac{1}{(q'-1)N_I\lambda_o} \ln (1/1-\lambda_{II}) \qquad (214)$$

where n_o is the holdup per stage (in the present case this would be the holdup for one of the first four hundred stages). For an arrangement like that of figure 43b, it can be shown that approximately

$$n_o = \frac{P_o\, n_A}{RT} \frac{f}{\ln (1/1-f)} V_o \ ,$$

where P_0 is the pressure in the stage, V_0 the volume of the stage, n_A Avogadro's number and f the fraction of heavy material diffusing through the barrier. With $P_0=1$ atm, $f=\frac{1}{2}$ and $V_0= 10^5$ cm^3 (see equation 213).

$$n_0 \cong 3 \text{ moles.}$$

Assuming $p'=1.006$, $\lambda_{II}=.9$ for 90 percent $U^{235}F_6$, $\lambda_o=1/140$ and $N_I \cong 10^5$ mole/day (see equation 212), one finds for the equilibrium time

$$\tau \cong 300 \text{ days} \ .$$

[73] R. Sherr, *J. Chem. Phys.* **6**, 251 (1938).
[74] J. R. Huffman and H. C. Urey, *J. Ind. Eng. Chem.* **29**, 531 (1937); H. C. Urey, *J. App. Phys.* **12**, 270 (1941).

If the holdup volume per stage (V_o) can be made any smaller, the equilibrium time is evidently decreased. $V_0=10^5$ cm^3 assumes that the barrier forms one wall of a channel of approximately 0.3 cm height. If the channel is made any smaller than this, one has to consider the viscous flow of gas along the channel and the calculations have to be modified accordingly.[75]

Countercurrent process; thermal diffusion.—We have shown that equilibrium single stage processes can be put in cascade by countercurrent flow. As the word countercurrent signifies, in this method the two mixtures into which the separation process divides the initial mixture of isotopes are made to flow one against the other. The flow is slow enough so that opposite parts of the mixtures have time to come almost to equilibrium.

The theory of countercurrent flow for thermal diffusion in gases has been worked out by Furry, Jones and Onsager[76] and for liquids it has been examined by Debye[77]. The theory of the countercurrent centrifuge however does not appear to have been published. The thermal diffusion method will be treated here only in an approximate fashion in order to stress the physical processes which occur. For more exact treatments the reader is referred to the original sources.

Cohen[78] has written a paper describing countercurrent flow in a fractionating tower, some results of which will be given here. A fractionating tower can be used to put in countercurrent any of the equilibrium processes similar to the heterogeneous chemical exchange process. Sometimes several of these processes are used together (Smyth 9.37 and 9.41).

When thermal diffusion is used in countercurrent, the separator is placed in an upright position as shown in figue 45. Only the plane separator will be considered here. From the previous discussion (page 201) it is known that on applying a temperature gradient, a flow of molecules will occur between the cold and hot walls. In the present arrangement, the resulting difference in density of the fluid at the two walls will cause a convection current to flow up the hot wall and down the cold wall. The average convection velocity v_g is approximately

$$v_g = \frac{\beta g \rho \Delta T}{100 \eta} a^2 , \tag{215}$$

[75] W. A. Nierenberg, *Bull. Am. Phys. Soc.* 21, 19 (Jan. 1946).
[76] Furry, Jones and Onsager, *Phys. Rev.* 55, 1083 (1939).
[77] P. Debye, *Ann. d. Physik* 36, 284 (1939).
[78] K. Cohen, *J. Chem. Phys.* 8, 588 (1940).

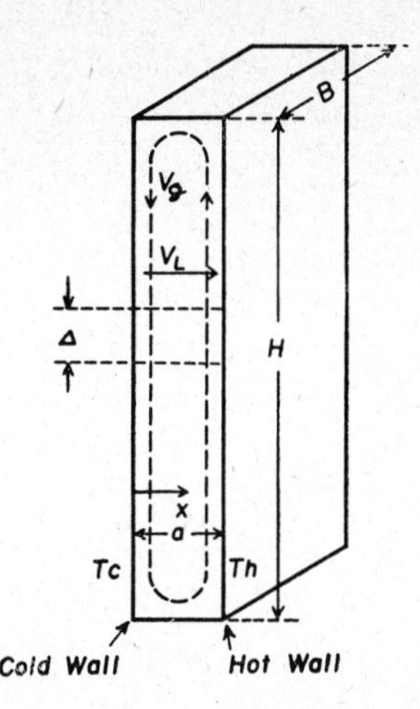

Fig. 45. Schematic diagram of a thermal diffusion column. The fluid containing the two isotopes to be separated is placed between two plane heat conducting walls which serve to establish a temperature gradient through the fluid. The direction of the convection flow (v_g) and the diffusion flow (v_L) is indicated. The end volumes at the top and bottom of the column are not shown.

where β is the coefficient of cubical expansion of the fluid, g is the acceleration due to gravity, ρ is the mean density of the fluid, η is the mean coefficient of viscosity, $\triangle T = T_h - T_c$ is the temperature difference and a is the distance between the walls. For opposite parts of the fluid to come to equilibrium, an approximate time τ_d is required. This is given by equation 191 as

$$\tau_d = \frac{a^2}{D} = \frac{a}{v_d}\ , \tag{216}$$

v_d being the diffusion velocity of the molecules. On the other hand, during this time opposite parts of the fluid will have moved a distance $\triangle$ with respect to each other where

$$\triangle = 2\ \tau_d \cdot v_g = 2a\,\frac{v_g}{v_d}\ . \tag{217}$$

The separator may therefore be considered to consist of a series of stages of length $\triangle$, each having a separation factor for the light isotope q', which is the same as that for the single stage equilibrium process. From equations 186 and 196 we recall

$$p' = 1 + \alpha\,\frac{\triangle T}{T} \tag{218}$$

where for gases $\alpha = .35 \dfrac{M_S - M_L}{M_S + M_L}$, $T = \frac{1}{2}(T_h + T_c)$ and

for liquids $\alpha = \dfrac{\epsilon}{kT}$, $T = \sqrt{T_h T_c}$, and $\Delta T = T_h - T_c$.

In these equations as well as in the ones following, it is always assumed that the initial concentration of the lighter isotope (λ_o) is much less than unity. If H is the overall length of the separator, the overall separation factor for the light isotope, Q', is

$$Q' = (q')^{H/\Delta} \quad . \qquad (219)$$

More exact calculations show that there is an optimum value for the "stage length" Δ, in order to obtain the maximum separation factor Q'. In fact if the convection velocity v_g is much larger than the diffusion velocity v_d, opposite parts of the liquid have no time to come to equilibrium and the separation does not proceed to its best value. If the convection velocity is much smaller than the diffusion velocity, back diffusion of molecules in a direction opposite to convection takes place because of the concentration difference existing between the top and bottom of the separator. This back diffusion decreases the overall separation factor Q'. It turns out that maximum separation is obtained if the convection velocity v_g is equal to the diffusion velocity v_d, so that

$$\Delta = 2a \qquad (220)$$

and

$$Q' = (q')^{H/2a} \qquad (221)$$

Using equations 215 and 217, it is seen that this requires

$$a^3 \doteq \frac{100\,\eta\, D}{\beta g \rho \Delta T} \quad . \qquad (222)$$

For gaseous uranium hexafluoride one can assume $D=1.4$, $\eta/\rho \cong 0.05$ cm^2/sec[79], $\beta \cong 1/273$, $\Delta T = 330$ so that $a \cong 0.05$ cm. If $H = 100$ cm, $Q' \cong 2.5$ (see equation 187). It may be remarked that this is for "no production" of isotope and neglects back diffusion along the convection flow. More exact calculations give $a \cong 0.13$ cm for gaseous uranium hexafluoride. For lighter gases $a \cong 0.5$ cm.

For liquids, assuming $\eta \cong 10^{-2}$, $D \cong 10^{-5}$ cm^2/sec, $\beta \cong 10^{-3}$, $g \cong 10^3$, $\rho \cong 1$, one finds $a^3 \cong 10^{-5}/\Delta T$, so that with $\Delta T = 100°$ C, $a \cong 0.005$ cm. If $H = 10$ cm and $q' = 1.001$, $Q' \cong 2.5$.

In any practical separator it is necessary to provide end volumes at the top and bottom of the separator. In the case of iso-

[79] W. Krasny-Ergen, *Nature* **145**, 742 (1940).

topes, the lighter isotope will concentrate in the top reservoir of the thermal diffusion apparatus and can be removed either continuously or discontinuously. Just as in the case of gas diffusion, it is often advantageous to use an infinite supply of the original mixture so as to have maximum enrichment.

An exact discussion of the most efficient production of isotopes (i. e., whether continuous or discontinuous, whether practically at equilibrium or not), will be found in the references previously mentioned. The equilibrium time τ_{tot} for the entire separator is approximately

$$\tau_{tot} \cong \frac{H^2}{D} , \tag{223}$$

where H is the length of the separator and D the coefficient of self diffusion of the isotope. For gases $H \cong 100$ cm and $D \cong 0.1$ cm^2/sec so that $\tau_{tot} \cong 1$ day and for liquids $H \cong 10$ cm and $D \cong 1$ cm/day so that $\tau_{tot} \cong 100$ days. In general therefore one operates the isotope separation for gases near equilibrium. In the case of liquids, however, this is not practical.

For a later comparison with the fractionating column it is interesting to see how the separation factor, Q', is affected when there is a continuous removal of light isotope from the top volume of the reservoir. Supposing that N moles/sec of material flow into the top reservoir, the amount of material which is removed from the top reservoir has been shown by Furry, Jones and Onsager to be approximately (assuming $\lambda_o < 1$)

$$\theta \, N(q' - 1) \tag{224}$$

where

$$N = \frac{a^3 B \rho g \triangle T}{720 \eta T} \frac{\rho}{M} \tag{225}$$

and

$$q' - 1 = \alpha \triangle T / T . \tag{226}$$

$N(q'-1)$, which is generally called the transport of material, is determined by experimental conditions so that θ can be calculated from the amount of material actually removed. We shall call θ the production factor. N, the flow of material into the top reservoir, is roughly given by

$$N \cong v_g \cdot \frac{a}{2} B \frac{\rho}{M} \text{ moles/sec} ,$$

where v_g is the convection velocity of the material given by equation 215, a is the distance between the plates, B is the breadth of the separator (see figure 45) and M/ρ is the molar volume of the gas. (In the case of gases the more exact calculation gives a different numerical factor in v_g and requires the substitution of $1/T$ for β).

Furry, Jones and Onsager show that the overall separation factor $Q'(\theta)$ depends on the production factor θ in the following way:

$$Q'(\theta) = \frac{Q'(0)}{1 + \theta\, Q'(0)}, \qquad (227)$$

where $Q'(0)$ is given by equation 221. Figure 46 gives a plot of equation 227. Evidently a low separation factor with no production ($Q'(0)$) is not much affected by production.

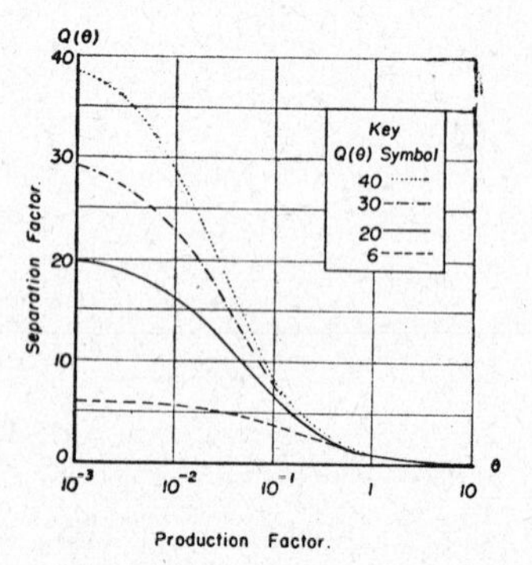

Fig. 46. Separation factor versus production factor for a thermal diffusion column. This is a plot of equation 227. The compromise which has to be made in practise between production and enrichment is clearly indicated. It is assumed that the isotope to be enriched is always present, is small concentration and that the production is continuous. Practically the same figure is obtained for the separation factor of a fractionating column, if the desired isotope is removed in gaseous form from the top reservoir. (K. Cohen, *Jour. Chem. Phys.* **8**, 588 (1940).)

In order to get an idea of the order of magnitude of possible production we apply equation 227 to the case of uranium hexafluoride assuming $a \cong 0.1$ cm. $B = 10$ cm. $\rho/\eta = 28$ sec/cm². $\alpha = 1.5 \times 10^{-3}$, $\Delta T/T = 1/1.5$, $M/\rho = 22 \times 10^3$ cm³. This gives $N(q'-1) \cong 10^{-3}$ moles/day and the actual production would be $\theta \cdot 10^{-3}$ moles/day of $Q(\theta)$ times enriched uranium hexafluoride gas.

Fractionating column.—A fractionating column is generally used to put into countercurrent flow those equilibrium processes which are similar to the heterogeneous chemical exchange method. Figure 47 gives a schematic diagram of a fractionating column. For fractional distillation, volume I is a condenser and volume II an evaporator. For chemical exchange volume I is a mixer for the gas and liquid, and volume II is a generator for the gas. In the column itself the gas travels upwards and the liquid downwards. The column is packed with plates or glass balls etc., as well as suitable catalysts, in order to have a large liquid-vapor interface.

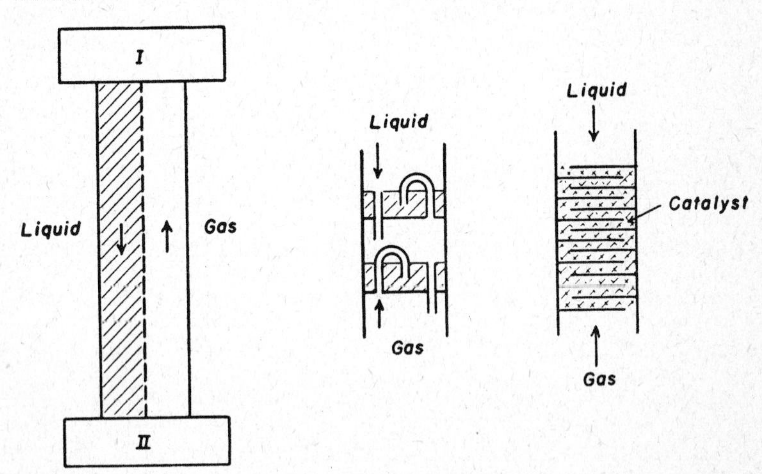

Fig. 47. Schematic diagram of a fractionating column. On the left the end volumes and the interface between the two phases in countercurrent flow are shown (K. Cohen, *Jour. Chem. Phys.* **8**, 588 (1940).) The role of the end volumes depends on the process for which the fractionating column is used. On the right possible constructions of the column are sketched (W. Walcher, *Erg. d. exakt. Naturwiss.* **18**, 155 (1939).)

Cohen's[80] theory of the fractionating column assumes constant holdup per unit length for the gas and liquid, which implies that back diffusion due to the pressure difference along the column is neglected. If the column is built with Z plates it can be seen that the overall separation factor Q for the heavy isotope is

$$Q = q^Z \tag{228}$$

where q for fractional distillation is given by equation 165 and for simple chemical exchange reactions by equation 195. In general, however, the experimentally determined Z is only a fraction of the actual number of plates. This is due mainly to the fact that the liquid and vapor do not have sufficient time or contact to realize complete equilibrium during their encounter on one plate. If the fractionating column is packed with glass balls or the like, Z may

[80] K. Cohen, *J. Chem. Phys.* 8, 588 (1940).

still be called the equivalent number of plates of the column but must be calculated from exact considerations of the transfer of molecules at the liquid-vapor interface. This will not be considered here. Generally $Z \leqq 100$ and $Q \cong 10$ for most fractionating columns. In the exchange of deuterium between hydrogen gas and water, though, much higher separation factors should be expected[81].

It is interesting to consider the case of continuous production. If N moles/sec of gas stream up the tower and δN moles/sec are removed, then figure 46 also gives the approximate variation of the separation factor $Q(\theta)$ in terms of the production factor θ (see equation 224)

$$\theta = \frac{\delta}{1-1/q} \quad . \tag{229}$$

If $q \cong 1$

$$\theta \cong \frac{\delta}{q-1} \tag{230}$$

From figure 46 it appears that $\theta = 10^{-1}$ will reduce a "no production" separation factor between 20 and 50 to about 10.

If $q = 1.01$, $\theta = 10^{-1}$ requires $\delta = 10^{-3}$, which means that a high separation factor in the present process is very sensitive to production. This is quite similar to the thermal diffusion case.

The time to equilibrium has been calculated in a very simple way by Huffmann and Urey[82] and is, if the original mole fraction of the heavy constituent $\sigma_0 \ll 1$,

$$\tau = \frac{n}{Z'} \frac{1}{(\ln q)\ (q-1)} \frac{1}{N'\sigma_0} \ln \frac{1}{1-\sigma_F} \quad . \tag{231}$$

n is the total holdup of the column in moles of material (liquid), N' is the flow of liquid down the column in moles/second and σ_F is the final mole fraction of the material. This equation assumes an infinite supply of the unenriched material. If $q \cong 1$ one can write in the above equation $\ln q = q - 1$. In general the equilibrium time is of the order of days.

Summary.—In this chapter various isotope separation methods were described with particular emphasis on methods which could, or have, served to produce U^{235} and deuterium.

The large scale separation of U^{235} has actually been accomplished with the gas diffusion method (Smyth 10.1) and the mass

[81] Crist and Dalin, *J. Chem. Phys.* **2**, 735 (1934).
[82] J. R. Huffmann and H. C. Urey, *J. Ind. Eng. Chem.* **29**, 531 (1937).

spectrometer (Smyth 11.1) method and has been partially successful using liquid thermal diffusion (Smyth 11.37), the centrifuge (Smyth 9.42) and the isotron (Smyth 11.24). The most successful preparation of deuterium is done in a fractionating tower and makes use of chemical exchange, fractional distillation and electrolysis (Smyth 9.37).

Disregarding these considerations, the separation methods can be classified into individual separation methods in which the separation depends on the motion of each molecule in the separator and into statistical separation processes in which the average behavior of many molecules determines the separation. There are only a few individual separation methods.

In the case of statistical processes, each separation method can be used in a single stage or in cascade. Also the separation methods can be classified into non-equilibrium (Rayleigh type) processes and equilibrium processes. This depends on the fact that in a statistical separation the separation divides the original mixture into two parts and the two parts are either not allowed to come to equilibrium (this could be called unidirectional division) or are allowed to come to equilibrium (in which case the process opposing the division comes to equilibrium with that favoring it). Any physical process is generally used only in one of the classes just described, although the discussion of evaporation illustrates the possibility of using one process in both the equilibrium and non-equilibrium methods.

From the discussion of enrichment at the beginning of the chapter it is clear that in statistical separation methods, a theoretical infinite supply of unenriched material should be used in order to obtain maximum enrichment. Furthermore, since the separation of most physical processes is not very large, the processes should be put in cascade. This increases the overall separation and permits a reasonable continuous or discontinuous production of isotopes, either of which can be the most efficient depending on the circumstances.

CHAPTER 13

CHEMICAL SEPARATION METHODS: ISOLATION OF PLUTONIUM

In this section we shall mention briefly the principles upon which common chemical separation methods depend and suggest applications to the case of the separation of plutonium from uranium and its fission products. It should be emphasized here that chemical separations depend upon the chemical properties of plutonium. Until such properties are determined by actual experiment with plutonium, specific separation procedures are to be regarded as examples of the chemical approach to the problem rather than as workable methods. The detailed chemical information upon which useful methods could be based has not been released for publication, and it is well known that separation procedures constructed in the absence of such data are often modified in development. It is interesting to note, however, that the Canadians have developed a plutonium separation method even better than that used at Hanford.[83]

The most popular chemical methods of separation depend upon the difference in solubility existing between different compounds. These differences are enormous in the more common cases and result in separations far more clean-cut than are to be expected from the physical methods described in the preceding chapter. It is because plutonium, being a different element than uranium, may be expected to show different chemical properties and hence be capable of separation by chemical means that this element is of especial importance.

In addition to differences in solubility, differences in volatility or varying ability to form complex ions or to undergo oxidation or reduction are characteristics extensively employed in chemical separation work. Since all of these properties depend in varying degrees upon the quantities generally grouped together under the title of chemical properties, it will be necessary to make some estimate of the chemical properties of plutonium before attempting specific applications.

[83] Part 2, pages 281-282, Atomic Energy Hearings, U. S. Senate.

Chemical properties and the periodic system.[84]—When an element has been located in the periodic system it is possible to approximate its chemical properties by the application of certain rules. These rules are in turn expressions of the periodicity of electrically similar states as atoms are built up from the simpler to the more complicated arrangements. It is, for example, interesting to be able to say whether a new element will have properties similar to a typical metal or to a non-metal. An element is regarded as metallic if it has a tendency to lose electrons and thus behave as a cation.

PERIODIC TABLE

O	I	II	III	IV	V	VI	VII	VIII
	H 1	He 2						
	Li 3	Be 4	B 5	C 6	N 7	O 8	F 9	
Ne 10	Na 11	Mg 12	Al 13	Si 14	P 15	S 16	Cl 17	
A 18	K 19	Ca 20	Sc 21	Ti 22	V 23	Cr 24	Mn 25	Fe Co Ni 26 27 28
	Cu 29	Zn 30	Ga 31	Ge 32	As 33	Se 34	Br 35	
Kr 36	Rb 37	Sr 38	Y 39	Zr 40	Cb 41	Mo 42	Ma 43	Ru Rh Pd 44 45 46
	Ag 47	Cd 48	In 49	Sn 50	Sb 51	Te 52	I 53	
Xe 54	Cs 55	Ba 56	La 57					
			rare earths	Hf 72	Ta 73	W 74	Re 75	Os Ir Pt 76 77 78
	Au 79	Hg 80	Tl 81	Pb 82	Bi 83	Po 84	85	
Rn 86	87	Ra 88	Ac 89	Th 90	Pa 91	U 92		
						Np 93		
						Pu 94		
						Am 95		
						Cm 96		

Fig. 48. Periodic table of the elements with arrows showing directions of increasing metallic properties.

It is easy to see to a first approximation how this property will vary with electronic configuration. Consider the effect of adding to lithium, in the first period of the periodic system (see fig. 48), one electron and one positive charge to give beryllium. The valence electrons are now under the influence of an increased positive charge so that their escape to yield the beryllium ion becomes more difficult. The operation of the same principle renders the escape of valence electrons progressively more difficult with rising group

[84] Ephraim, "Inorganic Chemistry", Gurney and Jackson, Edinburgh (1934); Morgan and Barstall. "Inorganic Chemistry", Heffer, Cambridge (1936).

numbers in the periodic table, so that elements become less metallic (i.e., less basic, from left to right in the table).

The situation is quite different when a comparison is made between lithium and sodium, for example. This change involves the addition of a complete shell of electrons below the single valence electron. The net effect is to diminish the ability of the sodium nucleus to hold electrons since the increased attraction of the nuclear charge is more than outweighed by the shielding effect of the complete electron shell added. Sodium is correspondingly more basic or more metallic than lithium and again this behavior can be observed throughout the table. The net result of these two effects is shown graphically in fig. 48 where the arrows show the directions of increase of basicity, and the dotted arrow represents their resultant or combined effect.

A consequence of these processes is that the metals tend to occupy the lower left corner and the non-metals the upper right corner of the periodic table, while the zone between is filled with elements partaking of both properties frequently referred to as semi-metals or metalloids. Arsenic and antimony are well known examples of this class. A rather obvious, but sometimes less appreciated corollary of these statements is the fact that positions in the table lying along lines at right angles to the dotted arrow of fig. 48 should exhibit a degree of chemical similarity. This fact is well illustrated by the marked resemblance between, for example, beryllium and aluminum, boron and silicon and vanadium and molybdenum. Indeed this resemblance is sometimes more pronounced than that between members of the same periodic group, and its existence is often useful in predicting the behavior of a little known element.

Closely related to this phenomena and especially interesting for our purposes, is the relationship observed between locations in the periodic system and the atomic volume. The atomic volume of an element which is defined as

$$\text{atomic volume} = \frac{\text{atomic weight}}{\text{density}},$$

is in a rough way a measure of the volume occupied by a single atom. This property, like metallic properties, is periodic and depends upon the electrical construction of the atom. Starting again with lithium and proceeding to the right in the first period, the atomic volume at first decreases due to the increased nuclear charge. Near the middle of the first period an increase of volume due to the mutual repulsion of valence electrons sets in and continues through the remainder of the first period and through the zero group of the second, reaching a maximum at sodium in the second period. If the atomic volume is plotted as a function of atomic number, the result is a series of sharp maxima occupied by the alkali metals with intervening broad minima made up of elements near the middle of the periods. Figure 49 is a plot of atomic volume vs. atomic number.[85] The five complete periods

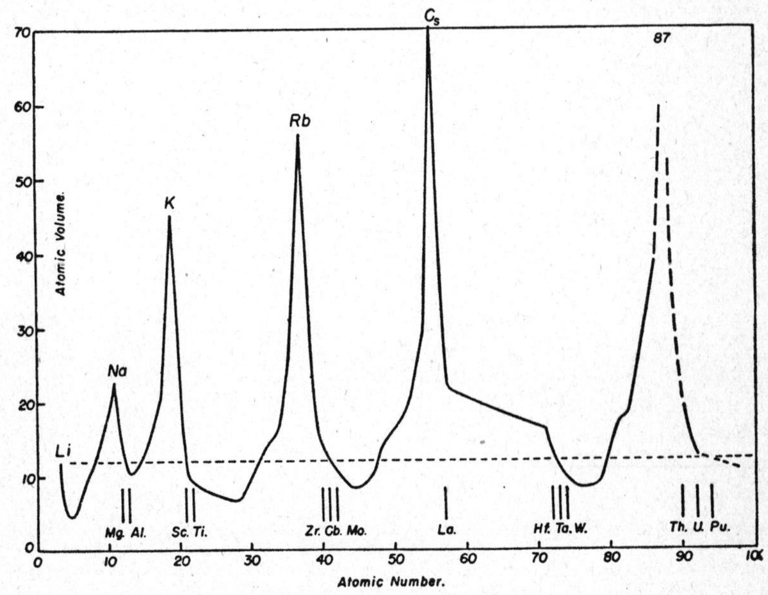

Fig. 49. Atomic volume as a function of atomic number.

are clearly shown here as is the increasing volume accompanying a descent in any given group. Note especially the increasing volume of the alkali metals associated with increasing metallic properties. Estimates of atomic volume are useful because, other things being equal, similar atomic volumes imply a degree of chemical similarity.

Near the middle of the last complete long period, fig. 49 shows a peculiar nearly linear portion in the position occupied by the so-called rare earths $_{57}$La to $_{71}$Lu. These metals, which lie in the third group, show such a similarity of properties that their separation by chemical methods is exceedingly difficult and cumbersome methods of fractional precipitation have to be applied. This resemblance is due to the fact that beyond lanthanum, the add-

[85] The data for this figure were taken partly from Ephraim, ''Inorganic Chemistry'', and partly from the ''Handbook of Chemistry and Physics'', 28th Ed. The curve has been slightly smoothed in the first two periods.

ed electrons do not enter the 6-shell (the valance shell), but the 4-shell, two shells below. For this reason the rare earths are identical in electron configuration so far as the two outer shells are concerned and of course their chemical properties reflect this similarity. It would be unreasonable to expect pronounced differences in behavior of these elements in processes which involve the relatively small energies usually entering chemical reactions. It is not true, however, that the chemical properties in these elements are precisely identical; they show slight differences in basicity, which decreases in the series from lanthanum to lutecium. The decrease in basicity is related to the decrease in atomic volume and may be pictured from an electrical point of view as arising from the fact that the single electron added between each element in this series is very strongly attracted by the nucleus because of its proximity to the positive charge. Since the total nuclear charge in the series is increasing, the net result is a decrease in volume. This *decrease* in volume, it will be observed, is quite the reverse of the *increase* usually found in progressing downward in a given group and is distinguished by the title of "lanthanide contraction".[86] Some of the separation methods employed for the rare earths depend upon the slight differences in basicity accompanying the contraction. It is worth noting that the effect of the lanthanide contraction extends beyond the rare earths. Normally there is a regularity in atomic volumes such that in any group the volume increases in descending through the group. But, the interpolation of the lanthanide contraction interrupts this regularity and in fact reverses it, so that elements following the rare earths have smaller atomic volumes than they would have otherwise. Some of the striking similarities between elements before and after the rare earths, are due to similarities in atomic volume produced by this effect. The striking chemical resemblances between zirconium and hafnium and between colombium and tantalum are well known examples. For the same reason, corresponding elements following the rare earths show closer chemical relationships than do those preceding them.

[86] If the shape of the atomic volume curve in the region containing the rare earths is compared with the equivalent portion in the second long period, it may be thought at first sight that the volumes shown by the rare earths are relatively *larger* than those of the corresponding elements in the preceding period and the term *contraction* may appear a misnomer. The confusion here arises from the custom of locating the rare earths, in printed forms of the periodic table, in a position separated from the third group in which they properly belong. If the rare earths are listed in a vertical column below lanthanum in group III it becomes apparent, according to the principles outlined above, that the atomic volume curve should rise beyond lanthanum, and the term "lanthanide contraction" becomes quite logical.

The chemical properties of plutonium.—In the absence of experimental data, we must be content with such deductions as can be made by the application of the rather qualitative principles outlined above. The first problem obviously is that of locating plutonium in the periodic table. This is made peculiarly difficult, as Seaborg[87] has pointed out, by the fact that plutonium lies beyond the confines of the hitherto known periodic system and thus requires an *extrapolation* of properties rather than the interpolation which can usually be made. The situation is complicated by the existing uncertainty as to the relationships among plutonium's near neighbors at the end of the known periodic system.

In 1941 Goeppert-Mayer[88] predicted, on the basis of calculated energy levels, that a new rare earth series should begin near uranium; the last added electrons entering the 5f shell rather than 6d. It was clearly recognized by Mayer, though, that these calculations were not sufficiently precise to determine the exact point at which the first electron would enter the 5f level. Smyth (6.35) implies that the first 5f electron appears in neptunium which would make uranium the first member of the series in the same sense that lanthanum is the first member of the first rare earth series. Chemically this would place neptunium and plutonium, 95 and 96 in the sixth group, below uranium. More recently Seaborg[89] has suggested that the new series begins with $_{89}$Ac, with the first 5f electron appearing in $_{90}$Th, and all the elements beyond actinium lying in the third group. There appears to be no doubt that neptunium and plutonium do not belong in groups seven and eight as would be the case if they were not members of some rare earth-like group, since they do not at all chemically resemble the corresponding seventh and eighth group elements, rhenium and osmium.[90] The alternative arrangements of Seaborg and Smyth for the last few elements of the periodic system are shown in table 20.

It is not possible, on the basis of available evidence, to decide between these schemes, and as a matter of fact the energies corresponding to the different possible configurations are probably so small that the decision, as Seaborg says, is largely academic. Nevertheless it is interesting to compare the situation here with that in the older rare earth series. In that series all the metals exhibit a valence of +3 only*, with the exception of the second, cerium, which has both +3 and +4. Evidently not more than one of the

[87] G. T. Seaborg, *Ind. Eng. Chem. News. Ed.* **23**, 2190 (1945).
[88] M. Goeppert Mayer, *Phys. Rev.* **60**, 184 (1941).
[89] G. T. Seaborg, *Ind. Eng. Chem. News* Ed. **23**, 2190 (1945).
[90] G. T. Seaborg, *Ind. Eng. Chem. News.* Ed. **23**, 2190 (1945); Smyth (6.35).
* A few have unstable lower valences.

Table 20

ELECTRONIC STRUCTURE OF THE HEAVY ELEMENTS

Element	n = 5 s	p	d	f	6 s	p	d	f	7 s
(After Smyth)									
88 Ra	2	6	10	0	2	6	0	0	2
89 Ac	2	6	10	0	2	6	1	0	2
90 Th	2	6	10	0	2	6	2	0	2
91 Pa	2	6	10	0	2	6	3	0	2
92 U	2	6	10	0	2	6	4	0	2
93 Np	2	6	10	1	2	6	4	0	2
94 Pu	2	6	10	2	2	6	4	0	2
*95	2	6	10	3	2	6	4	0	2
*96	2	6	10	4	2	6	4	0	2
(After Seaborg)									
88 Ra	2	6	10	0	2	6	0	0	2
89 Ac	2	6	10	0	2	6	1	0	2
90 Th	2	6	10	1	2	6	1	0	2
91 Pa	2	6	10	2	2	6	1	0	2
92 U	2	6	10	3	2	6	1	0	2
93 Np	2	6	10	4	2	6	1	0	2
94 Pu	2	6	10	5	2	6	1	0	2
95	2	6	10	6	2	6	1	0	2
96	2	6	10	7	2	6	1	0	2

*Not mentioned in the Smyth Report.

4f electrons is available for chemical reaction, and that only in the second member of the series. If the same rule were followed in the new series the maximum positive valences according to the two schemes of table 20 would be:

Element		Ra	Ac	Th	Pa	U	Np	Pu	95	96
Maximum positive valence	Smyth	2	3	4	5	6	7	6	6	6
	Seaborg	2	3	4	3	3	3	3	3	3

There is no good reason, of course, to believe that the rule is obeyed and we have the explicit statement of Smyth (6.35) that plutonium has valences of +3, +4, +5, and +6 and the suggestion that +4 and +6 are the most prominent. Seaborg infers that more than one electron in 5f can be involved in chemical reactions, but that the difficulty of removing these electrons becomes gradually greater with increasing atomic number, so that the +3 state becomes the most stable toward the end of the series. An examination of the atomic volume curve shows a much greater slope in the region thorium to uranium than that observed in the lanthanum to lutecium region. This might be interpreted to mean that the new rare earth series begins at uranium or later, but there is no assurance that slopes in the two rare earth series need be similar. From a chemical point of view it is fortunate that both Smyth and Seaborg agree that the chemical properties of the transuranic elements are very similar to those of uranium. It seems reasonable then, as a first approximation, to assign them to a position in the sixth group. It cannot be too strongly emphasized, however, that there is so much uncertainty here that the material which follows must be regarded not as a set of data but as a description of a possible line of chemical reasoning. It is offered as illustrative of the sort of outline that might serve as a guide for the study of some of the chemical problems associated with these new elements.

Separation of plutonium from the fission products of uranium.—It is appropriate to summarize here the chemical "facts" dealing with plutonium.

(a) Plutonium is primarily metallic in its properties. Whether placed in the "actinide series" of Seaborg, or in what might be called the "uranide series" following Smyth, the position of plutonium in the periodic table indicates that its properties will be chiefly metallic. If the third group position is chosen it is disturbing to have to attribute to it properties very like uranium. For this reason the sixth group position is more satisfactory from a chemical point of view. To avoid excessive discussion where so little can be known, the latter position is assumed in what follows.

(b) Plutonium has positive valences of +3, +4, +5, and +6. These valences would be expected on the basis of the Smyth form of table 20 by analogy with other transition elements and they are explicitly confirmed by him.

(c) Plutonium is somewhat less metallic than uranium. This follows from the decrease of atomic volume consequent upon either an "actinide" or "uranide' 'contraction.

(d) The atomic volume will be in the neighborhood of 12. This value is arrived at by extrapolating the atomic volume curve beyond uranium along a line parallel to that found for the first rare earth series. If the actinide series is assumed, so that thorium is part of the series, then the extrapolation would be carried out along a much steeper line, through thorium and uranium. It would yield an atomic volume of 7 or 8, appreciably lower than that of any other element in this portion of the periodic system. The higher value seems a more convenient choice.

(e) Plutonium will be in small concentration in the uranium Smyth states that the present production methods are operated to give guantities of the order of grams of plutonium in tons of total mixture. This fact is of great importance in choosing a separation method, since it makes the use of a carrier necessary.

The use of carriers in analytical procedures involving the separation of a small quantity of material from a much larger quantity, is a familiar process in work with radioactive elements, and occurs quite commonly in more ordinary types of analysis. For example, when barium is precipitated as the chromate in its separation from strontium, some strontium is partially precipitated along with the barium; the barium chromate thus acting as a carrier for strontium. This coprecipitation occurs in spite of the fact that strontium chromate is not precipitated under the conditions used for barium, if barium is not present. Many other examples of this type are known even where the chemical similarity of the metals concerned it not nearly as great as in the case of barium and strontium. Although the mechanisms entering here are not very well understood, there is reason to believe that at least two steps are involved. The first is probably an absorption of the carried substance on the carrier, followed by reaction in which the carried substance is actually built into the lattice of the carrier. If this explanation is accepted it follows that carrying will probably occur when the valence type and atomic dimensions of the two elements are sufficiently similar to make the latter step easy. A choice of suitable carrier might well be based upon these criteria, with perhaps less emphasis upon chemical similarity.

In devising a specific separation method for plutonium we have additional guidance from Smyth (8.23), who states that the separation method considered actually employed the precipitation of plutonium in the +4 state with a carrier, oxidation of plutonium to the +6 condition and reprecipitation of the carrier. By this means other elements not carried are separated in the first step, while those which are carried are separated in the second.

The decision as to a specific carrier can be assisted by reference to fig. 49 where the horizontal dotted line marks the estimated atomic volume of plutonium. The other elements, having approximately the same atomic volume and lying in corresponding positions in the periodic system, are seen to fall into five groups; one for each of the periods of the systems: (1) lithium, (2) magnesium and aluminum, (3) scandium and titanium, (4) zirconium, columbium and molybdenum and (5) hafnium, tantalum and tungsten. Of these, scandium, columbium, and hafnium can perhaps be eliminated for large scale purposes because of their rarity. Lithium, magnesium, aluminum, scandium, columbian and tantalum do not show prominent valences of either +4 or +6, and might be eliminated on that account—leaving only zirconium, molybdenum and tungsten to be considered. Of these, molybdenum and tungsten have valences of +4 and +6; zirconium having +4 only. Remembering that plutonium precipitated with the carrier will have to be separated from it later, it appears simpler to choose an element which will not be affected by the oxidation applied to the plutonium. Zirconium then will be the carrier and the working out of a separation scheme is now a matter of chemical detail; selecting a set of chemical reactions which might possibly give the desired separation, followed by trial and correction as defects in the procedure become apparent.

It is comforting to note that in spite of this quite approximate method of choice, the element picked as carrier lies above and to the left of plutonium in the periodic table and thus should show a degree of chemical kinship with it according to the general principles outlined above.

The procedure given in table 21 is based on data from Mellor,[91] Schoeller and Powell[92] and others. Unfortunately much of it is conflicting and even if it were not, the fact that separation is carried out while some of the elements are in a highly excited condition, is known to alter the normal behavior.[93] Smyth (8.54) notes that this condition of excitation necessitated some changes in procedure although not very profound ones. The elements listed at the left are those thought to be most prominent in the fission products where separation is concerned. They have been selected from Segré's chart by the elimination of elements with rather short half-lives and the elimination of all gases. Aluminum has been added to the list on the supposition that it may be used as a container for the uranium.

Even if this scheme should prove usable, it is quite probable that the cycle would have to be performed more than once to produce the required purity. It is also likely that a different scheme may have to be inserted somewhere in the process to deal with elements not efficiently handled by this procedure.

[91] Mellor, "A Comprehensive Treatise on Inorganic and Theoretical Chemistry," Longmans Green and Co., N.Y. (1928).

[92] Schoeller and Powell, "The Analysis of Minerals and Ores of the Rarer Elements," Griffin and Co., London (1919).

[93] Halford, Libby and DeVault, *J. Applied Phys.* 12, 312 (1941).

Table 21—PLUTONIUM SEPARATION SCHEME

U Pu Mo Ba La Ce Cs Al		Precipitate: $BaSO_4$ $La_2(C_2O_4)_3$ $Ce_2(C_2O_4)_3$ Mo, MoO_3 (c)		Precipitate: $ZrO(H_2PO_4)_2$ $PuO(H_2PO_4)_2$ (e)		Precipitate: $Zr(OH)_4$
	dissolve in H_2SO_4 (a) add $(NH_4)_2C_2O_4$ (b) in excess →	Filtrate: U^{++++} Pu^{++++} Cs^{+} Al^{+++}	add Zr^{++++} in excess; then a soluble phosphate and heat (d) →	Filtrate: Cs^{+} U^{++++} Al^{+++}	add excess KOH H_2O_2 →	Filtrate: K_2PuO_5 (f) (potassium per-plutonate)

(a) Sulphuric acid is chosen because $U(SO_4)_2$ is one of the most stable tetravalent uranium salts.

(b) Ammonium oxalate functions to: (1) keep U, Ce and Pu reduced (2) precipitate La and Ce as oxalates. It must be in excess to *prevent* the precipitation of U and Pu.

(c) Mo is very slightly soluble in sulphuric acid giving Mo^{++} which oxidizes in air to MoO_3.

(d) The solution should be hot and acid at this point for oxalates dissolve $ZrO(H_2PO_4)_2$ in cold dilate acid solution.

(e) Pu is assumed to be carried at this point, but not necessarily as the particular compound shown here.

(f) This formula corresponds to that of the U compound produced under the same conditions.

Preparation of metallic plutonium.—The similarity of plutonium to uranium encourages the belief that the procedures which have been worked out for uranium (Smyth 6.13) can also be used for plutonium. Starting with the perplutonate, the following reactions might be used:

(a) $K_2PuO_5 + 2\,Zn^{o} + 10H^{+} \rightarrow Pu^{++++} + 2\,Zn^{++} + 2K^{+} + 5H_2O$

(b) $Pu^{++++} + 4\,OH^{-} \rightarrow PuO_2 + 2\,H_2O$

(c) $PuO_2 + 4\,HF \rightarrow PuF_4 + 2\,H_2O$

(d) $PuF_4 \xrightarrow{\text{electrolysis}} Pu^{o} + 2\,F_2$

Unconverted uranium remaining after the separation of plutonium might be recovered by precipitation of U^{++++} as fluoride, followed by electrolysis. If separation from aluminum is required U^{++++} may be oxidized to U^{++++++} and dissolved in sodium carbonate, leaving aluminum as hydroxide. Metallic uranium would then be obtained by reduction of the uranate as in (a) (above) followed by reactions parallel to (b) (c) and (d).

Chemistry of the other transuranic elements.—All of these elements will be very similar to uranium since they apparently form part of a second rare earth series, although it seems that the chemical properties in this series vary more than in the first. Seaborg points out that the +3 valence becomes more important with increasing atomic numbers but no positive statements about any of the compounds of these metals have been forthcoming. Judging from the decreasing atomic volume it seems possible that the later elements may show an increased tendency to anion formation.

If the behavior of the first rare earth series is a reliable guide, the physical properties of these elements should not differ much from those of uranium. They should be dense metals (density 18+), resembling iron in appearance and readily tarnished in the air, have melting points in the neighborhood of 1150°C[94] and boiling points of the order of 3000°C.

[94] Smyth (2.27) gives the melting point of uranium metal as **1150°C** in contrast to the higher value usually quoted.

CHAPTER 14

POTENTIALITIES OF FISSION TECHNIQUES

Foreseeable uses.—It is interesting to discuss the predictable potentialities of our new knowledge. Just as fission itself is one of the unforeseen results of nuclear physics, so it can be expected that the most important potentialities of fission are not yet revealed. This is particularily true since the fundamental knowledge of the present is not yet assembled or available. The predictable uses, however, are so varied and important as to constitute an immediate challenge for their earliest and widest utilization.

One way of looking at the currently useful techniques would be to classify them according to their applicability to research, commerce and war. Our approach will be in this order of importance. The uses in scientific research are the most immediate, the most obvious and the most interesting. Commercial applications are less certain and so strongly dependent upon economic and political conditions that they are not amenable to simple evaluation. Much recent development has been applied to war use, the results of which are well known.

The primary achievement of nuclear engineers is the production of enormous numbers of slow neutrons by the construction of normal uranium piles. No less difficult, yet less spectacular, is the ability to separate or concentrate isotopes on a large scale. We shall first look into the potentialities of large scale isotope separation.

Increase of available nuclei.—The ability to change the natural relative concentration of isotopes of various elements increases the number of nuclei available for study and use. Heretofore, nuclear physicists have been limited to the approximately ninety-two naturally occurring mixtures which make up the normal elements. With the successful application of isotope concentration or separation methods to these elements, we can expect a large number of the known 274 stable isotopes to be made available in concentrated or pure form. These isolated nuclei can be useful both as objects of nuclear investigation and as stable isotope tracers. The methods described in chapter 12 should also prove useful for concentrating artificially produced nuclei when chemical methods do not suffice; in some cases chemical compounds are most easily separated by these methods.

The magnetic separators or high intensity mass spectrometers used at Oak Ridge can probably separate large quantities of isotopes of elements which can be introduced into the ion source as a gas. Presumably, the present arrangements are adapted to collecting the lighter of several isotopes and have a resolving power of the order of approximately one in eighty (since they must separate mass 235 from 238). It should also not be difficult to collect the heavier of several isotopes. The problems of making the ions stick in the collector have been solved for substances like potassium,[95] rubidium and lithium and have, no doubt, been solved for uranium. Consequently, the collecting problem may not be serious. Under these assumptions, most isotopes can probably be isolated in quantities sufficient for tracer work and nuclear research. Chemical exchange reactions in a counter current flow apparatus have been used to produce concentrations of the useful isotopes C^{13}, N^{15}, O^{18} and S^{34}.[96] These methods yield about one gram per day of the concentrated isotope and cost about fifteen dollars per gram for labor and chemicals. Gaseous thermal diffusion is conveniently used to concentrate C^{13} at the rate of milligrams per day at an approximate cost of two hundred dollars per gram according to Urey. A more recent large scale installation produces C^{13} at forty dollars per gram.[97] The large scale gaseous diffusion method and liquid thermal diffusion method developed at Oak Ridge for concentrating U^{235} in uranium hexafluoride may prove applicable to other isotopes. The liquid thermal diffusion plant is described by Condon[98] as relatively expensive to run due to the enormous consumption of steam for heating. The gas diffusion method may be more promising and possibly can be used without much adaptation for methane, oxygen, nitrogen and others.

The principal nuclear physics measurements on separated nuclei which should yield interesting information are: determination of mass, spin and magnetic moment (where not yet determined because of too small concentration), transmutation data such as thresholds, reaction energies, resonance energies, yields and cross sections both with charged particles and with neutrons and gamma rays. Separated isotopes should reduce the complexity and difficulty of interpreting nuclear experiments, especially those dealing with heavier elements, and should increase the amount of data available for theoretical interpretation. Particular nuclei should

[95] Hemmindinger and Smythe, *Phys. Rev.* **51**, 1052 (1937).
[96] H. C. Urey, *J. App. Phys.* **12**, 270 (1941).
[97] *Science News Letter*, Jan. 12, 1946.
[98] E. U. Condon, *Westinghouse Engineer*, November (1945).

be of especial interest for particular problems. Enrichment of specific isotopes could increase the yields of the transmutations derived therefrom. Enrichment of He^3 in ordinary helium would provide a valuable primary nucleus for study and possibly for use as a bombarding particle. The measurement of the magnetic moment of conjugate pairs of nuclei should be interesting. Values for such a pair as He^3 and H^3 would aid theoretical investigations according to Sachs (private communication).

Spectroscopic studies would be aided and simplified by use of separated nuclei since the isotope shift would be eliminated. Mass spectroscopic examination of concentrated isotopes might reveal stable isotopes previously missed because of their rarity. The existence of such stable isotopes as ${}_{27}Co^{57}$,${}_{53}I^{129}$ and ${}_{55}Cs^{135}$ could be checked in concentrated samples. The I and Cs isotopes should be either long-lived products of fission chains, or stable and rare; the existence of Co^{57} has been reported but not confirmed.

Stable isotope tracer production.—The most important contribution that nuclear physics has made to science in general is the technique of "tracers." The course of a chosen element or molecule can be traced through the most complicated chemical or physiological reaction by the use of "spy" atoms which reveal their presence by their radioactivity or by their difference in mass. Radioactive tracers have heretofore been the most popular. The detecting apparatus for measuring radioactive radiations (Geiger counters and recording circuits) is fairly simple and usually easy to use. The dilutions allowable with radioactive tracers is quite large since the detection is quite sensitive. But the major disadvantage is that the ionizing radiation from the radioactive isotopes must be kept below a level which disturbs the process investigated; thus the number of tracer atoms usable is often limited. Furthermore, only radioactive isotopes of long half-life and appreciable energy emission are convenient for practical use.

Stable isotopes have also been used for tracing. If the normal ratio of isotopes in an element is altered by concentration or separation of isotopes, this concentrated sample can be traced by the use of a mass spectrometer as detector. Heretofore, the paucity of supply and the few isotopes available has limited the use of stable isotope tracers. If these conditions are remedied by the large scale concentration of many isotopes, stable isotopes will take their place with radioactive tracers as a powerful tool of research. Their advantages are obvious: The stable isotopes give off no radiations to be guarded against. Because of their stability, the tracer experiments do not have to be hurried or corrected for half-lives. Iso-

Table 22-23
POSSIBLE STABLE ISOTOPE TRACERS

	Normal conc. %	Possible dilution as tracer	Possible suitable gas for analysis	Use as tracer	Competitive radioactive isotope tracer
He^3	10^{-5}	Depends on concentration achieved	He		
Li^6	7.9	1,000	LiC_2H_5		
B^{10}	18.4	500	BF_3		
C^{13}	1.1	10,000	CO_2	*	**C^{14}, difficult to produce**
N^{15}	0.38	30,000	N_2	*	
O^{18}	0.20	50,000	CO_2		
Ne^{22}	9.73	1,000	Ne		
Mg^{26}	11.1	1,000	$Mg(CH_3)_2$		
Si^{30}	4.2	2,000	$Si H_4$		
S^{34}	4.2	2,000	SO_2	*	S^{35}, weak Beta rays
K^{41}	0.307	30,000	A		
A^{36}	6.55	2,000		*	
Ca^{48}	0.19	50,000		*	Ca^{45}, hard to produce
Ti^{50}	5.34	2,000	$Ti Cl_4$		
Cr^{54}	2.3	5,000	CrO_2Cl_2		
Fe^{58}	0.28	30,000	$Fe(CO)_5$	*	Fe^{59}, hard to produce
Co^{57}	0.17	50,000	Carbonyl	*	
Ni^{64}	0.9	10,000	$Ni(CO)_4$		
Zn^{70}	0.5	20,000	$Zn(C_2H_5)_2$	*	Zn^{65}
Ge^{76}	6.5	2,000	$Ge H_4$		
Se^{74}	0.9	10,000	SeF_6, H_2Se		
Kr^{78}	0.35	30,000	Kr	*	
Sr^{84}	0.56	20,000			Sr^{89}
Zr^{96}	1.5	7,000			
Mo^{100}	9.25	1,000	$Mo F_6$		

Table 22-23 (*continued*)

	Normal conc.	Possible dilution as tracer	Possible suitable gas for analysis	Use as tracer	Competitive radioactive isotope tracer
Ru^{96}	5.68	2,000	RuO_4		
Pd^{102}	0.8	10,000			
Cd^{106}	1.4	15,000	$Cd(C_2H_5)_2$		
In^{113}	4.5	2,000			
Sn^{112}	4.5	2,000	SnH_4		
Sb^{123}	1.1	10,000	SbH_3		
Xe^{124}	.094	100,000	Xe		
Ba^{130}	0.1	100,000			
Sm^{144}	3	3,000			
Gd^{152}	0.2	50,000			
Dy^{158}	0.1	100,000			
W^{180}	0.2	50,000	WF_6		

topes should eventually be available for almost all even elements (except beryllium) as well as for some odd elements. The main disadvantage is that the allowable dilution is generally smaller than for radioactive tracers. Dilutions from about 1000 to 100,000 are possible with present mass spectrometric accuracies. The dilution allowable is approximately the reciprocal of the product of the relative accuracy of the mass spectrometer times the normal abundance of the tracer isotope. This assumes that the tracer isotope is initially separated completely. The measuring equipment (mass spectrometer) is more expensive than a counter system and the sample has to be in the form of a gas. Stable isotopes are of particular importance when suitable radioactive tracers are absent. The stable isotopes C^{13}, O^{18}, N^{15}, D^2 and S^{34} have already been used as tracers.[99] Table 22-23 is a list of possible stable isotope tracers. The asterisks indicate that these elements have been traced in actual experiments either by radioactive or stable isotopes. Previous tracer work has, of course, been influenced by the availability of tracers and of detecting equipment, consequently, we can expect many more elements to be used if the tracers become available. The wide range of problems which have already been at-

[99] Applied Nuclear Physics Conference, *J. App. Phys.* **12**, 259 (1941).

tacked with tracers is indicated in the reports given at the Applied Nuclear Physics Conference in 1941.[1]

Concentration of radioactive isotopes.—Long-lived radioactive isotopes can also be concentrated or separated by isotope separation methods. Concentration often can be accomplished by chemical methods but usually only with carriers which yield a mixture of the radioactive isotope with the normal isotopes of the element. Isotope separation methods will provide concentrations sufficient for determination of mass number and possibly for measurement of other nuclear data such as spin, magnetic moment, transmutation, absorption and scattering data. This should be of importance in several current problems. The interpretation of beta decay would be aided by a knowledge of the spin of the nucleus before and after the beta emission. Precision mass spectrometers can almost separate isomers. H^3 is a radioactive element which is of interest because of its simple composition; its concentration should enable it to be studied in appreciable detail. By concentration and mass analysis, many transmutation schemes leading to stable isotopes may be checked. Radioactive isotopes can also be detected or analyzed in a mass spectrometer. This technique should be especially useful for low energy, long-lived activities and may rival counters for detecting atoms such as H^3, Be^{10}, C^{14} etc.

Enormous supply of neutrons.—As we have mentioned before, the most striking achievement of nuclear engineering has been the production of such enormous numbers of neutrons as exist in piles. In order to utilize these neutron densities, it seems preferable to construct small, enriched, heavy water piles (page 167). Such a pile with its shielding can conveniently occupy a room of ordinary size. Extra shielding would be provided if the pile were buried, so that if it gets out of control protection will be afforded from the extra intense burst of neutrons. The neutron exposures can be made by lowering material into the pile to be irradiated or by using a channel to pipe the neutrons out. The total production of fast neutrons can be expected to be 10^{16} neutrons per sec for a 100 kw pile whose critical size may be less than 1 cubic meter. When these neutrons are slowed down, densities of 10^{10} per cc per second will be probable. Not all of these neutrons can be used, however. Dependent on the self-replenishment of the pile and the efficiency with which it runs (i.e. low losses to moderator, cooling, escape, etc.), anywhere from 0.01 to ½ of the neutrons will be available (page 168). For short exposures, however, much higher fractions can be utilized. If the pile is over-built, so that large amounts of boron are needed to control it, the boron can be re-

[1] *J. App. Phys.* **12**, 296 (1941).

moved as an equivalent absorber is inserted. This control is made easy by the lag due to the delayed neutrons.

In order to realize the enormity of these neutron intensities we may compare them with the production of neutrons by cyclotrons. The Berkeley cyclotron, using 100 microamps of 16 Mev deuterons on beryllium, gives a neutron production of about 10^{12} neutrons per second.[2] In this comparison then, the pile produces 10,000 times as many neutrons as a cyclotron. The cyclotron produces a beam of fast neutrons approximately four times as intense in the forward direction as the average value given above. The fast neutron beam of a pile is a little harder to predict; it certainly has a smaller average neutron energy. In intensity, however, the advantage is possibly still with the pile. In addition, piles are simpler to construct, to operate and to maintain. Their reliability is emphasized by Smyth (8.28).

Neutron beams.—There are several fields of usefulness of intense slow neutron beams from piles. In the first place they will provide strong neutron sources for purely nuclear experiments, such as neutron absorption, studies of nuclear resonances, study of neutron induced radioactivities, measurements of nuclear scattering of neutrons and measurement of the neutron magnetic moment. The large intensities will make it easy to carefully define the beam geometrically and to select neutron velocities exactly by a mechanical shutter and time of flight technique. The "cold" neutrons mentioned by Smyth (8.31) will afford a useful source of longer wave length neutrons. Smyth (8.29) mentions that comparative neutron absorption measurements can be readily made by observing the change in control bars (cadmium-iron) necessary to compensate for the introduction into the pile of a fixed amount of a particular substance.

Short pulses of neutrons produced by mechanically assembling and disassembling two subcritical masses of U^{235} or plutonium (see page 178) should provide a means of achieving very high neutron intensities. These pulses might also be useful in neutron experiments utilizing "time of flight" techniques.

Neutron beams will be useful for study of crystal lattices by transmission measurements. Order-disorder transitions,[3] crystal structure transitions and possibly Curie transitions can be investigated by neutron scattering. Bloch[4] has suggested that neutron beams will have two uses in the study of slow neutron scattering in ferromagnetic substances. The first is in the determina-

[2] *J. App. Phys,* **12,** 339 (1941); *Rev. Mod. Phys.* **9,** 330 (1937).
[3] Nix, Beyer and Dunning, *J. App. Phys.* **12,** 305 (1941).
[4] F. Bloch, *J. App. Phys.* **12,** 305 (1941).

tion of a "magnetic form factor" by the dependence of scattering on angle and energy. This will aid in clarifying the role of valency electrons in ferromagnetism. The second group of experiments deal with the magnetic scattering near magnetic saturation. This scattering should furnish information about the relation between the details of the ferromagnetic substance and magnetic saturation.

Intense neutron or gamma radiation may also "catalyze" industrially important or interesting chemical reactions. These radiations are known to cause changes in the electrical resistance, elasticity and heat conductivity of various materials (Smyth 7.24).

Bethe[5] has stated that piles will provide powerful sources of neutrons at a more constant rate than cyclotrons. He suggested that these large intensities of neutrons might make it more likely to detect neutrinos. He also pointed out the possibilities for research of neutron diffraction experiments in view of such large intensities. Using crystal diffraction, it may be possible to achieve monochromatic beams of neutrons, to investigate crystal structure, to locate hydrogen atoms (or preferably deuterium atoms) in crystals and to observe total reflection of neutrons by crystals.

Another important use for neutron beams is in cancer therapy and the study of physiological effects of neutrons. Fast neutrons have already been found to give encouraging results in cancer treatment[6] and small "piles" should be entirely adequate for such therapy. Slow neutron beams also have been considered.[7] Recent advances in the technique of localization of lithiated dyes in tumor tissue[8] hold promise of early clinical trial. The slow neutrons are expected to be selectively absorbed by the lithium and to release energy in the tumor region without appreciable effects elsewhere.

Neutron produced radioactive tracers.—One of the most important uses of piles is the production of new nuclei. This is, of course, the chief use to which they have already been put i.e., the production of plutonium. We shall treat the consequences of this later. Here we wish to point out that this plutonium production is accomplished by the absorption in U^{238} of approximately one neutron from every fission. If our purpose is to produce other nuclei than plutonium we can, in theory, manufacture any other neutron

[5] H. A. Bethe, "Scientific Aspects of Nuclear Energy", *Bull. Am. Phys. Soc.* **21**, no. 1, p. 12, January 24, 1946.
[6] Stone and Larkin, *J. App. Phys.* **12**, 332 (1941).
[7] P. G. Krieger, *Proc. Nat. Acad. Sci.* **26**, 181 (1940).
[8] Zahl and Cooper, *J. App. Phys.* **12**, 336 (1941).

producible nucleus at the same rate by properly proportioning the pile with fissionable material, moderator and source material. In practice this yield is hard to reach unless the pile is specifically designed for this single purpose. A multi-purpose pile will probably be close to self replenishing so that only about one hundredth of the neutrons produced by fission will be available for capture by added substances. One neutron goes to fission U^{235}, U^{233} or Pu making the pile self sustaining, and one produces Pu or U^{233} to make the pile self perpetuating. The remaining 0.3 of a neutron goes primarily into losses by absorption in cooler etc., leaving a small fraction available for use. Even then the number of neutrons absorbed depends upon the density of pile neutrons (slow neutrons considered primarily), the absorption coefficient and the number of nuclei added. We may, however, estimate the order of magnitude of yields by multiplying cyclotron yields by $10^4 (=10^{16}/10^{12})$ for a self sustaining but not self replenishing pile of 100 kw, or by 10^2 for a self sustaining and self replenishing pile or one used to make plutonium, also 100 kw. This is probably a good comparison for slow neutrons since the volume in which slow neutrons are produced in a cyclotron is similar to the volume of an enriched heavy water pile. The comparison for the Hanford pile is probably of a similar order of magnitude for although the total neutron production is much larger, the volume is also larger.

Table 24 lists slow neutron produced radioactive nuclei of long half-life suitable for tracers. The radioactive isotope is given in the first column; the second column gives the half-life. (Some half-lives of a few hours are included for their interest.) The third column gives the maximum energy radioactive particle emitted and its energy. The fourth column lists the maximum gamma ray energy emitted where known. The next column gives the slow neutron capture cross section for the stable isotope of one less mass number than the radioactive isotope, (except for C^{14} which is produced from N^{14} by a (n,p) reaction and element 43, Rh^{105}, I^{131}, Ag^{111}, Po^{210}, Pa^{233} and Np^{239} which are derived from radioactive elements in turn produced by neutron capture). Column six gives a few reported yields per hour in millicuries from the Berkeley cyclotron (one curie is an amount of radioactive substance which decays at a rate of 3.7×10^{10} disintegrations per second). The more important of these radioactive isotopes are probably C^{14}, Na^{24}, P^{32}, K^{42}, Br^{82} and I^{131} since they have already been used in many tracer experiments. The variety of problems which can be investigated with tracer techniques, however, will eventually encompass all the available isotopes.

Yields of radioactive materials.—In order to estimate the yields which might be achieved by practical piles we shall make a few arbitrary assumptions. Consider an enriched heavy water pile with a size of approximately one cubic meter. If this is run at 100 kw from U^{235}, Pu or U^{233} so that it is not self replenishing, then the neutrons available for capture to make radioactive substances will be approximately 10^{16} neutrons per second or 10^{10} neutrons per second per cc. If we insert 10 liters of the substance to be irradiated into the pile without changing the neutron distribution then we can estimate the yields of radioactive material. The dimensions of the containers in which the material is irradiated must, of course, be less than the mean free path of the thermal neutrons for capture in the substance and the effects of resonances should be included. Neglecting resonances, the yield Y in millicuries per hour is

$$Y = \frac{G\ d}{T\ A} \cdot 10^6 \text{ millicuries per hour} ,$$

where G is the cross section for thermal capture (in units of 10^{-24} cm^2) times the relative isotope abundance, d is the density (gms/cc), T is the half-life in hours and A is the effective atomic weight in atomic weight units. The results of such a calculation for some radioactive isotopes of interest are presented in table 25. It can be seen that in some cases the mean free path is small enough to result in appreciable shielding if we irradiate as much as 10 liters of material. In these cases the yields are too high. Where we can compare these estimates with observed cyclotron yields we get fair agreement (considering the fact that we do not know the exact disposition of material irradiated in the case of cyclotron yields). Bromine yields have been reported as 1 millicurie per hour for a large cyclotron with 100 microampere beam. We estimated that a non-self replenishing pile yield might be 10^4 times this or 10,000 millicuries per hour. Table 25 lists 1300 millicuries per hour which is quite a reasonable agreement. The inaccuracies in our calculation involve not only neglect of possible resonances but also neglect of possible destruction of the radioactive isotope by neutron capture. Obviously, yields vary markedly from isotope to isotope, but in many cases are quite large; usually of the order of millicuries per hour.

Table 24 — Slow Neutron Produced Radioactivities of Long Half-Life

Radioactive isotope	Half-life	Max. energy β particles emitted Mev	Maximum energy γ rays emitted Mev	Thermal neutron cross section in units of 10^{-24} cm²	Relative isotope abundance %	Cyclotron yields in millicuries per hour
$_1H^3$	25 yr	0.015		6.5×10^{-4}	0.02	
$_4Be^{10}$	10^5 yr	0.58	0.5	0.0085	100	
$_6C^{14}$	~6,000 yr	0.145		<1.7	99.6	5.10^{-3}
$_{11}Na^{24}$	14.8 hr	1.4	2.73	0.4	100	
$_{14}Si^{31}$	170 min	1.8	none	0.11	4.2	
$_{15}P^{32}$	14.3 d	1.72	none	0.23	100	
$_{17}Cl^{36}$	10^8 yr	0.64			75.4	
$_{19}K^{42}$	12.4 hr	3.5	1.5	1.0	6.6	
$_{20}Ca^{45}$	180 d	0.9	0.7	0.6	2.06	
$_{21}Sc^{46}$	85 d	0.26	1.25	2.8	100	
$_{22}Ti^{51}$	72 d	0.36	1.0		5.34	
$_{24}Cr^{51}$	26.5 d	K capture	1.0		4.49	
$_{26}Fe^{59}$	47 d	0.46	1.3	0.32	0.28	
$_{27}Co^{60}$	5.3 yr	0.3	1.3	0.73	100	
$_{29}Cu^{64}$	12.8 hr	0.66	none	3.1	68	
$_{30}Zn^{65}$	250 d	0.4	1.14	0.51	50.9	

Table 24 (*continued*)

Radio-active isotope	Half-life	Max. ener- β particles emitted Mev	Maximum energy γ rays emitted Mev	Thermal neutron cross section in units of 10^{-24} cm^2	Relative isotope abundance %	Cyclotron yields in millicuries per hour
$_{32}Ge^{71}$	40 hr	1.2		0.073	21.2	
$_{33}As^{76}$	26.8 hr	2.7	3.2	4.6	100	1
$_{35}Br^{82}$	34 hr	0.47	1.35	2.25	49.4	1
$_{37}Rb^{86}$	19.5 d	1.6		0.72	72.3	
$_{38}Sr^{89}$	55 d	1.5	none	0.005	82.56	
$_{39}Y^{90}$	60 hr	2.2	none	1.1	100	
$_{40}Zr^{95}$	65 d	1.0	0.92	0.05	17	
$_{42}Mo^{99}$	67 hr	1.5	0.4	0.37	25.4	
$_{43}\ ^{99}$	6.6 hr	I.T	.136	from Mo^{99} decay		
$_{44}Ru^{105}$	4 hr	1.5	.76	0.33	17	
$_{45}Rh^{105}$	35 hr	.6	.3	from Ru^{105} decay		
$_{46}Pd^{107}$	13 hr	1.1	none	12.1		
$_{47}Ag^{111}$	7.5 d	1.0	none	from Pd^{111} decay (26 min. 0.63)		
$_{47}Ag^{108}$	225 d	.59	1.4	48	52.5	
$_{48}Cd^{115}$	2.33 d	.25	.55	1.1	28	
$_{49}In^{114}$	48 d	I.T	.19	61	4.5	

Table 24 (*continued*)

Radio-active isotope	Half-life	Max. energy β particles emitted Mev	Maximum energy γ rays emitted Mev	Thermal neutron cross section in units of 10^{-24} cm^2	Relative isotope abundance %	Cyclotron yields in millicuries per hour
$_{50}Sn^{113}$	100 d		.085	1.1	1.1	
$_{51}Sb^{124}$	60 d	2.45	1.82		44	
$_{53}I^{131}$	8 d	.687	.4	from Te^{131} decay (25 min, 0.24)		
$_{55}Cs^{134}$	1.7 yr	.9		25.6	100	
$_{57}La^{140}$	40 hr	1.45	2.3		100	
$_{58}Ce^{141}$	30 d	.6	.2		90	
$_{59}Pr^{142}$	19.3 hr	2.14	1.9		100	
$_{60}Nd^{147}$	47 hr	.95			16.5	
$_{63}Eu^{154}$	7 yr	.9			50.9	
$_{64}Gd^{159}$	8 hr					
$_{65}Tb^{160}$	3.9 hr				100	
$_{66}Dy^{165}$	2.5 hr	1.2	1.1	2860	.6	
$_{67}Ho^{166}$	35 hr	1.6			100	
$_{68}Er^{169}$	12 hr				29.3	
$_{69}Tm^{170}$	105 d			100	100	
$_{70}Yb^{175}$	3.5 hr				37.2	

Table 24 (*concluded*)

Radio-active isotope	Half-life	Max. energy β particles emitted Mev	Maximum energy γ rays emitted Mev	Thermal neutron cross section in units of 10^{-24} cm^2	Relative isotope abundance %	Cyclotron yields in millicuries per hour
$_{71}Lu^{177}$	6.6 d	.44			2.5	
$_{72}Hf^{181}$	55 d				30	
$_{73}Ta^{182}$	120 d	.53	1.2	22.5	100	
$_{74}W^{185}$	77 d	.65			30.1	
$_{75}Re^{186}$	90 hr	1.05	none	109	38.2	
$_{76}Os^{193}$	17 d	.35		5.34	41	
$_{77}Ir^{192}$	60 d		.63		38.5	
$_{78}Pt^{197}$	3.3 d				7.2	
$_{79}Au^{198}$	2.7 d	.8	.44		100	
$_{80}Hg^{203}$	51.5 d	.46	.3		29.56	
$_{81}Tl^{206}$	3.5 yr	.87	none		70.9	
$_{83}Bi^{210}$	5 d	1.17		0.015	100	
$_{84}Po^{210}$	136 d	5.3		from Bi^{210} decay		
$_{90}Th^{233}$	26 m				100	
$_{91}Pa^{233}$	25 d	.23	.30	from Th^{233} decay		
$_{92}U^{239}$	23 m				100	
$_{93}Np^{239}$	2.3 d	.47	.27	from U^{239} decay		

References:
G. T. Seaborg, *Rev. Mod. Phys.* **16**, 1 (1944).
Segrè's chart, May 15, (1946).
J. G. Hamilton, *J. App. Phys.* **12**, 441 (1944).

Evans[9] gives a convenient estimate of maximum amounts of radioactive tracer which can be used and not produce biologic effects,

$$C = \left(\frac{5.7}{\text{max. energy of beta rays in Mev}} \right) \text{microcuries per kilogram}$$

Consequently, amounts of radioactive materials of the order of millicuries should be adequate for numerous biological tracer experiments.

Other uses of tracers require similar amounts of radioactive material, (i.e., microcuries). As examples of such uses and the amounts of tracers needed we will describe a few interesting applications.

Autoradiographic tracing is an extremely simple technique. With neutrons, the radioactive tracer can sometimes be produced in place by irradiating the whole system and then sectioning and placing on a photographic plate. Such a procedure was found useful in finding the distribution of aluminum present to a fraction of a percent in silicon.[10] The slab was irradiated with fast neutrons to produce sodium 24 from the aluminum in a (n, α) reaction. All neutron induced activities in silicon have short half-lives and die out in a day. Then, an autoradiograph exhibits the presence and distribution of the aluminum. (Figure 50 shows such an autoradiograph.) Other materials can be detected with slow neutron bombardment.[11] According to Hamilton,[12] it takes about 2×10^6 beta particles per square cm to affect a photographic plate. If the material is left on the plate for a time of the order of the half-life T of the radioactive material, then it will take at least $\frac{1}{2T(\text{sec})}$ microcuries per cubic millimeter concentration to produce an autoradiograph. In such experiments care must be taken to avoid the "Russell" effect[13] which is a darkening of the photographic plate exposed to freshly etched or ground surfaces of most metals.

In an extensive series of measurements of the self-diffusion of zinc in single and poly crystals, Miller and Banks[14] used a total of 0.01 of a microcurie of Zn^{65}.

[9] R. D. Evans, p. 657, *Medical Physics*, ed. O. Glasser, Year Book Publishers (1944).

[10] Lewis and Stephens, *Phys. Rev.*, January 1-15 (1946).

[11] Goodman and Thompson, *Am. Mineralogist* 28, 457 (1943).

[12] J. G. Hamilton, *J. App. Phys.* 12, 441 (1944).

[13] Lewis and Stephens, *Phys. Rev.*, January 1-15 (1946).

[14] Miller and Banks, *Phys. Rev.* 61, 648 (1942).

Table 25—PILE YIELDS OF SOME ISOTOPES
Calculated for 10 Liters of Material Exposed to 10^{10} Neutrons/cc/sec

Radioactive isotope	Cross section in units of 10^{-24} times relative isotope abundance	Density of material gms/cc	Half-life in hours	Atomic weight of material	Mean free path cm	Yields mc/hr
H^3	$13\cdot10^{-8}$	1	$2.2\cdot10^5$	9	$7\cdot10^7$	$6\cdot10^{-8}$
Be^{10}	0.0085	1.85	$8.75\cdot10^8$	9	570	$2\cdot10^{-6}$
C^{14}	1.7	1.6	$5\cdot10^7$	30	12	$2\cdot10^{-3}$
Na^{24}	0.4	0.97	14.8	23	60	1100
P^{32}	0.23	2.2	343	31	60	45
K^{42}	0.066	0.86	12.4	39	680	120
Ca^{45}	0.012	1.54	4310	40	2220	0.1
Fe^{59}	0.001	7.86	1130	56	7000	0.1
Zn^{65}	0.26	7.14	6000	65	65	4.5
As^{76}	4.6	5.7	26.8	75	2.86	1300
Br^{82}	1.12	3.12	34	80	22.8	1300
Rb^{86}	0.52	1.53	469	85	106	20
Sr^{89}	0.0041	2.6	1320	88	8000	0.1
Ag^{108}	25	10.5	5400	108	0.41	450
In^{114}	2.74	7.3	1150	115	5.7	150
Ta^{182}	22.5	16.6	2880	181	0.48	720
Bi^{210}	0.015	9.8	120	209	1420	6

Radioactive silver was used by Langer[15] in examining exchange rates. Less than a microcurie of the 45 day silver isotope was used in his investigations.

The use of radioactive materials in therapeutics often involves specific elements which will selectively deposit in particular regions and irradiate neoplasms. The selective deposition is traced first by radioactive tracers. Then, additional amounts of the suitable radioactive element are used to give appreciable radiation doses to the regions involved. Phosphorus has been found to be selectively absorbed in the nucleo protein fraction of transmitted leukemic cells and consequently P^{32} has been used in the treatment of leukemia.[16] Strontium has been found to deposit in the bones and Sr^{89} has been considered for irradiation of the skeleton.[17] Both these materials can be produced in piles in quantities of the order of millicuries. Evans[18] gives a formula for estimating the total tissue dose r in roentgens of a beta ray emitter of maximum beta ray energy E in Mev and half-life T in days when used in internal therapy as

$$r = 0.027\, C\, E\, T$$

where C is the concentration of the radioactive isotope in microcuries per kilogram of tissue. (C has to be related to the total amount to be used, by the results of previous tracer experiments). Since therapeutic doses are of the order of 10,000 roentgens, quantities of the order of 1 to 10 millicuries of the radioactive isotope are needed per patient.

Since carbon is such an importan element in organic and physiological chemistry, C^{14} is an especially desired tracer. Yet table 25 indicates that it is hard to produce under the conditions assumed. We may further consider the possibilities oi securing appreciable quantities. Smyth (8.43) mentions that the Clinton pile was run at 800 kw with air cooling and operated at about 150°C. Since it requires about 10^4 grams of air per second to keep such a pile down to 150°C, we may estimate the amount of air in the pile to be perhaps twenty times as much. If the air is recirculated and CO_2 is extracted, we can estimate a yield of C^{14}. If we accept our estimate of 10^{10} neutrons per cc per second as applicable to the Clinton pile, we can estimate the production of C^{14} as perhaps 0.4 mc/hour. Since this isotope has such a long half-life it will be usable many times.

[15] A. Langer, *J. Chem. Phys.* **10**, 321 (1942).
[16] J. H. Lawrence, Erf and Tuttle, *J. App. Phys.* **12**, 333 (1941).
[17] C. Pecher *J. App. Phys.* **12**, 318 (1941).
[18] R. D. Evans, p. 657, *Medical Physics*, ed. O. Glasser, Year Book Publishers (1944).

Production of radioactive sources.—Some of these radioactive substances should be particularly important for special purposes such as sources of gamma rays, electrons (monochromatic or continuous distribution in energy), positrons and alpha particles. The producible quantities of these radioactive substances are so large that these sources will constitute important and readily usable tools for research.

Many of the radioactive substances listed in table 24 emit gamma rays. While the intensities of these gamma rays are not often given, they are sometimes as numerous as the beta rays. Usually, when the energies are listed, the number of gamma rays may be taken to be at least of the order of one tenth of the number of disintegrations. The values of energy vary from 0.1 to 3.2 Mev. The highest energy Radium C gamma rays (the highest energy "radium" rays), it may be recalled, are 1.76 and 2.2 Mev and are 0.2 and 0.1, respectively, of the number of disintegrations. Consequently, we can prepare radioactive sources of gamma rays from the long-lived gamma emitters of table 24. The radioactive isotopes Ag^{108}, Sb^{124}, Ta^{182} and La^{140} may be useful for this purpose, but longer-lived substances would be more desirable and may be found with further research. It seems possible to secure curies of many of these isotopes (equivalent to grams of radium). Substitutes for radon as therapeutic gamma ray sources may be found in Na^{24}, K^{42} and As^{76}. It seems feasible to prepare these in compact form by concentration with the Szilard-Chalmers method and perhaps plating on thin foils which are then wrapped up to reduce size. Some of these radioactive isotopes may be used with beryllium or deuterium to produce photo-neutrons of various energies. These should be useful as ready sources of monochromatic neutrons. Monochromatic gamma ray sources of long life should be useful as cheap and handy standard sources to calibrate counters and other gamma ray detectors.

Beta ray sources should also have many uses. For example: standards for calibrating ionization chambers, sources for demonstrations, sources of positrons and sources of monochromatic internally converted electrons. Tl^{206} in addition to Radium D+E might prove useful as a calibrating standard source. Ge^{71} is a source of positrons which may have value in studies of positron effects. The possibility of a system of positive and negative electrons like the hydrogen atom may be studied with such sources[19] of positrons. Pool[20] gives a list of internally converted gamma

[19] A. E. Ruark, *Phys. Rev.* **68**, 278 (1945).
[20] M. L. Pool, *J. App. Phys.* **15**, 716 (1944).

rays which give monochromatic conversion electrons. A curie of electrons is almost 10^{-8} amperes and might prove useful as a source of electrons in scanning or transmission electron microscopes, special electronic tubes, etc. The internal conversion coefficient has not been measured in many substances and consequently more research will be necessary before it will be possible to select suitable sources from tables of the radioactive substances.

It should also be possible to prepare sources of monochromatic alpha particles. Polonium can be produced from bismuth by the following reactions:

$$_{83}Bi^{209} + n^{th} \longrightarrow {}_{83}Bi^{210} + \gamma \quad \sigma = .015 \times 10^{-24}\text{cm}^2 \quad [21]$$

$$_{83}Bi^{210} \longrightarrow {}_{84}Po^{210} + -\beta + 1.17 \text{ Mev} \quad [22]$$

$$_{84}Po^{210} \xrightarrow[139.5\text{d}]{} {}_{82}Pb^{206} + \alpha + 5.298 \text{ Mev} + \text{few gamma rays.} \quad [23]$$

Polonium is a well known source of alpha rays without appreciable gamma rays. It is used in calibrating ionization chambers, cloud chambers and as a reference standard for range and $H\rho$ measurements. It can also be used to produce neutrons by an (α, n) reaction. A mixture of beryllium and polonium gives neutrons of 11 Mev maximum energy with many neutrons of less than 2 Mev.[24] One curie of radon mixed with beryllium produces about 2.5×10^4 neutrons per second and Be (Po α, n) source gives somewhat less.

Production of Plutonium.—The production of plutonium in quantities of kilograms is the answer to the alchemists' fondest dreams. It might be pointed out that this production is made possible by the construction of enormous piles specifically designed for plutonium production, and by the large resonance capture cross section of uranium for neutrons initiating the series leading to the Pu isotope. This enormous production of plutonium is at the expense of many otherwise desirable features. The pile obviously is not self-replenishing and consequently produces plutonium at the expense of both U^{235} and U^{238}. This process depletes the available supply of U^{235} directly. Furthermore, the pile is water cooled and consequently would be an inefficient source of the by-product

[21] Segrè's chart.

[22] R. D. Evans, p. 656, *Medical Physics*, ed. O. Glasser, Year Book Publishers (1944).

[23] Rutherford, Chadwick and Ellis, p. 456, *Radiations from Radioactive Substances*, Cambridge University Press (1930).

[24] Bernardini, *Ricerca Scient.* 8, 33 (1937).

power which is now wasted. These wastes were obviously due to war haste but do not necessarily need to be continued. Our supply of uranium has been estimated to last 200 years, but Smyth (2.26) calls this optimistic. Other estimates are more pessimistic. Katzin calculates that there is not enough uranium available to supply all the power in the United States for one year.[25] Obviously, some policy of conservation and constructive use of available supplies of uranium and thorium are imperative.

The production of radioactive fission products is one useful by-product of the operation of a plutonium pile. The isotopes of Sr, Kr, Te, Xe and Ba produced as fission products should supplement the radioactive tracers already mentioned. The quantities produced should be enormous (page 150) and they are probably separated out either as gases or in the process of separation of plu-

Table 26

FISSION PRODUCTS OF LONG HALF-LIFE

Radio-active isotope	Half-life	Maximum β particle energy	Maximum gamma ray energy	Branching ratio in percent
Kr^{87}	>5 yr	0.8 Mev		
Sr^{90}	~30 yr	0.65	none	
Y^{91}	57 d	1.6	none	
Zr^{95}	65 d	1.0	0.92 Mev	
Cb or Nb^{95}	35 d	0.15	0.77	
Mo^{99}	67 hr	1.7	0.4	
43^{103}	62 d	*K* capture		
Ru^{103}	45 d	0.3	0.56	
Cd^{115}	43 d	1.7	0.5	
Sb^{127}	80 hr	1.2	0.72	0.18
Te^{127}	90 d	I.T.	0.8	0.18
I^{131}	8 d	0.6	0.37	
Xe^{133}	5.5 d	0.42	0.85	
Cs^{137}	27 yr	0.8	0.5	
Ba^{140}	300 hr	1.05	0.53	8.4
Ce^{141}	30 d	0.6	0.2	
Ce^{147}	275 d	0.348	1.25	
Pr^{143}	13.5 d	0.95	none	

[25] L. I. Katzin, *Postwar Industrial Uses of Atomic Energy*, Chicago (1945).

tonium or in recovery of uranium. Table 26 lists a few long-lived activities resulting from fission which might be useful as tracers or sources of radiation.

In connection with the fast neutron chain reaction for which plutonium has been used, some mention should be made of the possibility of enhancing the release of energy or, even further, of the possibility that such a release of energy could not be controlled. Since the temperatures and types of reactions involved are similar to those occurring in the stars, we can utilize the considerations summarized by Bethe.[26]

The danger of an uncontrolled release of energy is primarily due to the possibility of producing in the atmosphere a reaction such as

$${}_7N^{14}+{}_7N^{14} \longrightarrow {}_{14}Si^{28}+\gamma+29 \text{ mmu (milli mass units)}$$
$$\text{or} \longrightarrow {}_{12}Mg+\alpha+19 \text{ mmu}$$

At a temperature reached by a fast neutron induced chain reaction, the probability of such a reaction is estimated by Bethe's equation 7. Assuming a temperature of $T=50\cdot 10^6$ °A,[27] and a density of 10^{-3} gm/cm² we can calculate the number of reactions per gram per second. The particle reaction is more probable than the gamma emission reaction and can be calculated to occur at the rate of 10^{-80} per gram per second. Even at a pressure of 10^6 atmospheres which may accompany a shock wave in the atmosphere, the reaction rate is still 10^{-74}. This gives an energy release of 10^{-78} ergs /gm/sec which obviously won't propagate the reaction. Consequently, it seems impossible to achieve a chain reaction in the atmosphere at the estimated temperature of $50\cdot 10^6$ °A.[28]

However, on examining Bethe's Table V for more probable reactions, the D(*d,n*) reaction stands out as quite probable. That it does not occur in the energy cycle in the sun is attributed to the rarity of deuterium there. The rarity of deuterium on the earth would also eliminate this reaction in naturally occurring relative abundances. However, if pure deuterium or heavy water were added around the fissionable material in a fast neutron induced chain reaction, we can calculate the probability of the reaction:

$${}_1D^2+{}_1D^2 \longrightarrow {}_1He^3+{}_0n^1+3.5 \text{ mmu}$$

and its energy release. At a temperature of 50 million degrees

[26] H. A. Bethe, *Phys. Rev.* 55, 434 (1938).
[27] P. Morrison, Hearings before the Special Committee on Atomic Energy, United States Senate, Part 2, page 236, Dec. 5 (1945), testified that the explosion created a glowing ball one third of a mile across with a temperature about one hundred million degrees fahrenheit in the center.

and using the experimentally determined level width of $3 \cdot 10^5$ ev, the probability comes out $1.5 \cdot 10^{28}$ reactions per gm per sec. This seems adequate to propagate the chain reaction. The deuterium would be used up in 10^{-5} sec if the reaction goes to completion at the same rate. The release of all the available energy gives 10^{17} ergs per gm for deuterium. This compares with $2 \cdot 10^{20}$ ergs per gm released by a typical fission reaction. Consequently, (assuming ten percent efficiency) forty pounds of deuterium would double the energy release of the fast neutron induced chain reaction. The details of use, such as tamper, efficiency, deuterium compound, etc. require much more detailed consideration. However, it does seem feasible to add to a neutron induced fission chain reaction by a thermally induced chain reaction with deuterium.

The use of the fast neutron chain reaction seems primarily limited to war. As an explosive, the magnitude of the minimum energy release is so large and the attendant production of radioactive material so enormous that its use seems inadvisable for most purposes. The digging of a water level Panama Canal or melting the Polar ice cap would seem to be a wasteful and inefficient use of the precious fissionable material.

"Atomic" power production.—One of the tempting commercial applications of fission techniques is the production of power. Since the Hanford piles produce power at the rate of the Grand Coulee Dam (Smyth, 6.41), the use of piles as large central power stations is indicated. However, such installations will be most desirable for regions far from water, coal and oil.

The use of self replenishing piles which need U^{235}, U^{233} or Pu only to start it and which then uses U^{238} or Th is possibly the most economical way to run the power pile. From page 149 we can estimate that a large pile may contain twenty-five tons of uranium costing about $150,000.[29]

The energy produced by the Hanford pile could not be efficiently used since the temperature is not very high. A much more efficient conversion of this power would be achieved by running the pile at a higher temperature. Smyth (6.43) mentions the possibility of using bismuth as a coolant and heat interchanger. Since

[28] H. A. Bethe, Hearings before the Special Committee on Atomic Energy, United States Senate, Part 2, page 224, Dec. 5 (1945), testified that no temperature would "ignite" the earth's atmosphere and no achievable temperature would cause nuclear chain reactions in either the atmosphere or ocean.

[29] Improved methods of production have reduced the cost of metal uranium to $3 a pound according to a pamphlet on *Atomic Energy* by Robert E. Marshak, Eldred C. Nelson and Leonard I. Schiff, published by the University of New Mexico Press, March, 1946.

bismuth is a liquid between 271° and 1450°C, the pile could be operated at an elevated temperature and provide efficient conversion of heat to mechanical power.

On page 252, polonium was described as a product of exposing bismuth to neutrons. Estimating a minimum mass of bismuth for cooling 10^6 kw as 10^8 gms/sec or 10^{10} gms in the pile, we can estimate the rate of production of Po^{210} from a pile as $4 \cdot 10^{20}$ atoms per day or 0.1 gm. If this rate of production could be increased (by increasing the amount of coolant, etc.), the polonium itself could be used as an independent small power source. Since polonium emits 5.3 Mev alpha particles, it supplies power at the rate of 72 horse power per pound. Furthermore, since it does not emit appreciable gamma rays, the shielding necessary to protect individuals is nominal. However, such a power source could not be turned on or off and consequently would be useful only for continuously used special purpose power plants. The polonium has a half-life of 140 days and could therefore be separated from the end product, lead, and used again. Undoubtedly, piles specially constructed for polonium production would yield adequate amounts of polonium if its use proved desirable.

Considered purely as a fuel, uranium metal costing $20 a pound compares favorably with all other fuels for producing power. Costs of installation, upkeep and depreciation are difficult to estimate, however, and much development is necessary for economic use of fission as a power source.

Impress of fission on physics.—The most important impress on physics of fission work will be primarily in the number of physicists trained in nuclear techniques, the impetus and support given to nuclear research and the large number of tools of research developed and commercially manufactured. Such developments as mass spectroscopic analysis, leak hunting with a helium mass spectrometer, counting circuits which collect and record the electron pulse from ionizing particles in an ionization chamber, convenient vacuum gauges such as the Phillips ionization gauge, neutron counters utilizing thin films of U^{235} for slow neutrons and U^{238} for fast neutrons, and numerous other experimental and theoretical techniques will be of considerable use in nuclear physics. The increased medical knowledge of the dangers of radiations will also be of the utmost importance to nuclear science when it is released.

Fission has not affected fundamental knowledge appreciably. In fact, it may not long remain the most important field of nuclear research. The swing to the ultra-nuclear field of 100 Mev energy has already begun. The development of electron acceleration techniques will certainly profit from physicists' acquaintance with large scale and commercial development. Consequently, although fission research may not have revealed the fundamentals of nuclear forces, it may benefit the future of such research.

The impress of fission on the nation and the world is more difficult to assess. The most important aspect concerning physicists has been the universality with which the physicists, especially the atomic physicists, have awakened to the serious failure of political institutions to provide for the impact of newly discovered scientific facts. It is the physicists' faith that it is good to discover scientific facts. This is not yet the faith of our country nor the world. This conflict can be only partially resolved by the activities of the newly formed Federation of American Scientists. Obviously, we face a critical situation because of the threat of continued world conflict now made suicidal by the atom bomb. Consequently, the major effort at present must be to achieve some stable or developing world control for the atom bomb and, of course, for all war.

However, physicists must still remember that they are primarily scientists. If physicists believe that it is good to uncover scientific fact, then they must strive to keep conditions under which research is carried on as conducive as possible to intelligent action. It is only if research is absolutely unhampered by *any* restriction that it is truly scientific. It must *not* be the continuing responsibility of physicists as scientists to provide for the moral and intelligent use of scientific fact. It is the responsibility of physicists to pursue research in as an intelligent and scientific fashion as possible and to support other scientific endeavors by the most widespread and efficient dissemination of knowledge.

Partial Bibliography on Nuclear Fission and Transuranic Elements

(Not all these references were consulted by the authors and the references in the text are not all included in this bibliography.)

M. Ageno, E. Amaldi, D. Bocciarelli, B. Cacciapuoti and G. Trabacchi, *Ricerca Sci.* **11**, 302 (1940), "Fission of heavy elements."

M. Ageno, E. Amaldi, D. Bocciarelli and G. Trabacchi, *Ricerca Sci.* **11**, 413 (1940, "The decomposition of U with fast neutrons."

M. Ageno, E. Amaldi, D. Bocciarelli, B. Cacciapuoti and G. Trabacchi, *Phys. Rev.* **60**, 67 (1941), "Fission yield by fast neutrons."

E. Amaldi, D. Bocciarelli and G. Trabacchi, *Ricerca Sci.* **12**, 134 (1941), "The fission of thorium and protoactinium."

H. L. Anderson, *Phys. Rev.* **57**, 566 (1940), "Resonance capture of neutrons by uranium."

H. L. Anderson, E. Fermi and A. V. Grosse, *Phys. Rev.* **59**, 52 (1941), "Branching ratios in the fission of U."

B. Arakatsu, Y. Uemura, M. Sonoda, S. Shimizu, K. Kimura and K. Muraoka, *Proc. Phys. Math. Soc. Japan* **23**, 440 (1941), "Photo-fission of uranium and thorium produced by the γ-rays of lithium and fluorine bombarded with high-speed protons."

B. Arakatsu, M. Sonoda, Y. Uemura and S. Shimizu, *Proc. Phys. Math. Soc. Japan* **23**, 633 (1941), "Range of the photo-fission fragments of uranium produced by the γ-ray of lithium bombarded with protons."

C. J. Bakker, *Nederland. Tijdschr. Natuurkunde* **6**, 333 (1939), "Splitting of heavy atom nuclei by neutrons."

C. G. Bedreag, *Bull. Sect. Sci. Acad. Rowmaine* **25**, 410 (1943); *Chem. Zentr.* 1943, II, 2037, "Analytical tests for the uranides 93^{239}U; the f-electrons of the uranides 93, 94."

V. Berestetskiĭ and A. Migdal, *Compt. Rend. Acad. Sci. U.S.S.R.* **30**, 706 (1941) (in English), "Mechanism of nuclear fission."

J. K. Bøggild, *Phys. Rev.* **60**, 827 (1941), "Range/velocity relation for fission fragments in helium."

J. K. Bøggild, K. J. Brostrøm and E. Lauritsen, *Kgl. Danske Vide Euskab, Seleskab, Math. fus. Medd.* **18**, No. 4, 32 pp (1940), "Cloud chamber studies of fission-fragment tracks."

J. K. Bøggild, K. J. Brostrøm and T. Lauritsen, *Phys. Rev.* **59**, 275 (1941), "Range and straggling of fission fragments."

N. Bohr, *Uspekhi. Khim.* **8**, 544 (1939), "Disintegration of heavy nuclei."

N. Bohr, *Phys. Rev.* **58**, 654 (1940), "Scattering and stopping of fission fragments."

N. Bohr, *Phys. Rev.* **58**, 864 (1940), "Successive transformations in nuclear fission."

N. Bohr, *Tek. Ukeblad* **87**, 199 (1940), "Utilization of atomic energy."

N. Bohr, *Phys. Rev.* **59**, 270 (1941), "Velocity/range relation for fission fragments."

N. Bohr, *Phys. Rev.* **59**, 1042 (1941), "Mechanism of deuteron-induced fission."

N. Bohr and J. A. Wheeler, *Phys. Rev.* **55**, 1124 (1939), "Mechanism of nuclear fission."

N. Bohr, J. K. Bøggild, K. J. Brostrøm and T. Lauritsen, *Phys. Rev.* **58**, 839 (1940), "Velocity/range relation for fission fragments."

N. Bohr and F. Kul'kar, *Uspekhi Fiz. Nauk* **20**, 319 (1938), "Transformations of atomic nuclei produced by collisions with material particles."

E. T. Booth, J. R. Dunning and F. G. Slack, *Phys. Rev.* **55**, 1124, "Fission of uranium and production of delayed emission by slow neutron bombardment."

E. T. Booth, J. R. Dunning, A. V. Grosse, and A. O. Nier, *Phys. Rev.* **58**, 476, (1940), "Neutron capture by uranium (238)."

H. J. Born and W. Seelman-Eggebert, *Naturwiss.* **31**, 86 (1943), "Identification of uranium decomposition products with corresponding isotopes produced by (n) and (np) processes."

H. J. Born and W. Seelmann-Eggebert, *Naturwiss.* **31**, 201 (1943); C. A. **38**, 195, "Maximum energy of the β-rays of isotopes obtained in uranium disintegration."

H. J. Born and W. Seelmann-Eggebert, *Naturwiss.* **31**, 420 (1943), "New decomposition products from irradiation of U with fast neutrons (Ru, Rh)."

W. Bothe and W. Gentner, *Z. Phys.* **119**, 9-10, 568 (1942), "The energy limit of the neutrons produced by the fission of uranium."

W. Bothe and A. Flammersfeld, *Naturwiss.* **29**, 194 (1941), "The 18-minute-molybdenum from uranium."

H. Bradt, *Helv. Phys. Acta* **12**, 553 (1939), "Neutrons from U."

K. J. Brøstrom, J. K. Bøggild and T. Lauritsen, *Phys. Rev.* **58**, 651 (1940), "Cloud-chamber studies of fission fragment tracks."

G. Cocconi, *Nuovo Cimento* **16**, 417 (1939); *Chem. Zentr.* **1940**, I, 979, "Fission of heavy atomic nuclei."

A. W. Coven, *Phys. Rev.* **68**, 279 (1945), "Evidence of increased radio-activity of the atmosphere after the atomic bomb test in New Mexico."

W. Czulias and J. Shintlmeiter, *Nature* **143**, 911 (1939), Note in *Nature* from Academy of Sciences of Vienna, February 2.

K. K. Darrow, *Bell System Tech. Jour.* **19**, 267 (1940), "Nuclear fission."

K. K. Darrow, *Proc. Am. Phil. Soc.* **82**, 351 (1940), "Status of (atomic) nuclear theory."

K. K. Darrow, *Science* **91**, 514 (1940), "Nuclear fission."

G. Dessauer and M. Hafner, *Phys. Rev.* **59**, 840 (1941), "Proton-induced fission."

R. W. Dodson and R. D. Fowler, *Phys. Rev.* **57**, 966 (1940), "Products of U fission, radioactive isotopes of I and Xe."

N. Feather, *Nature* **143**, 877 (1939), "Fission of heavy nuclei: a new type of nuclear disintegration."

E. Fermi, *Nature* **146**, 640 (1940), "Reactions produced by neutrons in heavy elements."

E. Fermi and E. Segrè, *Phys. Rev.* **59**, 680 (1941), "Fission of uranium by alpha particles."

B. Ferretti, *Ricerca Sci.* **10**, 471 (1939), "Cleavage of uranium."

A. Flammersfeld, P. Jensen and W. Gentner, *Zeits. für Physik.* **120**, 450 (1943), "Partition relations and energy changes for uranium fission."

Flerov and Petrjak, *Phys. Rev.* **58**, 89 (1940), "Spontaneous fission of uranium."

S. Flügge, *Naturwiss.* **29**, 462 (1941), "The production of naturally radioactive elements in an artificial way."

S. Flügge, *Zeits. für Physik* **121**, 298 (1943), "Spontaneous fission of U and its neighboring elements."

L. S. Foster, *J. Chem. Education* **17**, 448 (1940), "Bombardment of uranium with fast and slow neutrons."

I. M. Frank, *Priroda* **9**, 20 (1939); *Khim Referat. Zhur.* **5**, 1 (1940), "A new type of nuclear reaction."

T. Franzini and M. Galli, *Ricerca Sci.* **12**, 1157 (1941); *Chem. Zentr.* **1**, 1595 (1942), "Nuclear cleavage of uranium."

T. Franzini and V. Ricca, *Atti. Accad. Italia, Rend. Classe Sci. Fis. Mat. Nat.* **3**, 247 (1941); *Chem. Zentr.* **1**, 6 (1943), "The production of Xe by the bombardment of U with neutrons."

Ya. I. Frenkel, *J. Phys.* (*USSR*) **1**, 125 (1939), "Electrocapillary theory of the splitting of heavy elements by slow neutrons."

Ya. L. Frenkel, *J. Exptl. Theoret. Phys.* (*U.S.S.R.*), **9**, 641 (1939) "Electrocapillary theory of the splitting of heavy nuclei by slow neutrons."

D. H. T. Gant and R. S. Krishnan, *Proc. Roy. Soc. A.* **178**, 474 (1941), "Deuteron-induced fission in uranium and thorium."

G. N. Glasoe and J. Steigman, *Phys. Rev.* 58, 1 (1940), "Radioactive products from gases produced in U fission."

L. Goldstein, A. Rogozinski and R. J. Walen, *de Physique et le Radium* **10**, 477 (1939), "Interaction of fast neutrons with U nuclei."

I. N. Golovin, *Uspekhi Khim.* 8, 529 (1939), "Splitting of the uranium and thorium nuclei into two complex nuclei."

I. N. Golovin, *Uspekhi Khim.* 8, 1923 (1939), "Splitting of the nuclei of uranium and thorium."

Hans Götte, *Naturwiss.* **28**, 449 (1940), "A nuclear isomer of xenon appearing in U disintegration."

Hans Götte, *Naturwiss.* **29**, 496 (1941), "New strontium and ytterbium isotopes from uranium."

A. Grosse, *J. Amer. Chem. Soc.* **57**, 440 (1935), "The chemical properties of elements 93 and 94."

A. Grosse et al., *Phys. Rev.* **46**, 241 (1934), "The chemistry of element 93 and the Fermi effect."

A. V. Grosse and E. T. Booth, *Phys. Rev.* **57**, 664 (1940), "Radioactive zirconium and columbium from uranium fission."

A. V. Grosse, E. T. Booth and J. R. Dunning, *Phys. Rev.* **59**, 322 (1941), "The fourth (4n+1) radioactive series."

E. Haggstrom, *Phys. Rev.* **59**, 322 (1941), "The β-ray spectrum of „Ekatantalum[233]."

T. Hagiwara, *Rev. Phys. Chem. Japan* **13**, 145 (1939); *Mem. Coll. Sci., Kyoto Imp. Univ.* A**23**, 19 (1940), "Liberation of neutrons in the nuclear explosion of uranium irradiated by thermal neutrons."

O. Hahn, *Ann. Physik* **36**, 368 (1939), "Some peculiarities of the atoms from the nuclear fission of uranium and thorium."

O. Hahn, *Scientia* **68**, 8 (1941), "Intranuclear processes and the splitting of the uranium atom."

O. Hahn, *Chem. Ztg.* **66**, 317 (1942), "The experimental unraveling of the elements and kinds of atoms arising from the fission of U."

O. Hahn, *Jena Z. Med. Naturw.* **76**, 36 (1943); *Chem. Zentr.* **1**, 2175 (1943), "Artificial atom transformation and the splitting of heavy nuclei."

O. Hahn and F. Strassmann, *Abhandl. Preuss. Akad. Wiss., Physik. Math. Klasse,* 1939, No. **12**, 3-20; *Chem. Zentr.* 1940, **1**, 8, "The 'exploding' of the uranium nucleus by the action of slow neutrons."

O. Hahn and F. Strassmann, *Phys. Zeits.* **40**, 673 (1939), "The collapse of U and Th nuclei into lighter atoms."

O. Hahn and F. Strassmann, *Naturwiss.* 28, 54 (1940), "New method for separation of active fission-products from U and Th."

O. Hahn and F. Strassmann, *Naturwiss.* 28, 61 (1940), "Use of the 'emanating ability' of thorium hydroxide for recovery of decomposition products of thorium."

O. Hahn and F. Strassmann, *Naturwiss.* **28**, 455 (1940), "Separate recovery of the krypton and xenon isotope fractions formed in U disintegrations."

O. Hahn and F. Strassmann, *Naturwiss.* **28**, 543 (1940), "Further products of uranium disintegration."

O. Hahn and F. Strassmann, *Forschungen u. Fortschr.* **16**, 31 (1940); *Chem. Zentr.* **1**, 1795 (1940), "The splitting of uranium into lighter atoms through the action of neutrons."

O. Hahn and F. Strassmann, *Naturwiss.* **29**, 285 (1941), "Formation of zirconium and protactinium by bombardment of thorium with neutrons."

O. Hahn and F. Strassmann, *Naturwiss.* **29**, 369 (1941), "On the molybdenum isotopes from the fission of U."

O. Hahn and F. Strassmann, *Z. Physik* **117**, 789 (1941), "Molybdenum isotopes from uranium fission."

O. Hahn and F. Strassmann, *Naturwiss.* **30**, 324 (1942), "The short-lived barium and lonthanum isotopes from uranium fission."

O. Hahn and F. Strassmann, *Z. Physik* **121**, 729 (1943), "The strontium and yttrium isotopes appearing in uranium fission."

O. Hahn and F. Strassmann, *Naturwiss.* **31**, 249 (1943), "Active Sr and Y isotopes from U disintegration."

O. Hahn and F. Strassmann, *Naturwiss.* **31**, 499 (1943), "New disintegration products of U."

H. Halban Jr., F. Joliot, L. Kowarski and F. Perrin, *J. de Physique et le Radium* **10**, 428 (1939), "Evidence for a chain nuclear reaction in the middle of a uraniferous mass."

P. Havas, *J. de Physique et le Radium* **1**, 146 (1940), "Slowing down of heavy ions by matter. Application to the rupture of U."

S. Haynes, *Phys. Rev.* **59**, 834 (1941), "Effect of neutron energy on the total decay curves of fission products."

R. O. Haxby, W. E. Shoupp, W. E. Stephens and W. H. Wells, *Phys. Rev.* **57**, 1088 (1940), "Thorium fission threshold."

R. O. Haxby, W. E. Shoupp, W. E. Stephens and W. H. Wells, *Phys. Rev.* **58**, 92 (1940), "Photo-fission of uranium and thorium."

R. O. Haxby, W. E. Shoupp, W. E. Stephens and W. H. Wells, *Phys. Rev.* **58**, 199 (1940), "Fast neutron threshold for uranium fission."

R. O. Haxby, W. E. Shoupp, W. E. Stephens and W. H. Wells, *Phys. Rev.* **59**, 57 (1941), "Photo-fission of U and Th."

Malcolm C. Henderson, *Phys. Rev.* **58**, 200 (1940), "The heat of fission of uranium."

Malcolm C. Henderson, *Phys. Rev.* **58**, 774 (1940), "Heat of fission of U."

P. Huber, *Schweiz. Arch. angew. Wiss. Techn.* **10**, 97 (1944), "Drop model of the atomic nucleus."

J. C. Jacobsen and N. O. Lassen, *Phys. Rev.* **58**, 867 (1940), "Deuteron-induced fission in U and Th."

J. C. Jacobsen and N. O. Lassen, *Phys. Rev.* **59**, 1043 (1941), "Fission cross section in uranium and thorium for deuteron impact."

W. Jentschke and F. Prankl, *Phys. Zeits.* **40**, 706 (1939), "Nuclear disintegration products of U bombarded with neutrons."

W. Jentschke, F. Prankl and F. Hernegger, *Naturwiss.* **28**, 315 (1940), "Disintegration of ionium by neutron irradiation."

W. Jentschke and F. Prankl. *Sitzber. Akad. Wiss. Wien, Math. Naturw. Klasse,* Abt. IIa **148**, 237 (1940), "Further investigations on the decomposition of uranium and thorium on neutron irradiation."

W. Jentschke and F. Prankl, *Z. Physik* **119**, 696 (1942), "The energies and masses of the U nuclear fragments produced by irradiation with preponderantly thermal neutrons."

W. Jentschke, F. Prankl and F. Hernegger, *Sitzber. Akad. Wiss. Wien, Math. Naturwiss. Klasse* Abt. IIa **151**, 147 (1942); *Chem. Zentr.* **1**, 1547 (1943), "Proof of nuclear fission of ionium under neutron irradiation."

W. Jentschke and F. Prankl, *Wien. Chem. Ztg.* **45**, 267 (1942), "Nuclear fission of the heaviest elements by irradiation with neutrons."

W. Jentschke, *Z. Physik* **120**, 165 (1943), "Energies and masses of the uranium fission products on irradiation with neutrons."

I. Joliot-Curie and Tsien San-Tsiang, *J. de Physique et le Radium* **10**, 495 (1939), "Comparison of radiation of radioactive isotopes formed from U and Th."

I. Joliot-Curie and P. Savich, *Uspekhi Khim.* **8**, 540 (1939), "Radioelements formed in uranium and thorium irradiated by neutrons."

I. Joliot-Curie and F. Joliot, *Ann. Phys.* **19**, 107 (1944), "The fission of ionium by neutrons."

F. Joliot, *Uspekhi Khim.* **8**, 537 (1939), "Experimental evidence for the explosive disintegration of uranium and thorium nuclei under neutron bombardment."

F. Joliot, *Bull. Soc. Franc. Elec.* **2**, 13 (1942), "Neutrons and artificial radioactivity."

F. Joliot, *Compt. Rend.* **218**, 488 (1944), "A method of measuring the ranges of radioelements of known chemical nature coming from the fission of uranium.

M. H. Kanner and H. H. Barschall, *Phys. Rev.* **57**, 372 (1940), "Distribution in energy of fragments from U fission."

V. G. Khlopin, M. A. Pasvik-Khlopina and N. F. Volkov, *Nature* **144**, 595 (1939), "A particular mode of fission of the uranium nucleus."

V. G. Khlopin, M. A. Pasvik-Khlopina and N. F. Volkov, *Compt. Rend. Acad. Sci. U.S.S.R.* **24,** 118 (1939), "Division of uranium nuclei under the action of neutrons and the problem of the existence of trans-uraniums."

V. G. Khlopin, *Bull. Acad. Sci. U.S.S.R., Ser. Phys.* **4,** 305 (in English, 309) (1940), "Chemical nature of uranium fission products."

K. H. Kingdon, H. C. Pollock, E. T. Booth and J. R. Dunning, *Phys. Rev.* **57,** 749 (1940), "Fission of the separated isotopes of uranium."

I. V. Kurchatow, *Uspekhi Fiz. Nauk* **25,** 159 (1941), "Fission of uranium and thorium under deuteron bombardment."

R. S. Krishman and E. A. Nahum, *Proc. Roy. Soc.* (*London*) *A***180,** 321 (1942), "Cross section measurements for disintegrations produced by deuterons in the heavy elements."

R. S. Krishman and E. A. Nahum, *Proc. Roy. Soc.* *A***180,** 333 (1942), "Excitation function measurements for disintegrations produced by deuterons in the heavy elements."

I. V. Kurchatow, *Uspekhi Fiz. Nauk* **25,** 159 (1941). "Fission of heavy nuclei."

W. E. Lamb, Jr., *Phys. Rev.* **58,** 696 (1940), "Passage of uranium fission fragments through matter."

W. E. Lamb, Jr., *Phys. Rev.* **59,** 687 (1941), "Range of Fission Fragments."

Alois Langer and W. E. Stephens, *Phys. Rev.* **58,** 759 (1940), "Radioactive barium and strontium from photo-fission of uranium."

A. Langsdorf, Jr., and A. Segrè, *Phys. Rev.* **57,** 105 (1940), "Nuclear isomerism in Se and Kr."

K. Lark-Horovitz and W. A. Miller, *Phys. Rev.* **59,** 941 (1941), "Fission tracks on the photographic plate."

K. Lark-Horovitz and R. E. Schreiber, *Phys. Rev.* **60,** 156 (1941), "Uranius fission with Li-D neutrons: energy distribution of the fission fragments."

N. O. Lassen, *Phys. Rev.* **68,** 142 (1945), "Hρ—distribution of fission fragments."

N. O. Lassen, *Phys. Rev.* **68,** 230 (1945), "Ionization of fission fragments in nitrogen, argon and xenon."

A. I. Leipunskii, *Bull. Acad. Sci. U.S.S.R., Ser. Phys.* **4,** 291 (1940), "Fission of the nuclei."

A. I. Leipunskii and Maslov, *Compt. Rend. Acad. Sci. U.S.S.R.* **27,** 783 (1940), "The fission of uranium nuclei caused by the capture of slow neutrons."

A. I. Leipunskii, *Visti Akad. Nauk U.S.S.R.* **8,** 61 (1941); *Chem. Zentr.* 1942, 1, 158, "The disintegration of uranium."

W. F. Libby, *Phys. Rev.* **55,** 1269 (1939), "Stability of uranium and thorium for natural fission."

C. Maghan, *Compt. Rend.* **214,** 110 (1942), "Energy balances for different modes of fission of the uranium nucleus under the action of neutrons."

V. Majer, *Chem. Listy* **35,** 23 (1941); *Chem. Zentr.* **1941,** 2, 305, "Emanation capacity and the active precipitate of U."

W. Maurer and H. Pose, *Zeits. f. Physik,* **121,** 285 (1943), "Neutron emission of the U nucleus as a result of its spontaneous fission."

M. G. Mayer, *Phys. Rev.* **60,** 184 (1941), "Rare-earth and transuranic elements."

McMillan and Abelson, *Phys. Rev.* **57,** 1185 (1940), "Radioactive Element 93".

E. McMillan, *Phys. Rev.* **58,** 178 (1940), "The seven day uranium activity."

L. Meitner and O. R. Frisch, *Kgl. Danske Videnskab. Selskab. Math.-fys. Medd.* **17,** 14 (1939), "Products of the fission of uranium and thorium under neutron bombardment."

L. Meitner, *Nature* **145,** 422 (1940), "Capture cross sections for thermal neutrons in thorium, lead and uranium 238."

L. Meitner, *Phys. Rev.* **60**, 58 (1941), "The resonance energy of the thorium capture process."

J. Mende, *Pôtfüzetek Természettudományi Közlönghöz* **70**, 43 (1938), *Chem. Zentr.* 1939, **1**, 328, "Artificial shattering of the atom of uranium."

A Moussa and L. Goldstein, *Compt. Rend.* **212**, 986 (1941), Radioactive bromine isotopes produced by the fission (forced disintegration) of uranium."

A. Moussa and L. Goldstein, *Phys. Rev.* **60**, 534 (1941), "Radioactive bromine isotopes from uranium fission."

A. O. Nier, E. T. Booth, J. R. Dunning and A. V. Grosse, *Phys. Rev.* **57**, 546 (1940), "Nuclear fission of separated uranium isotopes."

A. O. Nier, E. T. Booth, J. R. Dunning and A. V. Grosse, *Phys. Rev.* **57**, 748 (1940), "Further experiments on fission of separated uranium isotopes."

Y. Nishina, T. Yasani, H. Ezoe, K. Kimura and M. Ikawa, *Nature* **146**, 24 (1940), "Fission products of uranium produced by fast neutrons."

Y. Nishina, T. Yasaki, H. Ezoe, K. Kimura and M. Ikawa, *Phys. Rev.* **57**, 1182 (1940), "Induced β - activity of uranium by fast neutrons."

Y. Nishina, T. Yasaki, K. Kimura and M. Ikawa, *Phys. Rev.* **58**, 660 (1940), "Fission products of uranium by fast neutrons."

Y. Nishina, T. Yasaki, K. Kimura and M. Ikawa, *Phys. Rev.* **59**, 323 (1941); *Phys. Rev.* **59**, 677 (1941), "Fission products of uranium by fast neutrons."

Y. Nishina, K. Kimura, T. Yasaki and M. Ikawa, *Z. Phys.* **119**, 195 (1941); *Phys. Rev.* **59**, 677 (1941), "Fission products of uranium by fast neutrons."

I. Noddack, *Naturwiss.* **27**, 212 (1939), "Remarks on the work of O. Hahn, L. Meitner and F. Strassmann on the products formed in irradiation of uranium with neutrons."

I. S. Panasyuk and G. N. Flerov, *Compt. Rend. Acad. Sci. U.S.S.R.* **30**, 704 (1941), "Spontaneous fission of thorium."

M. A. Pasvik-Khlopina and V. G. Khlopin, *Akad. V. I. Vernadskomu Pyatidessyatiletiyu Nauch. Dey atelnosti* **1**, 539 (1936); *Chem. Zentr.* 1938, **1**, 1533, "The chemical nature of artificial transformation products of uranium."

N. A. Perfilov, *Compt. Rend. Acad. Sci. U.S.S.R.* **23**, 896-8 (1939) (in French), "Disintegration of uranium by neutrons."

N. A. Perfilov, *Compt. Rend. Acad. Sci. U.S.S.R.* **28**, 426 (1940, "Relation mv/E for products of uranium fission."

N. A. Perfilov, *Bull. Acad. Sci. U.S.S.R., Ser. Phys.* **4**, 300 (1940), "Observation of tracks of recoiling nuclei a."

N. A. Perfilov, *Compt. Rend. Acad. Sci. U.S.S.R.* **33**, 485 (1941) (in English), "The spectrum of β-rays from I^{133}."

K. A. Petrzhak, *Compt. Rend. Acad. Sci. U.S.S.R.* **27**, 209 (1940), "Ranges and energies of fragments in fission of uranium by fast neutrons."

K. A. Petrzhak and G. N. Flerov, *J. Exper. Theoret. Phys. (U.S.S.R.)*, **10**, 1013 (1940), "Spontaneous fission of uranium."

K. A. Petrzhak and G. N. Flerov, *Uspekhi Fiz. Nauk* **25**, 171 (1941), "Spontaneous fission of uranium."

G. Placzek, *Phys. Rev.* **60**, 166 (1941), "Diffusion of thermal neutrons."

M. S. Plesset, *Amer. J. Phys.* **9**, 1 (1941), "On the classical model of nuclear fission."

A. Polesitskiĭ and K. A. Petrzhak, *Compt. Rend. Acad. Sci. U.S.S.R.* **24**, 854 (1939) (in English), "An attempt to detect the formation and fission of uranium nuclei."

I. A. Polesitskiĭ and M. Orkeli, *Compt. Rend. Acad. Sci. U.S.S.R.* **28**, 215 (1940), "Chemical nature of the radioactive fragments of thorium fission. Radioactive halogens."

L. B. Ponizovskiĭ, *Doklady. Akad. Nauk. U.S.S.R.* **40**, 59 (1943); *Compt. Rend. Acad. Sci. U.S.S.R.* **40**, 50 (1943), "Radioactive families."

L. Ponsovsky, *Nature (London)* **152**, 187 (1943), "Radioactivity and the completion of the periodic system."

M. L. Pool, *Phys. Rev.* **65**, 353 (1944), "Potential nuclear monochromatic electron sources."

M. L. Pool, *Phys. Rev.* **65**, 357 (1944), "Spontaneous neutron emission." (Title only.)

H. Pose, *Zeits. f. Physik* **121**, 293 (1943), "Spontaneous neutron emission of U. and Th."

R. D. Present and J. K. Knipp, *Phys. Rev.* **57**, 751 (1940); **57**, 1188 (1940), "On the dynamics of complex fission."

R. D. Present, *Phys. Rev.* **59**, 466 (1941), "Possibility of ternary fission."

M. Piocopio, *Chim. Ind. Agr. Biol.* **15**, 803 (1939), "Trans-uranium elements."

L. I. Rusinov and G. N. Flerov, *Bull. Acad. Sci. U.S.S.R., Ser. Phys.* **4**, 310 (1940), "Experiments on fission of uranium."

L. I. Rusinov, *Uspekhi Khim.* **10**, 662 (1941), "Spontaneous disintegration of uranium."

G. T. Seaborg, J. W. Gofman and J. W. Kennedy, *Phys. Rev.* **59**, 321 (1941), "Radioactive isotope of protoactinium."

G. T. Seaborg. *Rev. Mod. Phys.* **16**, 1 (1944), "Table of isotopes."

W. Seelmann-Eggebert, *Naturwiss.* **31**, 491 (1943), "Active Xe isotopes."

E. Segrè and C. S. Wu, *Phys. Rev.* **57**, 552 (1940), "Some fission products of uranium."

E. Segrè and G. T. Seaborg, *Phys. Rev.* **59**, 212 (1941), "Fission products of uranium and thorium produced by high energy neutrons."

E. V. Shpolskiĭ, *Uspekhi Fiz. Nauk.* **21**, 253 (1939), "Artificial disintegration of heavy nuclei."

Kurt Starke, *Naturwiss.* **30**, 107 (1942), "Separation of element 93."

Kurt Starke, *Naturwiss.* **30**, 577 (1942), "Enrichment of the artificial radioactive uranium isotope $_{92}U^{239}$ and its product $_{93}$ (element $93)^{239}$."

A. Stettbacher, *Tech. Ind. Schweiz. Chem. Ztg.* **25**, 115 (1942), "Power of atomic disintegrations nuclear-transformation explosions."

F. Strassmann and O. Hahn, *Naturwiss.* **28**, 817 (1940), "Short-lived bromine and iodine isotopes from uranium disintegration."

F. Strassmann, *Angew. Chem.* **54**, 249 (1941), "The splitting of heavy nuclei."

F. Strassmann and O. Hahn, *Naturwiss.* **30**, 256 (1942), "The isolation and some of the properties of element 93."

T. Takenti, *Bull. Tokyo Univ. Eng.* **8**, 285-6 (1939); *Rev. Phys. Chem. Japan* **13**, 189, "Photographic demonstration of uranium fission."

J. Thibaud, *Scientia* **66**, 11 (1939), "A new aspect of atomic transmutation; splitting of the nucleus of uranium."

L. A. Turner, *Phys. Rev.* **57**, 157 (1940), "The nonexistence of transuranic elements."

L. A. Turner, *Phys. Rev.* **57**, 334 (1940), "Secondary neutrons from uranium."

L. A. Turner, *Phys. Rev.* **58**, 181 (1940), "Regularities among the heavy nuclei."

L. A. Turner, *Rev. Mod. Phys.* **12**, 1 (1940), "Nuclear fission."

L. A. Turner, *Phys. Rev.* **57**, 950 (1940), "The missing heavy nuclei."

L. A. Turner, *Rev. Mod. Phys.* **17**, 292 (1945), "The missing heavy nuclei."

Kwai Umeda, *Proc. Phys. Math. Soc. Japan* **22**, 257 (1940), "Nuclear fission."

W. D. Urry, *Amer. J. Sci.* **239**, 191 (1941), "Radioactive determination of small amounts of U."

N. F. Volkov, *Compt. Rend. Acad. Sci. U.S.S.R.* **24**, 528 (1939) (in English), "Fission of uranium nuclei."

H. Volz, *Zeit. f. Phys.* **121**, 201 (1943), "Absorption cross sections for slow neutrons."

T. F. Wall, *Engineering* **160**, No. 4153, 134 (1945), "Atomic fission and the cyclotron."

L. Wertenstein, *Nature* **144,** 1045 (1939), "Radioactive gases evolved in uranium fission."

G. L. Weil, *Phys. Rev.* **60,** 167 (1941), "β-spectra of some uranium fission products."

C. S. Wu, *Phys. Rev.* 58, 926 (1940), "Identification of two radioactive xenons from uranium fission."

Chien-Shiung Wu and Emileo Segrè, *Phys. Rev.* **67,** 142 (1945), "Radioactive xenons."

T. Yasaki, *Inst. Phys. and Chem. Research, Tokyo, Sci. Papers* **980,** 457 (1940), "Fission products and induced β-ray radioactivity of uranium by fast neutrons."

Zel'dovich and Khariton, *J. Exptl. Theoret. Phys. U.S.S.R.* **9,** 1425 (1939), "Chain disintegration of the abundant uranium isotope."

Ya. B. Zel'dovich and Yu. B. Khariton, *J. Exptl. Theoret. Phys. U.S.S.R.* **10,** 29 (1940), "Chain-forming disintegration of uranium under the action of slow neutrons."

Ya. B. Zel'dovich and Yu. B. Khariton, *J. Exptl. Theoret. Phys. U.S.S.R.* **10,** 477 (1940), "Kinetics of chain disintegration of uranium."

Ya. B. Zel'dovich and Yu. A. Zysin, *J. Exptl. Theoret. Phys. U.S.S.R.* **10,** 831 (1940), "Theory of nuclear fission."

Ya. B. Zel'dovich and Yu. B. Khariton, *Uspckhi Fiz. Nauk* **23,** 329 (1940), "Fission and chain decomposition of uranium."

A. P. Zhdanov and L. V. Mysovskiĭ, *Compt. Rend. Acad. Sci. U.S.S.R.* **23,** 135 (1939), "The observation of recoil nuclei from the neutron bombardment of uranium."

A. P. Zhdanov, L. V. Mysovskiĭ and M. Mysovskaya, *Compt. Rend. Acad. Sci. U.S.S.R.* **23,** 388 (1939), "Tracks of the recoil nuclei of the disintegration of uranium by neutrons."

A. P. Zhdanov and L. V. Mysovskiĭ, *Compt. Rend. Acad. Sci. U.S.S.R.* **25,** 9 (1939), "The observation of recoil nuclei from the disintegration of uranium."

SYMPOSIUM

Symposium on Atomic Energy and Its Implications, Proceedings of the American Philosophical Society, vol. 90, no. 1, Jan. 29, 1946.

H. D. Smyth, "Fifty Years of Atomic Physics."
J. R. Oppenheimer, "Atomic Weapons."
R. S. Stone, "Health Protection Activities of Plutonium Project."
E. Fermi, "Development of the First Chain Reacting Pile."
E. P. Wigner, "Resonance Reactions."
H. C. Urey, "Methods and Objectives in Separation of Isotopes."
J. A. Wheeler, "Problems and Prospects in Elementary Particle Research."
J. H. Willits, "Social Adjustments to Atomic Energy."
J. Viner, "The Implications of the Atomic Bomb for International Relations."
J. T. Shotwell, "The Control of Atomic Energy under the Charter."
I. Langmuir, "World Control of Atomic Energy."
A. H. Compton, "Atomic Energy as a Human Asset."

BOOKS

H. D. Smyth, "Atomic Energy for Military Purposes," Princeton University Press, 1945.

J. Dement and H. C. Dake, "Uranium and Atomic Power," Chemical Publishing Co., Inc., Brooklyn, N.Y., 1945.

D. Dietz, "Atomic Energy in the Coming Era," Dodd, Mead and Co., 1945.

E. Fermi, et al, "Nuclear Physics," University of Pennsylvania Press, Philadelphia, 1941.

J. T. O'Neill, "Almighty Atom, The Real Story of Atomic Energy," Ives Washburn, Inc., New York, 1945.

Supplemental Bibliography

(The references in this list were added after the manuscript of this book was first submitted but include nothing published after October, 1947)

Ageno, M., *Nuovo Cim.* **3**, 3-14 (1946), The albedo of slow neutrons.

Ageno, M., *Phys. Rev.* **69**, 241-2 (1946), The albedo of thermal neutrons.

Ageno, M., Amaldi, E., Bocciarelli, D., Trabacchi, G. C., *Gazz. Chim. Ital.* **75**, 3-33 (1945), Collision of neutrons with protons and deuterons.

Ageno, M., Amaldi, E., Bocciarelli, D., Trabacchi, G. C., *Phys. Rev.* **71**, 20-31 (1947), On the scattering of fast neutrons by protons and deuterons.

Akers, W., *Nature* **160**, 182 (1947), Metallurgical problems involved in the generation of useful power from atomic energy.

Akhieser, A., Pomeranchuk, I., *Journ. of Phys. U.S.S.R.* **11**, 167-178 (1947), On the scattering of slow neutrons in crystals.

Akhieser, A., Pomeranchuk, I., *Journ. of Phys. U.S.S.R.* **9**, 471-476 (1945), On the elastic scattering of fast charged particles by nuclei.

Akhieser, A., Pomeranchuk, I., *Journ. of Phys. U.S.S.R.* **9**, 461 (1945), On the scattering of low energy neutrons in helium II.

Alexander, Jerome, *Colloid Chemistry* **6**, 1157-73 (1946), Potential nuclear energy and some consequences of its release, including the atomic bomb.

Allen, K. W., Burchan, W. F., and Wilkinson, D. H., *Nature*, London **159**, 473-4 (1947), Interaction of fast neutrons with beryllium and aluminum.

Amaldi, E., Bocciarelli, D., Cacciapuoti, B. N., Trabacchi, G. C., *Nuovo Cim.* ser. 9, **3**, 203-34 (1946), On the elastic scattering of fast neutrons by medium and heavy nuclei.

Amaldi, E., *Ric. Sci.* **15**, 213-18 (1945), Utilization of nuclear energy.

Amaldi, E., Bocciarelli, D., Trabacchi, G. C., *Ric. Sci.* **12**, 830-42 (1941), On proton-neutron collision. I.

Amaldi, E., Bocciarelli, D., Cacciapuoti, B. N., Trabacchi, G. C., *Nuovo Cim.* **3**, 15-21 (1946), Diffraction effects in the scattering of fast neutrons.

Amaldi, E., Bocciarelli, D., Ferretti, B., Trabacchi, G. C., *Ric. Sci.* **13**, 502-31 (1944), On proton-neutron collision. II.

Amaldi, E., Bocciarelli, D., Trabacchi, G. C., *Ric. Sci.* **12**, 134-8 (1941), On the fission of thorium and protoactinium.

Anderson, H. L., Fermi, E., Marshall, L., *Phys. Rev.* **70**, 815-17 (1946), Production of low energy neutrons by filtering through graphite.

Anderson, H. L., Fermi, E., Wattenberg, A., Weil, G. L., Zinn, W. H., *Phys. Rev.* **72**, 16-23 (1947), Method for measuring neutron absorption cross sections by the effect on the reactivity of a chain reacting pile.

Anderson, J. S., *Australian Chem. Inst. J., and Proc.* **13**, 182-98 (1946), Atomic energy and the chemist.

Anderson, O. E., White, H. E., *Phys. Rev.* **71**, 911 (1947), Hyperfine structure and the nuclear spin of U^{235}.

Atomic Scientists of Chicago, "The Atomic Bomb", 62pp. (1946).

Atomic Energy Act—1946, A bill for the control and development of atomic energy. Wash., D. C., U. S. Gov't. Printing Office.

Atomic Power, McGraw Hill, Vol. 1, No. 1, Aug. 1945. Theory—Design—Application.

Auger, P., Munn, A. M., Pontecorvo, B., *Canad. J. Res. A.* **25**, 143-56 (1947), [The transport mean free path of thermal neutrons in heavy water.]

Auluck, F. C., Kothari, D. S., *Nature*, London **157**, 662 (1946), Distribution of energy levels for the liquid drop nuclear model.

Autonoff, G., *Phys. Rev.* **68**, 288 (1945), Position of uranium Y in radioactive series.

Bailey, C. L., Bennett, W. E., Bergstralh, T., Nuckolls, R. G., Richards, H. T., Williams, J. H., *Phys. Rev.* **70**, 583-9 (1946), The neutron-proton and neutron-carbon scattering cross section for fast neutrons.

Baker, A. L., *Mech. Eng.* **68**, 775-8 (1946), Nuclear energy. Problems of the engineer and the A.S.M.E. in its application for beneficent purposes.

Baldwin, G. C., Klaiber, G. S., *Phys. Rev.* **70**, 259-70 (1946), Multiple nuclear disintegrations induced by 100 Mev Xrays.

Baldwin, G. C., Klaiber G. S., *Phys. Rev.* 71, 3-10 (1947), Photofission in heavy elements.

Barnard, C. I., Oppenheimer, J. R., Thomas, C. A., Winn, H. A., Lilienthal, D. E., *U. S. Department of State, Publication* 2498, 61pp. (1946), A report on the international control of atomic energy.

Barschall, H. H., Battat, M. E., *Phys. Rev.* **70**, 245-8 (1946), On the disintegration of nitrogen by fast neutrons.

Barschall, H. H., Battat, M. E., Bright, W. C., *Phys. Rev.* **70**, 458-9 (1946), Scattering of fast neutrons by boron.

Beck, J. S. P., Meissner, W. A., *Am. J. Clin. Path.* **16**, 586-92 (1946), Radiation effects of the atomic bomb among the natives of Nagasaki, Kyushu.

Belgium Symposium, Dec. 27-28, 1946. *Tech. Wetenschap. Tijdschr.* **16**, 47-51 (1947).

van Itterbeek, A., Physics of the atom.

Rutgers, A. J., Nuclear structure and nuclear energy.

van Paemil, O., Separation and physics of isotopes.

Sizoo, G. J., The uranium pile as a source of energy.

Bernstein, S., Preston, W. M., Wolfe, G., Slattery, R. E., *Phys. Rev.* **71**, 573-81 (1947), Yield of photoneutrons from U^{235} fission products in heavy water.

Beyden, J., *Comptes Rendus* **224**, 1715-16 (1947), Study of a rapid method of separating La. and Yt. by the use of radioactive indicators.

Bleksley, A. E. H., *S. African Mining Eng. J.* **56**, 383-5 (1945), The radium-uranium minerals.

Bless, A. A., *Proc. Florida Acad. Sci.* **8**, 267-79 (1945), Atomic energy.

Bøggild, J. K., Brostrøm, K. J., Lauristen, T. K., *Danske Vidensk. Selsk. Mat.-fys Medd.* **18**, (1940), Cloud chamber studies of fission fragment tracks.

Bøggild, J. K., Arrøe, O. H., Sigurgeirsson, T., *Phys. Rev.* **71**, 281-87 (1947), Cloud chamber studies of electronic and nuclear stopping of fission fragments in different gases.

Bohm, D., Richman, C., *Phys. Rev.* **71**, 567-72 (1947), On the neutron-proton scattering cross section.

Born, Max, ''Atomic Physics'', Blackie & Sons, London (1944).

Borst, L. B., Ulrich, A. J., Osborn, C. L., Hasbrouck, B., *Phys. Rev.* **70**, 557 (1946), Neutron diffraction and nuclear resonance structure.

Bothe, W., Jensen, P., *Z Phys.* **122**, 749-55 (1944), Absorption of thermal neutrons in carbon.

Bowen, H. G., *Mech. Eng.* **68**, 779-81 (1946), Nuclear energy as a power source. U. S. Navy's interest in nuclear power as distinct from use in nuclear ordinance.

Bradt, H., Heine, N.G., Scherrer, P., *Helv. Phys. Acta* **16**, 455-70 (1943), Conversion lines in the β-spectrum of UX.

Breit, G., Zilsel, P. R., *Phys. Rev.* **71**, 232-7 (1947), The scattering of slow neutrons by bound protons. II. Harmonic binding.

Breit, G., *Phys. Rev.* **71**, 215-31 (1947), The scattering of slow neutrons by bound protons. I. Methods of calculation.

Breusch, F. L., *Experientia* **2**, 363 (1946), Rule concerning the missing isotone numbers.

British Information Service, *Rev. Mod. Phys.* **17**, 472-490 (1945), Statements relating to the atomic bomb.

British Reports, *Nature* **159**, 411 (1947), List of British declassified reports.

Broda, E., *Nature*, London, **158**, 872-3 (1946), Determination of the upper limits of fission cross sections of lead and bismuth for Li-D neutrons by a track count method.

Broda, E., Wright, P. K., *Nature*, London, **158**, 871-2 (1946), Determination of the upper limits of the fission cross sections of lead and bismuth for Li-D neutrons by a chemical method.

Broda, E., Feather, N., *Proc. Roy. Soc.* A**190**, 20-30 (1947), Assignment of the slow-neutron-produced activities of thallium and the dual disintegration of radium E.

Broda, E., *Nature* **157**, 307 (1946), Discovery of elements 95 and 96 and the chemical properties of the transuranic elements.

Bunge, M., *Nature* **156**, 301(1945), Neutron-proton scattering at 8.8 and 13 Mev.

Burton, M., *J. Phys. Coll. Chem.* **51**, 611-25 (1947), Radiation chemistry.

Canadian Reports, *Nature* **159**, 719 (1947), List of Canadian declassified reports.

Carnegie Endowment Comm. on Atomic Energy, *Chem. Eng.* **53**, #10, 125-33 (1946), Atomic energy. Its future in power production.

Chadwick, J., *Nature* **159**, 421 (1947), Atomic energy and its applications.

Chadwick J., *J. Inst. Fuel* **20**, #3, 34-41 (1946), Atomic energy and its applications.

Chamberlain, W. E., *Chem. Eng. News* **24**, 1352-6 (1946), Future of atomic power. Biological phase.

Chamberlain, O., Gofman, J. W., Segrè, E., Wahl, A. C., *Phys. Rev.* **71**, 529-30 (1947), Range measurements of alpha particles from 94^{239} and 94^{238}.

Chamberlain, O., Williams, D., Yuster, P., *Phys. Rev.* **70**, 580-2 (1946), Half-life of uranium (234).

Chatterjee, S. D., *Indian J. Physics* **19**, 211-16 (1945), The spontaneous emission of neutrons from the uranium nucleus.

Clark, C. H., "The Story of the Atomic Bomb," Brighton, England, Machinery Pub. Co. (1945).

Compton, A. H., *Gen. Elec. Rev.* **50**, #7, 9-15 (1947), The birth of atomic energy and its human meaning.

Compton, A. H., *Am. J. Phys.* **14**, 173-8 (1946), The social implications of atomic energy.

Coon, J. H., Barschall, H. H., *Phys. Rev.* **70**, 592-6 (1946), Angular distribution of 2.5 Mev neutrons scattered by deuterium.

Corinaldesi, E., *Nuovo Cim.* **3**, 131-41 (1946), On density measurements of thermal neutrons.

Cork, J. M., "Radioactivity and Nuclear Physics", New York, Van Nostrand Co. (1947).

Corvalen, M.I., *Phys. Rev.* **71**, 132 (1947), Concentration of plutonium in pitchblende.

Coster, D., de Vries, Hl., Groendijk, H., *Nature* **159**, 569 (1947), Neutron experiments of physical laboratory of Groningen.

Coven, A. W., *Phys. Rev.* **68**, 279 (1945), Evidence of increased radioactivity of the atmosphere after the atomic bomb test in New Mexico.

Curie, I., Joliot, F., *Ann. Phys.* (Paris) **19**, 107-9 (1944), The fission of ionium by neutrons.

Daniel, F., *Chem. Eng. News* **24**, 1514-17; 1588-9 (1946), Peacetime uses of atomic power.

Darwin, C., *Science Progress* **34**, 449-65 (1946), Atomic energy.

DaSilveira, M., *Portugaliae Physica*, **1**, 167-74 (1945), Natural radioactivity by neutron emission.

Debierne, A., *Comptes Rendus* 224, 1220-1222 (1947), About the Bikini cloud.

DeMent, J. A., Dake, H. C., "Handbook of Uranium Minerals", Portland, Oreg., Mineralogist Pub. Co., 78 pp. (1947).

Demers, P., *Phys. Rev.* **70**, 974-5 (1946), Pairs of fission fragments from U^{235}.

Demers, P., *Can. J. Res.* **24A**, 117-41 (1946), Slow neutron studies.

Demontvignier, M., *Rev. gén. élec.* **55**, 85-98 (1946), Nuclear energy and its recent applications.

Dempster, A. J., *Phys. Rev.* **72**, 431 (1947), Yield of nuclei formed by fission.

Dewey, B., *Chem. Eng. News* **24**, 2478-80 (1946), The atomic bomb and common sense.

Diemer, G., deVries, H., *Physica's Grav.* **11**, 345-52 (1946), On the activities caused by nearly thermal neutrons.

Dietz, D., ''Atomic Energy Now and Tomorrow'', London, Westhouse, 172pp. (1946).

Dostrovsky, I., Hughes, E. D., *Nature* **158**, 164-5 (1946), A convenient and efficient fractionating column and its use in the separation of the heavy isotopes of hydrogen and oxygen.

Dubs, W. R., *Schweiz. Bauztg.* **128**, 107-11, 123-7 (1946), The fundamental physics of the atomic energy power plant.

Duckworth, H. E., Hogg, B. G., *Phys. Rev.* **71**, 212 (1947), Relative abundance of the copper isotopes and the suitability of the photometric method of detecting small variations in isotopic abundance.

Duffuld & Calvin, *Journ. Am. Chem. Soc.* **68**, 1129.

Eidinoff, M. L., *J. Chem. Education* **23**, 60-5 (1946), Uranium fission. Discoveries leading to a chain reaction.

Eidinoff, M. L., Ruchlis, H., ''Atomics for the Millions'', New York, Whittlesey House (1947).

Eklund, S., *Tek. Tiel.* **75**, 1425-33 (1945), The release of atomic energy.

Ellis, C. D., *J. Roy, Soc. Arts* **94**, 235-41 (1946), Nuclear energy and its peacetime applications.

Ellis, C. D., *J. Incorp. Brewers' Guild* **32**, 233-45 (1946), Nuclear energy and its peacetime applications.

Eriksson, H. A., *Ark. Mat. Astr. Fys.* **33**B **#2**, paper 5, (1946), On the distribution of neutron energies in a moderator of infinite size.

Eriksson, H. A., *Ark. Mat. Astr. Fys.* **34**B **#1**, paper 2, (1947), On the stationary energy and spatial distribution of neutrons in a medium of infinite size.

Farwell, G., Segrè, E., Wiegand, C., *Phys. Rev.* **71**, 327-30 (1947), Long-range alpha particles emitted in connection with fission. Preliminary report.

Fearsen, R. E., Engle, A. W., *Phys. Rev.* **70**, 564 (1946), Atmospheric analyses done at Tulsa, Oklahoma, during Bikini atomic bomb test.

Feather, N.. *Nature*, London, **159**, 607-8 (1947), Emission of secondary charged particles in fission.

Feather, N., Krishnan, R. S., *Proc. Camb. Phil. Soc.* **43**, 267-73 (1947), The radiations emitted by U_{92}^{239} and its formation in the deuteron bombardment of uranium.

Feather, N., *J. Roy Aeronaut. Soc.* **50**, 553-67 (1946), Atomic disintegration.

Feather, N., *Chem. Products* **9**, 31-4 (1946), Atomic energy.

Feenberg, E., *Rev. Mod. Phys.* **19**, 239-58 (1947), Semi-empirical theory of the nuclear energy surface.

Feeny, H., *Canad. J. Res. A* **23**, 73-76 (1945), Absorption of thermal neutrons in indium.

Feinfer, E., Bothe, W., *Z. Phys.* **122**, 769-77 (1944), Interactions of neutrons and gamma rays with beryllium.

Feldenkrais, M., *Nature* **157**, 481 (1946), ''Industrial'' separation of isotopes.

Fenning, F. W., Graham, G. A. R., Seligman, H., *Canad. J. Res. A* **25**, 73-76 (1947), The ratio of the capture cross section of lithium and boron for thermal neutrons.

Fermi, E., Sturm, W. J., Sachs, R. G., *Phys. Rev.* **71**, 589-94 (1947), The transmission of slow neutrons through microcrystalline materials.

Fermi, E., *Science* **105**, 27-32 (1947), Elementary theory of the chain reacting pile.

Fermi, E., Marshall, L., *Phys. Rev.* **71**, 915 (1947), Phase of scattering of thermal neutrons by aluminum and strontium.

Fermi, E., *Chem. Eng. News,* **24**, 1357-9 (1946), Future of atomic power. Power phase.

Fermi, E., Marshall, L., *Phys. Rev.* **71**, 666-77 (1947), Interference phenomena of slow neutrons.

Ferretti, B., *Ric. Sci.* **12**, 843-52; 993-1019 (1941), On the theory of collisions of protons and fast neutrons, I--II.

Ferretti, B., *Ric. Sci.* **12**, 512-15 (1941), On the collision of slow neutrons and protons.

Feshback, H., Peaslee, D. C., Weisskopf, V. F., *Phys. Rev.* **71**, 145-58 (1947), On the scattering and absorption of particles by atomic nuclei.

Fields, R., Russell, B., Sachs, D., Wattenberg, A., *Phys. Rev.* **71**, 508-10 (1947), Total cross sections measured with photoneutrons.

Fischer, C., *Phys. Z.* **43**, 507-15 (1942), Nuclear transformation in nitrogen by means of fast neutrons.

Fisher, J. W., Flint, H. T., *Nature* **159**, 741 (1947), X-ray spectra of transuranic elements.

Foster, G., *Am. Rev. Soviet Union* **7**, ✻2, 47-50 (1946), Research on atomic energy in the U.S.S.R.

Foster, L. F., *J. Chem. Education* **22**, 619-23 (1945), Synthesis of the new elements neptunium and plutonium.

Fox, M. C., *Chem. & Met. Eng.* **52**, ✻12, 102-3 (1945)), Thermal diffusion as an adjunct of electromagnetic process.

Frank, F. C., *Proc. Phys. Soc.* **57**, 408-412 (1945), Mass of the neutrino.

Franzini, T., Galli, M. G., *Ric. Sci.* **12**, 1157-60 (1941), On the disintegration of uranium.

Frenkl, J., *J. Phys. U.S.S.R.* **10**, 533 (1946), On some features of the progress of fission of heavy nuclei.

Frisch, D. H., *Phys. Rev.* **70**, 589-92 (1946), The total cross sections of carbon and hydrogen for neutrons of energies from 35-490 Kev.

Frisch, O. R., "Meet the Atoms", New York, L. B. Fischer Pub. Corp. (1946).

Fryer, E. M., *Phys. Rev.* **70**, 235-44 (1946), Transmission of velocity selected neutrons through magnetized iron.

Furry, W. H., Jones, R. C., *Phys. Rev.* **69**, 459-71 (1946), Isotope separation by thermal diffusion: the cylindrical case.

Gamow, G., "Birth and Death of the Sun", New York, Penguin Books (1946).

Gamow, G., "Atomic Energy in Cosmic and Human Life", New York, Macmillan Co. (1945).

Gehlen, J., *Z. Phys.* **121**, 268-84 (1943), Determination of the effective cross section of commercial aluminum for the capture of slow neutrons.

Gilbert, A., Keller, R., Rossel, J., *Helv. Phys. Acta* **19**, 493-502 (1946), Influence of molecular bonding on the scattering of thermal neutrons by nitrogen.

Gilbert, A., Rossel, J., *Helv. Phys. Acta* **19**, 285-306 (1946), Effect of temperature on neutron-proton scattering.

Giles, C., *J. Am. Rocket Soc.* ✻63, 5, 20 (1945), Atomic powered rockets. Energy from atoms for propulsive purposes.

Goldberger, M. L., Seitz, F., *Phys. Rev.* **71**, 294-310 (1947), Theory of the refraction and the diffraction of neutrons by crystals.

Goldhaber, M., Yalow, A. A., *Phys. Rev.* **69**, 47 (1945), Resonance scattering of group neutrons.

Goldschmidt, V. M., *Norsk. Fys. Tids.* **3**, ✻3, 179-90 (1941-42), Transuranic elements (elements with nuclear charge greater than 92).

Goloborodko, T., *Journ. of Phys. U.S.S.R.* **11**, 44-48 (1947), Resonance phenomena in the elastic scattering of photoneutrons with energies of 0.1, 0.2, 0.3, and 0.4 Mev by atomic nuclei.

Goodman, C., *Petroleum Refiner* **24**, ✻11, 427-35 (1945), Also *Petroleum World* **43**, ✻4, 72-3 (1946), Petroleum vs. plutonium.

Goodman, C., *Automotive and Aviation Inds.* **93**, #11, 18-19; 62; 66 (1945), Utilization of atomic energy.

Goodman, C. (editor), "Notes from Seminar in Nuclear Science and Engineering", Mass. Inst. of Tech. (1947).

Götte, H., *Z Naturforsch.* **1**, 377-82 (1946), Separation of the rare earths formed in uranium fission by the method of Szillard and Chalmers.

Goudsmit, S. A., "ALSOS", New York, Henry Schuman, 259pp. (1947).

Grahame, D. C., *Phys. Rev.* **69**, 369 (L) (1946), Resonance scattering of fast neutrons.

Green, L. L., Livesey, D. L., *Nature,* London **158**, 272 (1946), Fission fragment tracks in photographic plates.

Green, L. L., Livesey, D. L., *Nature* **159**, 332 (1947), Emission of light charged particles in the fission of uranium.

Grew, K. E., *J. Incorp. Brewers' Guild* **31**, 301-13 (1945), Atoms and matter.

Grew, K. E., *J. Incorp. Brewers' Guild* **32**, 71-6 (1946), Atomic energy.

Grummitt, W. E., Wilkinson, G., *Nature,* London **158**, 163 (1946), Fission products of U^{235}.

Hagemann, F., Katzin, L. I., Studier, M. H., Ghiorso, A., Seaborg, G. T., *Phys. Rev.* **72**, 252L (1947), The decay products of U^{235}: $(4n+1)$ Radioactive Series.

Hahn, O., *Trans. Chalmers Univ. Technol.*, Gothenburg, Sweden #28, 9 (1944), Transformations of the chemical elements and the fission of uranium.

Hahn, O., Strassmann, F., *Z. Phys.* **121**, 729-45 (1943), On the strontium and yttrium isotopes resulting from the fission of uranium.

Hahn, O., *Angew. Chem.* A**19**, 2-6 (1947), Working with radioactive atoms.

Hahn, O., Strassmann, F., Götte, H., *Abhandl. press. Akad. Wiss. Math.-Naturw. Klasse* #3, 30pp (1942), The experimental unraveling of the elements and atoms formed in uranium fission.

Hahn, O., *Veroffentl. deut. wiss. Inst.*, Stockholm, Reche III, 1-20 (1944), Artificial nuclear changes and the splitting of heavy nuclei.

Hahn, O., Strassmann, F., *Abhandle. preuss. Akad. Wiss. Math.-Naturw. Klasse* #12, 14pp (1944), The chemical separation of the elements and atoms formed in uranium fission.

Hahn, O., Strassman, F., Seelmann-Eggebert, W., *Z. Naturforsch* **1**, 545-56 (1946), The chemical separation of elements originating in uranium fission. I.

Hamermesh, M., Schwinger, J., *Phys. Rev.* **69**, 145-54 (1946), The scattering of slow neutrons by ortho- and para-deuterium.

Hamermesh, M., Schwinger, J., *Phys. Rev.* **71**, 678-80 (1947), Neutron scattering in ortho- and para-hydrogen.

Harkins, Wm. D., *Science* **103**, 289-302 (1946), The neutron, the intermediate or compound nucleus, and the atomic bomb.

Havers, W. W., Jr., Rainwater, J., *Phys. Rev.* **70**, 154-173 (1946), Slow neutron cross sections of indium, gold, silver, antimony, lithium, and mercury as measured with a neutron beam spectrometer.

Haxel, O., Volz, H., *Z. Phys.* **120**, 493-512 (1943), On the absorption of neutrons in aqueous solutions.

Heale, H. G., *Nature* **157**, 225 (1946), Electrochemistry of uranium.

Heisenberg, W., *Nature* **160**, 211 (1947), Research in Germany on the technical application of atomic energy.

Hereward, H. G., Lawrence G. C., Paneth, H. R., Sargent, B. W., *Canad. J. Res.* A**25**, 15-41 (1947), Measurement of the diffusion length of thermal neutrons in graphite.

Herzog, G., *Phys. Rev.* **70**, 227-8 (1946), Gamma ray anomaly following the atomic bomb test of July 1, 1946.

Hess, V. F., Luger, P., *Phys. Rev.* **70**, 564-5 (1946), The ionization of the atmosphere in the New York area before and after the Bikini atom bomb test.

Ho, Zah-Wei, Tsien San Tsiang, Vigneron, L., Chastel, R., *Comptes Rendus* **223**, 1119-21 (1946), Experimental proof of the disintegration of the uranium nucleus into four particles.

Hoed, D. den, *Nederland. Tijdschr. Geneeskunde* **90**, 161-4 (1946), Biological aspects of the atomic bomb.

Hoffman, F. de, Feld, B. T., *Phys. Rev.* **72**, 567-9 (1947), Delayed neutrons from Pu^{239}.

Hogerton, J. F., *Chem. & Met. Eng.* **52**, 98-101 (1945), Oak Ridge gives industry a unit operation—gas diffusion.

Hulubei, H., Cauchois, Y., *Comptes Rendus* **224**, 1265-6 (1947), Researches on a stable isotope of atomic number 84.

Hulubei, H., *Phys. Rev.* **71**, 740-1 (1947), Search for element 87.

Houtemans, F. G., *Phys. Zeitsch.* **45**, 258 (1945), Regarding a semi-empirical relationship between the strength of a neutron source and the maximum attainable density of slow neutrons in a material containing hydrogen.

Hughes, A. L., *Mining Congr. J.* **32**, ※6, 22-25 (1946), Atomic energy. I.

Hughes, A. L., *Mining Congr. J.* **32**, ※7, 45-7 (1946), Atomic energy. II.

Hursley, F., Hursley, D., "Atomic Bombs", Syracuse, N. Y., Syracuse Univ. Press (1945).

Hutchison, D. A., *J. Chem. Phys.* **14**, 401-8 (1946), Efficiency of the electrolytic separation of potassium isotopes.

Inghram, M. G., Hayden, R. J., Hess, D. C., Jr., *Phys. Rev.* **71**, 643 (1947), Activities induced by pile neutron bombardment of samarium.

Inghram, M. G., Hess, D. C., Jr., Hayden, R. J. Parker, G. W., *Phys. Rev.* **71**, 743 (1947), 55 hour element [61] formed in fission.

Inghram, M. G., Hess, D. C., Jr., Hayden, R. J., *Phys. Rev.* **71**, 561-2 (1947), Neutron cross sections for Hg. isotopes.

Jacobsen, T. C., Lassen, N. D., *K. Danske Vidensk. Selsk. Mat.-fys. Medd.* **19**, ※6 (1941), Deuteron induced fission of uranium and thorium.

Janser, A. M. U., *Chem. Products* **9**, ※1-2, 2-10; 14 (1945), Space flight and atomic power.

Jeffries, Z., *Chem. Eng. News* **24**, 186 (1946), The new term "nucleonics", generic name for atomic energy and related subject matter.

Jensen, P., *Z. Phys.* **122**, 756-68 (1944), Retardation of neutrons.

Jentschke, W., Prankl, F., Hernegger, F., *S. B. Akad. Wiss. Wien* **151**, Abd. IIa, 149-157 (1942), Proof of the fission of ionium under neutron bombardment.

Jesse, W. P., *Chem. Eng. News* **24**, 2906-9 (1946), Role of instruments in the atomic bomb project.

Joliot, F., *Rev. gén. sci.* **53** ※8/9 172-83 (1946), Atomic disintegration.

Joliot, F., *Comptes Rendus* **218**, 733-5 (1944), Physical method for extraction of radioelements from fission of heavy atoms and evidence for a radio praseodymium of a 13 day period.

Joliot, F., *Comptes Rendus* **218**, 488-91 (1944), A method of measuring the ranges of radioelements of known chemical nature coming from the fission of uranium.

Jones, R. C., Furry, W. H., *Rev. Mod. Phys.* **18**, 151-224 (1946), Separation of isotopes by thermal diffusion.

Karlik, B., Bernert T., *Naturwiss.* **32**, 44 (1944), The element 85 in the actinium series.

Katcoff, S., *Phys. Rev.* **71**, 826 (1947), Preliminary determination of the neutron absorption cross section of long-lived I^{129}.

Keiss, C. C., Humphreys, C. J., Lauer, D. D., *J. Research Nat'l. Bur. Standards* **37**, 57-72 (1946), Preliminary description and analysis of the first spectrum of uranium.

Keith, P. C., *Chem. & Met. Eng.* **53**, ※2, 112-22 (1946), The role of the process engineer in the atomic bomb project.

Kelly, J. E., *Mining J.* **29**, 7-8 (1945), Raw materials of atomic power.

Kendall, J. T., *Engineering* **161**, 75-6;100 (1946), Production of atomic energy.

Kennedy, J. W., Wahl, A. C., *Phys. Rev.* **69**, 367-8 (1946), Search for spontaneous fission in 94^{239}.

Kennedy, J. W., Seaborg, G. T., Segrè, E., Wahl, A. C., *Phys. Rev.* **70**, 555-6 (1946), Properties of 94^{239}.

Kidd, G. E., *Petroleum Engr.* **17**, #6, 240-6 (1946), Uranium-source of atomic power.

Kjellson, H., *Tek. Tid.* **75**, 1417-24 (1945), Atomic theory and atomic bombs.

Klema, E. D., *Phys. Rev.* **72**, 87L (1947), Fission cross section of $N\rho^{237}$.

Klemm, A., *Z. Naturforsch.* **1**, 252-7 (1946), The phenomenology of two procedures for separating isotopes.

Koch, J., Bendt-Nielsen, B., *Kgl. Danske Videnskab. Selskab. Math. Fys. Medd.* **21**, #8, 28pp (1944), A high-intensity mass spectograph for experiments on the separation of isotopes.

Korsching, H., Wirtz, K., *Abhandl. preuss. Akad. Wiss. Math.-Naturw. Klasse* #3, 3pp (1939), Separation of zinc isotopes by thermal diffusion in the liquid phase.

Kothari, D. S., *J. Sci. & Ind. Research* (India) **4**, 324-8 (1945), Nuclear energy and its utilization.

Kothari, D. S., Auluck, F. C., *Nature*, London, **159**, 204-5 (1947), Surface tension of nuclear matter and the enumeration of eigenfunctions of an enclosed particle.

Kowarski, L., *J. Phys. Radium* **7**, 253-8 (1946), Neutrons from fission and the chain reaction.

Kramer, A. W., *Power Plant Eng.* **49**, #9, 78-84; #12, 99-104 (1945); **50**, #2, 41-43; #1, 87-92 (1946), The development of atomic energy.

Kwal, B., *Comptes Rendus* **224**, 563-4 (1947), On several phenomena which may accompany the nuclear fission of uranium.

Lassen, N. O., *K. Danske Vidensk. Selsk. Mat. fys. Medd.* **23**, #2 (1945), On the effective charge of fission fragments.

Lassen, N. O., *Phys. Rev.* **68**, 142-3 (1945), $H\rho$-distribution of fission fragments.

Lassen, N. O., *Phys. Rev.* **68**, 230 (1945), (L) Ionization by fission fragments in nitrogen, argon and xenon.

Lassen, N. O., *Phys. Rev.* **69**, 137-9 (1946), The variation along range of the $H\rho$-distribution and the charge of the fission fragments of the light group.

Lassen, N. O., *Phys. Rev.* **70**, 577-9 (1946), Specific ionization by fission fragments.

Langevin, P., *Ann. Phys.* **17**, 303-17 (1942), Collision between fast neutrons and nuclei of any mass.

Laughlin, J. S., Kruger, P. G., *Phys. Rev.* **71**, 736 (1947), Scattering of neutrons of energy between 12 Mev and 13 Mev by protons.

Laurence W. L., "The Story of the Atomic Bomb" (Off. War Department Release) The New York Times, New York; 40pp (1946).

Lapoite, M., "L'energie atomique", Paris, H. Dunod, 44pp (1946).

Le Caine, J., *Phys. Rev.* **72**, 564-66 (1947), Application of a variational method to Milne's problem.

Lilienthal, D. E., *Chem. Eng. News* **24**, 2483-5 (1946), How can atomic energy be controlled?

Lilienthal, D. E., Bacher, R. F., Pike, S. T., Strauss, L. L., Waymack, W. W., *Science* **105**, 199-204 (1947), First report of the U. S. Atomic Energy Commission.

Lind, S. C., *Trans. Electrochem. Soc.* **89**, 4pp. (1946), Uranium and radium in the postwar period.

Los Alamos Laboratory—Lecture Notes on Nuclear Physics.

Luhr, Overton, "Physics Tells Why; Atomic Energy", Rev. Ed. Lancaster, Pa. Jaques Cattell Press, 27pp. (1946).

McBrady, J. J., Livingston, R., *J. Phys. Chem.* **50**, 176-90 (1946), The formation of quadrivalent uranium during the uranyl-sensitized photochemical decomposition of oxalic acid.

McKolls, R. G., Bailey, C. L., Bennett, W. E., Bergstralh, T., Richards, H. T., Williams, J. H., *Phys. Rev.* **70**, 805-7 (1946), The total scattering cross sections of deuterium and oxygen for fast neutrons.

Macklin, R. L., Knight, G. B., *Phys. Rev.* **72**, 435 (1947) (L), X-rays associated with U^{234}.

McMahon, B., *Mech. Eng.* 68, 782-4 (1946), Effect of McMahon bill on nuclear-energy applications.

McMahon, B., *Chem. Corps. J.* **1**, 13-15 (1947), Chemistry's contribution to the development of atomic energy.

Magnan, C., Chanson, P., Ertand, A., *Comptes Rendus* **218**, 712-14 (1944), Electrostatic spectograph for evaluating the charge of the nuclear fragments of uranium.

Mallet, L., *J. Radiol. Electrol.* **24**, 247-54 (1941), Biological effects and possible therapeutic effects of neutrons.

Manley, J. H., Agnew, H. M., Barschall, H. H., *Phys. Rev.* **70**, 602-5 (1946), Elastic backscattering of d-d neutrons.

Mark, C., *Phys. Rev.* **72**, 558-563 (1947), Neutron density near a plane surface.

Marshak, R. E., *Phys. Rev.* **72**, 47-50 (1947), Milne problem for a large plane slab with constant source and anisotropic scattering.

Marshak, R. E., *Rev. Mod. Phys.* **19**, 185-238 (1947), Theory of the slowing down of neutrons by elastic collision with atomic nuclei.

Marshak, R. E., *Phys. Rev.* **71**, 688-93 (1947), The variational method for asymptotic neutron densities.

Massey, H. S. W., Buckingham, R. A., *Phys. Rev.* **71**, 558 (1947), The collision of neutrons with deuterons and the reality of exchange forces.

Masters, D., Way, K., "One World or None", New York, McGraw-Hill (1946).

Mattauch, J., "Nuclear Physics Tables", Translated by E. P. Gross, New York, Interscience Publishers, 175pp, (1946).

Matthias, F. R., *Wisconsin Engr.* **50**, #6, 7-9, 26 (1946), The atomic bomb.

Maurer, R. J., *Science Counselor* **9**, 41-2, 61 (1946), The atomic bomb and world peace.

Maurer, W., Pose, H., *Z. Phys.* **121**, 285-92 (1943), Neutron emission of the uranium nucleus as the result of its spontaneous fission.

Mebus Brezansky, R., "Transmutación y desintegración de la materia, la bomba atómica", Conferencia dada el 13 de agosto de 1945, en la Escuela de ingeniería de la Universidad de Chile, Santiago de Chile, Talleres gráficos. "Horizonte" 16pp (1945).

Meggers, W. F., *Science* **105**, 514-16 (1947), Electron configurations of "rare-earth" elements.

Meitner, L., *Fortune* **33**, 137-44, 185-6, 188 (1946), The nature of the atom. A review.

Meitner, L., *Rev. Mod. Phys.* **17**, 287-91 (1945), Attempt to single out some fission processes of uranium by using the differences in their energy release.

Merle, J. M., *U. S.* **2**, 395, 286, Feb. 19, 1946. Aluminothermic reduction of metal oxides in a centrifuge.

Migdel, A., *J. Phys. U.S.S.R.* **9**, #1, 45-51 (1945), Pair creation in nuclear collisions.

Miro Quesada, O., "La bomba atómica, explicación de su potencia prodigiosa", Lima, Imprenta Torres Aguirre, s.a. 51pp. (1945).

Mode, H. G., Graham, R. L., *Canad. J. Res. A* **25**, 1-14 (1947), A mass spectrometer investigation of the isotopes of xenon and krypton resulting from the fission of U^{235} by thermal neutrons.

Moerkerk, J. J., "Atoomkrachten, Hun Ontdekking en Beteekenis", Antwerp, Belgium, Helicon, 91pp (1945).

Morrison, P., *J. Applied Phys.* **18**, 133-52 (1947). Physics in 1946.

Muehlhause, C. O., Goldhaber, M., *Phys. Rev.* **66**, 36 (1944), Slowing down of R neutrons into C neutrons in lead.

Muehlhause, C. O., Goldhaber, M., *Phys. Rev.* **70**, 85 (1946) (L), Capture cross sections for slow neutrons. II. Small capture cross sections.

Munn, A. M., Pontecorvo, B., *Canad. J. Res. A* **25**, 157-67 (1947), Spatial distribution of neutrons in hydrogenous media containing bismuth, lead and iron.

Muraour, H., *Comptes Rendus* **221**, 200-1 (1945), The luminous and thermal phenomena which accompany the detonation of the atomic bomb.

Murphy, E. J., *Chem. Eng. News* **24**, 182-6 (1946), Development of atomic energy.

Murphy, W. J., *Chem. Eng. News* **24**, 1782-6 (1946), Operation Crossroads.

Niini, R., *Suomen Kemistilehti* 18A, 153-7 (1945), The principle of atomic bombs.

Nix, F. C., Clement, G. C., *Phys. Rev.* **68**, 159-62 (1945), Thermal neutron scattering studies in metals.

Occhialini, G. P. S., Powell, C. F., *Nature*, London, **159**, 186-90 (1947), Nuclear disintegrations produced by slow charged particles of small mass.

Occhialini, G. P. S., Powell, C. F., *Nature*, London, **159**, 93-4 (1947), Multiple disintegration processes produced by cosmic rays.

Oliphant, M. L., *Proc. Roy. Inst. Great Brit.* Dec. 14, 1945 - 9pp. Utilization of nuclear energy.

Oliphant, M. L., *Nature*, London **157**, 5-7 (1946), The release of atomic energy.

O'Neall, R. D., *Phys. Rev.* **70**, 1-4 (1946), Slowing down of low energy neutrons in water. II. Determination of photoneutron energies.

O'Neill, J. T., "Almighty Atom, the Real Story of Atomic Energy", New York, I. Washburn (1945).

Oppenheimer, J. R., *Chem. Eng. News* **24**, 1350-2 (1946), Future of atomic power. Atomic explosives.

Pais, A., *Proc. Cambridge Phil. Soc.* **42**, pt. 1, 45-54 (1946), Scattering of fast neutrons by protons.

Paneth, F. A., *Nature* **159**, 8-10 (1947), The making of the missing chemical elements.

Peaceman, D., *City Coll. Vector* **9**, 71- 2 (1946), The atomic pile.

Pearson, H., *J. Inst. Production Engr.* **25**, 41-58 (1946), Atomic energy.

Pendergrass, E. P., *Trans. & Studies Coll. Physicians, Phila.* E47, **14**, 107-12 (1946), Atomic bomb experiments in the Pacific, with a consideration of the relationship of atomic energy to the future of medicine.

Perey, M., *J. Chem. Phys.* **43**, 155-68 (1946), Element 87.

Perfilov, N. A., *Comptes Rendus Acad. Sci., U.S.S.R.* **47**, 623-6; *Doklady Akad. Nauk U.S.S.R.* **47**, 648-51 (1945), Registration of uranium fragments with removal of background due to alpha particles emitted by uranium.

Perfilov, N. A., *Journ. of Phys. of U.S.S.R.* **10**, 1-12 (1946), A new method of recording particles of the type of uranium fragments by means of a photographic plate.

Peierls, R., *Nature, London*, 158, 773-5 (1946), Fundamental particles.

Perkins, D. H., *Nature*, London, **159**, 126-7 (1947), Nuclear disintegration by meson capture.

Perlman, I., Goeckermann, R. H., Templeton, D. H., Howland, J. J. *Phys. Rev.* **72**, 352 (1947) (L), Fission of bismuth, lead, thallium, plantinum, and tantalum with high energy particles.

Peryman, I., *J. Chem. Education* **24**, #3, 115-20 (1947), Atomic energy in industry.

Persico, E., *Scientia* **79**, 83-92 (1946), The new fire.

Peter, O., *Z. Naturforsch* **1**, 557-559 (1946), Neutron "radiography".

Philipp, K., Riedhammer, J., *Naturforsch.* **1**, 372-6 (1946), Energy states of 44 hr. lanthanum and of element 93.

Philipp, K., Riedhammer, J., Wiedemann, M., *Naturwiss.* **32**, 216-30 (1944), The electron line spectrum of element 93. II.

Placzek, G., *Phsy. Rev.* **69**, 423-38 (1946), On the theory of slowing down of neutrons in heavy substances.

Placzek, G., Seidel, W., *Phys. Rev.* **72**, 550-55 (1947), Milne's problem in transport theory.

Placzek, G., *Phys. Rev.* **72**, 556-7 (1947), Angular distribution of neutrons emerging from a plane surface.

Plutonium Project, The, *J. A. Chem. Soc.* **68**, 2411-42 (1946), Nuclei formed in fission. Decay characteristics, fission yields, and chain relationships.

Potter, R. D., "The Atomic Revolution", New York, R. M. McBride and Co. (1946).

Poole, J. H., *Sci. Proc. Roy. Dublin Soc.* **24**, #7, 71-6 (1946), Possibility of initiating thermo-nuclear reactions under terrestial conditions.

Pose, H., *Z Phys.* **121**, 293-7 (1943), Spontaneous neutron emission of uranium and thorium.

Prata, E., "La Seissione Nucleare Dell'uranio", Milan, U. Hoepli, 160pp. (1946).

Present, R. D., *Phys. Rev.* **72**, 7-15 (1947), On the division of nuclear charge in fission.

Present, R. D., Reines, F., Knipp, J. K., *Phys. Rev.* **70**, 557-8 (1946), The Liquid Drop Model for nuclear fission.

Preston, M. A., *Phys. Rev.* **71**, 865-77 (1947), The theory of áα radio activity.

Rainwater, J., Havens, W. W., Jr., *Phys. Rev.* **70**, 136-53 (1946), Neutron beam spectrometer studies of boron, cadmium and the energy distribution from paraffin.

Ravalico, D. E., "Da Volta all-energia Atomica", Milan, U. Hoepli, 274pp. (1946).

Redman, W. C., Saxon, D., *Phys. Rev.* **72**, 570-75 (1947), Delayed neutrons in plutonium and uranium fission.

Robertson, Solomon, Wendt, 422 C. A. 1945, "British Statement Relating to the Atomic Bomb", London, H. M. Stationery Office (1945).

de Roos, J. L., "*De Atoombom de Bom met Atoomsplitsende Werking, een Populair-Wetenschappelijke Verhandeling*", Haarlem, Netherlands, H. Stam, 1041pp. (1945).

Rose, K., *Materials and Methods* **22**, 1054-7 (1945), Materials for producing the atomic bomb.

Rosenfeld, L., *Nature*, London **156**, 141-2 141-2 (1945), Penetration of fast nucleous into heavy atomic nuclei.

Rossel, J., *Helv. Phys Acta* **19**, 421-2 (1946), Study of intermolecular forces by scattering of slow neurtons.

Ryzhanov, S., *Zhurnal Experiment alnoi y teoreticheskoi physik* **17**, 554 (1947), The radiative transitions of the heavy atomic nuclei.

Saha, M., *Proc. Phys. Soc.* **57**, 271-86 (1945), A physical theory of the solar corona.

Salvetti, C., *Chimico e industrio* **28**, 78-86 (1946), The fission of uranium and the transuraniums.

Sargent, B. W., *Eng. J.* (*Canada*) **28**, 752-7 (1945), Atomic power.

Sargent, B. W., Booker, D. V., Cavanagh, P. E., Hereward, H. G., Niemi, N. J., *Canad. J. Res.* **A25**, 134-42 (1947), The diffusion length of thermal neutrons in heavy water.

Sauerwein, K., *Z. Naturforsch.* **2a**, 73-9 (1947). Neutron capture of uranium in various energy regions.

Sawyer, R. B., Wollan, E. O., Bernstein, S., Peterson, K. C., *Phys. Rev.* **72**, 109-14 (1947), Bent crystal neutron spectrometer and its application to neutron cross section measurements.

Scharff-Goldhaber, G. Klaiber, G. S., *Phys. Rev.* **70**, 229 (1946), (L) Spontaneous emission of neutrons from uranium.

Schiff, L. I., *Phys. Rev.* **70**, 562 (1946), Thresholds for slow neutron induced reactions.

Schrödinger, E., *Proc. R. Irish Acad.* A **51**, 1-8 (1945), Probability problems in nuclear chemistry.

Schulz, L. Goldhaber, M., *Phys. Rev.* **67**, 202 (1945), Capture cross section of hydrogen for slow neutrons.

Seaborg, G. T., Gofman, J. W., Stronghton, R. W., *Phys. Rev.* **71**, 378 (1947), Nuclear properties of U^{233}: a new fissionable isotope of uranium.

Seaborg, G. T., McMillan, E. M., Kennedy, J. W., Wahl, A. C., *Phys. Rev.* **69**, 366 (1946), Radioactive element 94 from deuterons on uranium.

Seaborg, G. T., *Science* **104**, 379-86 (1946), The transuranium elements.

Seaborg, G. T., *Chem. Eng. News* **23**, 2190-3 (1945), The chemical and radioactive properties of the heavy elements.

Seaborg, G. T., *Chem. Eng. News* **24**, 1193-8 (1946), The heavy elements.

Seaborg, G. T., *Chem. Eng. News* **24**, 1192-3 (1946), The impact of nuclear chemistry.

Seaborg, G T., *Chem. Eng. News* **24**, 3160-1 (1946), Plutonium and other transuranium elements.

Seaborg, G. T., *Chem. Eng. News* **25**, 358-60, 397 (1947), Plutonium and other transuranium elements.

Seaborg, G. T., Segrè, E., *Nature*, London, **159**, 863-5 (1947), The transuranium elements.

Seaborg, G. T., *Industr. Eng. Chem.* (*News Edit.*) 2085, 2190 (1945), *Nature*, London, **157**, 307-8 (1946), Discovery of the elements 95 and 96 and the chemical properties of the transuranic elements.

Seaborg, G. T., *Record Chem. Progress* **7**, 1-12 (1946), Chemical processes in plutonium production.

Seelmann-Eggebert, W., *Naturwiss.* **33**, 279-80 (1946), On some new fission products of uranium (ruthenium, rhodium and silver).

Seelmann-Eggebert, W., Strassmann, F., *Z. Naturforsch* **2a**, 80-6 (Feb. 1947), On the fragments to be expected in the fission of uranium.

Segrè, E., Wiegand, C., *Phys. Rev.* **70**, 808-11 (1946), Stopping power of various substances for fission fragments.

Semenchenko, V. K., Korobov, V. V., *Uspekhi Khim.* **15**, 657-84 (1946), Present state of the periodic law and the new elements.

Sherr, R., *Phys. Rev.* **68**, 240-5 (1945), Collision cross sections for 25 Mev neutrons.

Shoenberg, D., *Nature* **159**, 303 (1947), Uranium not a superconductor.

Shotwell, J. T., *Am. J. Phys.* **14**, 179-85 (1946), The international implications of nuclear energy

Shroder, E. F., *Phys. Rev.* **69**, 439-42 (1946), Partial separation of the isotopes of chlorine by thermal diffusion.

Siegel, J. M., *Rev. Mod. Phys.* **18**, 513-44 (1946), *J. Am. Chem. Soc.* **68**, 2411-42 (1946), Nuclei formed in fission: Decay characteristics, fission yields and chain relationships.

Silveira, M. de, *Portugaliǻe Phys.* **1**, 167-74 (1945), Natural radioactivity with neutron emission.

Simon, Ralph, *Phys. Rev.* **69**, 596-603 (1946), Performance of a hot-wire Clusius and Dickel Column.

Simons, L. K., *Danske Vidensk. Selsk.* **17**, #7 (1940), Measurements of the neutron-proton scattering cross section.

Smith, E. S. T., Fox, A. H., Sawyer, T., Austin, H. R., "Applied Atomic Power", New York, Prentice-Hall, 240pp. (1946).

Smorodinsky, J., *J. Phys. U.S.S.R.* **8**, 219-24 (1944), On the scattering of neutrons by protons.

Smythe, H. D., *Rev. Mod. Phys.* **17**, 351-471 (1945), Atomic energy for military purposes.

Snell, A. H., Levinger, J. S., Meiners, E. P., Jr., Sampon, M. B., Wilkinson, R. G., *Phys. Rev.* **72**, 545-49 (1947), Studies of the delayed neutrons. II. Chemical Isolation of the 56 - Second and the 23 - Second Activities.

Snell, A. H., Nedzel, V. A., Ibse, H. W., Levinger, J. S , Wilkinson, R. G., Sampon, M. B., *Phys. Rev.* **72**, 541-44 (1947), Studies of the delayed neutrons. I. The decay curve and the intensity of the delayed neutrons.

Solomon, A. K., *Fortune* **33**, #5, 115-22, 173-4, 176 (1946), The physics of the bomb.

Smorodinsky, J., *Journ. of Phys. of U.S.S.R.* **11**, 195-6 (1947), On the scattering of neutrons by protons.

Spencer, H. M., *J. Chem. Education* **24**, 19-20 (1947), Charts of isotopes and of disintegration and transmutation reactions.

Squires, A. M., *J. Chem. Education* **23**, 538-41 (1946), The fractionation of isotopes.

Stanton, H. E., *J. Western Soc. Engrs.* **51**, 176-80 (1946), Industrial applications of atomic energy.

Stephenson, R. J., *Am. J. Phys.* **14**, 30-5 (1946), A brief account of the physics of the atomic bomb.

Stout, J. W., Jones, W. M., *Phys. Rev.* **71**, 582-5 (1947), Colorimetric determination of the energy produced by plutonium (239).

Sturm, W. J., *Phys. Rev.* **71**, 757-76 (1947), Measurement of neutron cross sections with a crystal spectrometer.

Sugerman, N., *Journ. of Chem. Phys.* **15**, 544-551 (1947), Determination of the ranges of fission fragments emitting delayed neutrons.

Sundarachar, C. K., Venkatanarasimiah, C. K., *Current Sci.* **15**, 15-16 (1946), Techniques in the study of nuclear fission.

Sutton, R. B., McDaniel, B. D., Anderson, E. E., Lavatelli, L. S., *Phys. Rev.* **71**, **272** (1947), The capture cross section of boron for neutrons of energies from .01 ev to 1000 ev.

Suzor, F., *Comptes Rendus* **224**, 1155-6 (1947), Course in aluminum and stopping power of gold for a uranium fission fragment.

Swain, P. W., *Atomic Power* **1**, #3, 2-4(1946), The economics of atomic power.

Swain, P. W., *Atomic Power* **1**, #1, 3-9 (1945), Uranium 235: power fuel of the future.

Swann, W. F. G., *Aero Digest* **51**, #5, 33-7, 113-14 (1945), Nature and portent of atomic power.

Taylor, H. S., *Chem. Eng. News* **24**, 1360-2 (1946), Future of atomic power. Chemical phase.

Thibaud, J., "Energie Atomique et Univers," 2nd. Ed., Lyons, France, M. Audin, 304pp. (1945).

Thode, H. G., Graham, R. L., *Can. J. Research* **25A**, 1-14 (1947), Mass spectrometer investigation of the isotopes of xenon and krypton resulting from the fission of U235 by thermal neutrons.

Thomas, C. A., *Chem. Eng. News* **24**, 2480-3 (1946), Non-military uses of atomic energy.

Thomson, G. P., *Machinery Market* (London) #2356, 13 (1945); *J. Junior Inst. Engrs.* **56**, 89- 92 (1946), Atomic energy.

Timmes, J. J., *U. S. Naval Med. Bull.* **46**, 219-24 (1946), Radiation sickness in Nagasaki. A preliminary report.

Tsien, H. S., *J. Aeronaut. Sci.* **13**, 171-80 (1946), Atomic energy.

Tsien San-Tsiang, Ho Zah-Wei, Farragir, *Comptes Rendus* **224**, 825-6 (1947), On the fission energy of thorium.

Tsien San Tsiang, Ho Zah-Wei, Chastel, R., Vigneron, L., *Comptes Rendus* **224**, 272-3 (1947), Energies and frequency of occurrence of the phenomena of tripartition and quadripartition of uranium.

Tsien San-Tsiang, Marty, C., *Comptes Rendus* **221**, 177-8 (1945), Existence of very low energy gamma rays from radium D.

Tsien San-Tsiang, Chastel, Ho Zah-Wei, Vigneron, *Comptes Rendus* **223**, 986-7 (1946), Tripartition of uranium caused by neutron capture.

Tsien San-Tsiang, *Comptes Rendus* **224**, 1056-8 (1947), On the mechanism of the tripartition of uranium.

Tsien San-Tsiang, Ho Zah-Wei, Vigneron, L., Chastel, R., *Nature*, London, **159**, 773-5 (1947), Ternary and quaternary fission of uranium nuclei.

Tsien San-Tsiang, Ho Zah-Wei, Chastel, R., Vigneron, L., *Phys. Rev.* **71**, 382-3 (1947), The new fission processes of uranium nuclei.

Turner, L. A., *Rev. Mod. Phys.* **17**, 292-6 (1945), The missing heavy nuclei.

Turner, L. A., *Phys. Rev.* **69**, 366 (1946), Atomic energy from U^{238}.

Ubbelohde, A. R., *Proc. Phy. Soc.* **59**, 139-44 (1947), Rates of nuclear transformations under conditions of freezing in.

U. S. Library of Congress, "The Social Impact of Science", A select bibliography, with a section on atomic energy, Wash., D. C., U. S. Gov't Printing Office (1945).

U. S. Secretary of State Committee (on atomic energy). ''A Report on the International Control of Atomic Energy,'' Wash., D. C., U. S. Gov't Printing Office (1946).

Valente, F. A., Zagor, H. I., *Phys. Rev.* **69**, 55-9 (1946), Fast neutron resonance with nitrogen.

Velasso de Pando, M., ''La Bomba Atomica'', Madrid, Dossat, 54pp. (1946).

Verde, M., Wick, G. C., *Phys. Rev.* **71**, 852 (1947), Some stationary distributions of neutrons in an infinite medium.

Verde, M., *Nuovo Cim.* **3**, 116-29 (1946), Distribution of neutrons from a point source of fast neutrons in an infinite hydrogenous medium. I. Resonance neutrons.

Volz, H., *Z. Phys.* **121**, 201-35 (1943), Effective cross section for the absorption of slow neutrons.

Vowles, H. P., *Indian Eng.* **120**, 33-5 (1946), Particle accelerators for atomic bombardment.

Wagner, C. F., Hutcheson, J. A., *Westinghouse Engr.* **6**, 125-7 (1946), Nuclear energy potentialities.

Wall, T. F., *Engineering* **160**, #4153, 134-5(1945), Atomic fission and the cyclotron.

Waller, I., *Ark. Mat. Astr. Fys.* **34**A (#1) Paper 3-5 (1947), On the theory of the diffusion and the slowing down of neutrons.

Ward, F. A. B., Proc. Phy. Soc. **57**, 113-17 (1945), A mechanical model illustrating the uranium chain reaction.

Warren, Shields, *Cancer Research* **6**, 449-53 (1946), The pathological effects of an instantaneous dose of radiation.

Warren, Shields., Draeger, R. H., *U. S. Naval Med. Bull.* **46**, 1349-53 (1946), The pattern of injuries produced by the atomic bomb at Hiroshima and Nagasaki.

Weekes, E. D., Weekes, D. F., *Phys. Rev.* **70**, 565 (1946), Effort to observe anomalous gamma rays connected with atomic bomb test of July 1, 1946.

Welles, S. B., *Phys. Rev.* **69**, 586-9 (1946), Partial separation of the oxygen isotopes by thermal diffusion and the deuteron bombardment of 0^{17}.

White, J. R., Cameron, A. E., *Phys. Rev.* **71**, 907 (1947), The critical ionization potentials of uranium hexafluoride and hydrogen fluoride.

Wigner, E. P., *J. Applied Phys.* **17**, 857 (1946), Theoretical physics in the Metallurgical Laboratory of Chicago.

Wigner, E. P., *Proc. Nat. Acad. Sci., Wash.* **32**, 302-6 (1946), Reaction and scattering cross sections.

Wigner, E. P., *Phys. Rev.* **70**, 606-18 (1946), Resonance reactions.

Williams, R. R., Jr., *J. Chem. Education* **23**, 423-33 (1946), Nuclear reactions.

Wilson, R. R., *Phys. Rev.* **72**, 98-100 (1947), On the time required for the fission process.

Wilson, R. R.,*Phys. Rev.* **72**, 189-92 (1947), Directional properties of fission neutrons.

Wilson, R. R., *Phys. Rev.* **71**, 560 (1947), Delayed neutrons from Pu^{239}.

Winans, J. G., *Phys. Rev.* **71**, 379 (1947), A classical model for the nucleus.

Wollan, E. D., Moak, C. D., Sawyer, R. B., *Phys. Rev.* **72**, 447 (1947), Alpha particles associated with fission.

Wu, C. S., Rainwater, L. J., Havens, W. W., Jr., Dunning, J. R., *Phys. Rev.* **69**, 236-7 (1946), Neutron scattering in ortho- and para-hydrogen and the range of nuclear forces.

Wytzes, S. A., Van der Mass, G. J., *Physica's Grav.* **13**, 49-61 (1947), A new determination of the mean ranges of the alpha particles from UI and U II.

Yates, R. F., ''Atom Smashers'' (Revised), New York, Didier Publishers (1946).

Zammattio, C., ''L'energia atomica'', Milan, Cavallotti, 157pp. (1946).

Zinn, W. H., *Phys. Rev.* **71**, 752-7 (1947), Diffraction of neutrons by a single crystal.

AUTHOR INDEX

(The following abbreviations are used: p, page; fig., figure)

SUBJECT INDEX